John T. Gresser

A Mathematica Approach to Calculus

SECOND EDITION

Gerald L. Bradley ▲ Karl J. Smith

PRENTICE HALL, Upper Saddle River, NJ 07458

Executive Editor: George Lobell
Editorial Assistant: Gale A. Epps
Special Projects Manager: Barbara A. Murray
Production Editor: Michele Wells
Supplement Cover Manager: Paul Gourhan
Supplement Cover Designer: Liz Nemeth
Manufacturing Buyer: Alan Fischer
Cover Photo: Louvre Museum and Pyramid,
Tom Craig/FPG International

Simon & Schuster / A Viacom Company
Upper Saddle River, NJ 07458

Printed in the United States of America

10 9 8 7 6 5 4 3 2

ISBN 0-13-010586-4

Prentice-Hall International (UK) Limited, *London*
Prentice-Hall of Australia Pty. Limited, *Sydney*
Prentice-Hall Canada, Inc., *London*
Prentice-Hall Hispanoamericana, S.A., *Mexico*
Prentice-Hall of India Private Limited, *New Delhi*
Prentice-Hall of Japan, Inc., *Tokyo*
Simon & Schuster Asia Pte. Ltd., *Singapore*
Editora Prentice-Hall do Brazil, Ltda., *Rio de Janeiro*

To Pamela, Morgan, and Nathan
for their patience
during this writing project

and

to Captain Ralph
off on another adventure.

Preface

The last decade of the twentieth century has been an exciting and interesting time for the study of mathematics. At the dawn of the new millennium, mathematics is undergoing a technological change which will have a profound effect on the way mathematics is studied, understood and used in the future. What is driving all this change is the introduction and widespread availability of computer software programs which allow one to compute and manipulate mathematical symbols on a computer screen in much the same way that these symbols are computed or manipulated with pencil and paper. Several such programs, generally referred to as Computer Algebra Systems (CAS) or symbol processors are commercially available. One of the most elaborate of these is the program called Mathematica, which will be used in this manual.

One of the strengths of Mathematica is that it is so easy to use. To be sure, there are some special language peculiarities that one must learn before Mathematica can be used to solve problems, but these are held to a minimum. To a large extent, Mathematica mimics the ordinary language of mathematics. With no more than an hour or so worth of effort, one can begin to use Mathematica to solve mathematical problems. Virtually all of the words used in mathematics are commands or data structures in the Mathematica language. Think of a mathematical word, and you will likely find it in Mathematica, with the first letter of the word capitalized, but otherwise spelled exactly as it is in English. If not, you will certainly be close enough to find something in Mathematica's great on-line help, which will be discussed shortly.

With Mathematica, there is no longer a need to focus attention on the difficulties of carrying out a mathematical computation. The user becomes the director of the production, the creative force which drives the solution, while Mathematica does all the work. Problems which were previously inaccessible because of computational complexity, can now be solved using Mathematica. Just the freedom to concentrate on the creative aspects of a problem rather than the computational aspects, makes some problems much more approachable.

Mathematicians, indeed, all scientists, are naturally skeptics. Their first reaction to any claim is to ask for a proof, and then they are inclined to look for mistakes and counter examples when they study the work of their colleagues. It is exceedingly important that students of science and mathematics acquire this attitude of skepticism. Is the answer reasonable? Is it correct? Is it unique? Are there easier or alternate ways to get the answer? Is there a counter example? These are questions we should always ask, but in the past there was such a heavy price to pay to answer these questions, that students typically became accustomed to accepting as correct the outcome of any computation or argument. Understandably, few of us were willing to verify the correctness of an answer with yet another lengthy calculation.

With Mathematica, on the other hand, it is usually easy to establish the correctness of a solution by an alternate calculation of some kind. Consequently it is now easier to be a skeptic, and it is just as important to question the validity of a calculation now as it ever was before. Mathematica, computers, and human beings are all capable of giving inappropriate answers, incomplete answers, misunderstood answers and even wrong answers. Mathematica provides an excellent opportunity for students of science and mathematics to question their work and to acquire, in the process, this all important skeptical attitude at an early stage in their careers.

Computer Preliminaries

Mathematica is available for Macintosh, Windows, and Unix operating systems. Generally speaking, Mathematica needs at least 9 MB RAM to operate, but 12 MB RAM should be set aside for its operation, if it is available. Even more RAM is clearly beneficial. As you begin to demand more of Mathematica, consider the benefits of giving Mathematica more RAM to operate, if it is available on your computer. A color monitor is obviously nice to have, especially for graphic displays, but a color monitor is not necessary.

This manual is written for **Mathematica 3.0 for the MacIntosh**. Mathematica for the **Windows operating system is similar enough** that very little difficulty should be experienced in modifying this manual to the Windows operating system. Rather than risk confusion, however, by discussing both operating systems, this manual is written entirely in terms of just one operating system.

We assume a passing familiarity with a computer using one of these operating systems. Excellent on-line tutorials are available on most computers. Two hours or so spent on such a tutorial should be sufficient preparation for anyone with no computing experience. In particular, we assume that the reader knows how to: 1) open up applications and files, 2) create folders and files and save files, 3) select, click, double click, and drag, 4) use menu bars.

Calculus and Mathematica

This manual is not meant to be a self-contained calculus text, but rather a supplement to a regular calculus text. Topics are covered in a way which is meant to parallel the sequence of topics in a fairly typical calculus text. New Mathematica commands are discussed as they arise in the process of solving problems in calculus.

In chapter 2, the primary focus of attention is to learn basic Mathematica, by using it to do problems in algebra, trigonometry, and introductory differential calculus. Starting with the integral in Chapter 3, emphasis shifts back to the study of calculus, where it remains for the rest of the manual. As you will see, Mathematica, as involved as it may be, is also straightforward enough that it will come to be understood quite naturally, while attention is paid primarily to calculus. It takes some time, however, to adjust to this merger of mathematics, computers, and Mathematica. Chapter 2 is meant to nudge us gently in this direction. Introductory differentiation is well suited for more intense Mathematica activity, but the opportunity to use Mathematica in this way will come soon enough.

A practice emerging at many colleges and universities is to postpone the use of a Computer Algebra System until the second calculus course, and this manual is ideally suited for such a program. Chapter 2 provides a quick tour through the usual topics of elementary differentiation, and so it can be used as a review while learning basic Mathematica. This chapter is important as a vehicle to learn the fundamentals, and it should be covered by everyone with no previous Mathematica experience. This chapter also provides an opportunity to get used to the practice of thinking about mathematics with a keyboard and monitor rather than with pencil and paper.

Of course, the best way to learn Mathematica is not to read about it, but to use it, and so, to a certain extent, this manual is a list of problems. Many of the problems at the end of each section are meant to provide routine experience in using Mathematica commands. **Routine problems from your main calculus text can also be used as Mathematica practice problems**. You can, with such problems, **look up their answers in the appendix of your main text** to verify that Mathematica commands are being used correctly.

Problems are, however, also designed to foster an attitude of skepticism and experimentation. The importance of adopting a skeptical scientific attitude has already been discussed.

Answers are, by design, not provided in the back. **How do you know** that you have the correct answer? **Is there another** solution to that equation which is being solved? Is there an interesting feature to a graph under consideration which is **too small to be seen** in a window or which occurs outside the window being used? Problems in this manual sometimes create **unexpected, incomplete or occasionally wrong answers**. In the first few problem sets, you will usually be alerted to look for unexpected results, but eventually, such warnings will not be supplied. **Whenever possible or appropriate, you should supply evidence that your answer is correct**. One of the goals of this manual is to develop a good, skeptical, scientific attitude.

Mathematica can be used to gain a deeper understanding of calculus, by focusing complete attention on an issue of calculus rather than on a computation. Some of the problems are designed with this in mind.

Without question, Mathematica can be used to enhance problem solving skills. Just imagine the creative freedom that you will have, when you can focus all of your energy on ideas rather than computations. Mathematica makes more substantial and interesting mathematical problems accessible, and this may be its most important contribution to mathematical education. One should strive to get as much experience as possible in doing problems of this sort. Problems designed to improved problem solving skills are included in the exercise sets along with all of the other problems.

Many of the exercise sets have additional problems, labeled **"Projects"**., which are somewhat more involved. They range in difficulty from being just longer and more interesting versions of ordinary problems, to being quite difficult. They should be accessible without outside background reading. These problems are designed to enhance problem solving skills, by making use of not only current topics under discussion, but, occasionally, a wide variety of previously discussed topics as well. At least some of them are presented in a playful way and are meant to be enjoyed as well as to be instructive.

These projects, however, should be tackled with some discretion as well. Using Mathematica for the first few times is an interesting, but very different way of doing mathematics, and it takes some time to get accustomed to it and to take advantage of all the opportunities Mathematica presents to the user. When you are ready to take on the issue of "putting it all together," you are encouraged to work on the projects that appear after the exercise sets at the ends of the chapters.

Some Closing Thoughts

It is the nature of computer languages that certain words, phrases, sentences, and punctuations have to be read quite carefully, symbol by symbol. We have made an attempt to emphasize all such structures with **bold face** type. **When you encounter these items, read them with great care**. In addition, whenever a Mathematica command is introduced for the first time, it also appears in bold face type.

Explanations of new commands are given just once, when they are first introduced, so if chapters or sections are skipped, it would help to skim over the material that was skipped, looking for, and reading the material on the introduction of new commands. Since new commands appear in bold face type when they are first introduced, this should be relatively painless.

While Mathematica is fairly easy to learn, it is a wide ranging language with hundreds of commands effecting many subjects of mathematics. No attempt has been made to make this manual a complete study of Mathematica. **You are encouraged to explore on your own and to seek help frequently with Mathematica's on-line help**. Adopt the attitude that **any "mathematical or logical word" under consideration is a word that can be found in some form, somewhere in Mathematica**.

Exercises should be presented in an organized, thoughtful and readable manner. Remember that **Mathematica is a very good word processor**, and it should be used frequently in the exercises. Projects, in particular, should be done with great care paid to presentation. Think of a project as a term paper or as a report which is going to your supervisor at work. The same matters of presentation should play a role in creating a paper on any subject—including mathematics.

A pencil and paper strategy session can be a useful way to start a Mathematica work session. It should be a strategy session, however, and not a complete pencil and paper solution, unless, of course, such a solution is desired. Remember that Mathematica will do all of the computations for you. Think of a pencil and paper strategy session as a flow chart, where steps are organized, notation is devised, etc.

And finally, above all, remember that **Mathematica was created by human beings**. Human beings make mistakes. Mathematica makes mistakes, infrequently, perhaps, but mistakes nevertheless. The same is true for all other software packages. Mathematica is under continuous improvement, but it will always have some potential for error. Human beings are, of course, a more common source of mistakes. A small mistake in an input statement can have enormous consequences. Even without an input line mistake, a Mathematica computation can be misinterpreted or used improperly by us with grave consequences.

Work with Mathematica should be a partnership between a human being and a machine, not a thoughtless ride on a machine. The way to avoid wrong answers is to know mathematics well enough to see the warning signs when mistakes have been made or when Mathematica is misbehaving. **Make it a practice to verify that answers are correct, especially if an answer "looks" doubtful**. Frequently, answers can be verified very quickly.

In order to keep the main story line simple, brief, and readable, this manual will not always participate in this practice of verifying answers. Make it a habit, when reading this manual, and when doing your own work to **check Mathematica's performance**.

——— ⋆ ⋆ ⋆ ———

Solving problems with Mathematica, especially more involved problems, can be a rich and rewarding mathematical experience. I hope that this manual and its problems meet with your approval. My e-mail address is included below, because good text books are a community effort, and your comments and suggestions for improvements would be greatly appreciated.

John T. Gresser
Department of Mathematics and Statistics
Bowling Green State University
Bowling Green, OH 43403
jgresse@math.bgsu.edu

Contents

To The Teacher

A Mathematica Approach to Calculus is intended to be used as a supplement to a main text. There is a significant change in philosophy between Chapter 2 (basic Mathematica learned by a quick tour through introductory differentiation) and the rest of the manual where the focus is more on calculus than on software. This matter was discussed in the Preface as well.

Whether Mathematica is introduced in the first course or the second, students still need time to adjust to the rigors of merging mathematics, computers, and Mathematica. There are certain advantages to postponing the introduction of Mathematica until the second course (a more mature audience is one reason), and this supplement is ideally suited for such courses. By working their way through Chapter 2, students can quickly learn basic Mathematica and adjust to this computer environment, by reviewing familiar topics from the previous course. While students work on Chapter 2, somewhat on their own, a regular second calculus course can proceed almost at a normal pace. In a fairly short amount of time, students will be ready to use Mathematica in their second course activity.

If Mathematica is introduced in the first course, then this adjustment period must still be built into a course. One way to do this is to depend initially on a more traditional development of calculus, and then slowly introduce Mathematica into the course as it progresses. For such a program, Chapter 2 may well be an acceptable way to start a Mathematica based calculus program. Certainly it lacks an in depth study of the elementary properties of the derivative, but it allows calculus to proceed at a fairly normal pace, and it gives students time to adapt to a computer based program.

The mathematical community has responded to the issue of calculus reform in a variety of ways. The reform process is still evolving, but it does not appear to be evolving towards one uniform way to teach calculus. Nevertheless, it is hard to deny that computers could (or should) play at least some minimal role in calculus, if not a maximal role, in all of these schemes. Time, unfortunately, is a major player in this game.

Calculus classes have typically been packed with so much material that there has been little or no room left to add new material. If a computer laboratory component is added to a calculus class, certainly classroom time has to be made available to deal with this addition. It is probably true that some topics have to be dropped, or the level of expectation has to be relaxed on some topics, in order to make room for the additional demands made by computer activity. Therein lies the problem. How do we make time for computer work in calculus?

Time plays a role in another way as well. Mathematicians are extremely busy people. Adding a computer component to a calculus class means taking time out of an already packed work schedule to learn new software, and more important, to learn how to use it in a class room setting. The influence of a computer can produce an overwhelming change in the way mathematics is presented, and this can create a demand for more time than a teacher is able to provide for a course. How are mathematicians suppose to make time for all of this change?

A software manual may supply all of the necessary computer background, and a main text may supply all of the pertinent mathematical material. If, however, there is nothing connecting the two books, then a teacher has to spend significant class room time and energy to explain the connection. A student may ask, "Why did I do this problem on a computer?

I pushed a button and got this answer. Now, what's the point?"

This manual will certainly not eliminate these problems, but, hopefully, it can soften them somewhat. I believe that it is possible to include a fairly significant computer component to a calculus class with only minor alterations to whatever method is being used to teach calculus. More dramatic reform might be encouraged by some, but then again, each strategy has its advantages and disadvantages.

This manual is mainly about calculus. It contains a fair amount of intuitive—hopefully readable, and student oriented—mathematical material in addition to Mathematica material. These same ideas may well appear in a main calculus text, but they also appear in the manual to provide a connection between mathematics and Mathematica. Student questions such as those raised above may be answered by reading the manual. If so, then a teacher can spend most of the class room time on strictly mathematical ideas. (Mathematica, however, frequently offers a slick way to present an idea.) The computer component of the course can then be handled with a command to, "Read the manual, and do problems,...."

Some classroom lab time is probably essential. At my home institution, Maple is used in calculus, rather than Mathematica, but the situation would be much the same, if Mathematica were used instead. In my 5 credit (introductory Maple) second calculus course, the first 3 days of the semester are spent in the lab devoted exclusively to Maple. This is a very intense time for my students, but they are fresh, they are not yet burdened by the demands of their other courses, and a great deal of Maple is learned in three days. After that, classroom time is devoted almost entirely to mathematics, with slightly less than one day per week spent in the lab for the rest of the semester. Maple frequently enters into classroom discussion, but usually in a mathematical way.

In my 3 credit follow-up Maple based, multivariate calculus course, no class room time is spent in the lab. My classroom activity, is almost exclusively mathematical. Maple issues are almost always discussed in some mathematical context. Students are told to read the manual, and to do certain problems. I make myself available on a regular basis for lab activity outside of class.

I confess that my students get frustrated and need human encouragement. Much of this frustration is a communication problem they have with computers. Human beings can understand content, sometimes in spite of what is said to them. A computer (lacking a human ability to interpret) can only understand exactly what it is given. But this is a plus, is it not? What a great way to enforce precise thinking and communication!

Except for the computer component, calculus at my home institution is taught in a very traditional way. A traditional book (read big) is used, and this approach has left a clear impact on the structure of this manual. As you can see, topics are covered in much the same way they would be covered in a traditional calculus course.

In spite of the traditional bent, this manual should also be usable in courses which are more reform oriented. Computers are, after all, one of the principal features of this movement. As a supplement, this manual can simply be regarded as a list of topics and problems, and so the choice of topics and their order of presentation does not have to be strictly adhered to.

In fact, regardless of the main text you use, and whether your approach to teaching calculus is traditional or more reform oriented, you may wish to avoid a cover to cover study of this manual. A cover to cover study would be nice, but it would require somewhat of a commitment to the Mathematica based approach. If material is skipped, new Mathematica commands that are introduced in the skipped material will be missed. It is reasonably important to become acquainted with most of these commands, even if material is skipped, because they are introduced, in detail, only once. New Mathematica commands appear in bold face type, when they are first introduced, so it should be easy to skim through the skipped material and pick up the new commands. If a command is missed in this way, it will be noticed later on, when the command is used again. If the command use is not clear from context, it should be a simple matter to find the introduction of the command in the missed section, or look it up in Mathematica's on-line help file.

In my courses, students need to spend at least two or three hours per week outside of class in the lab in order to finish lab assignments. They are permitted (almost encouraged) to discuss mathematics with fellow class members. So much learning takes place during these sessions, that I am only slightly concerned about how much of their work is entirely their own. In order to make their hard work a bit more tolerable, I assign, whenever possible, the problems in this manual that have been written in a playful way. Students are quick to pick up the playful spirit, and grading their lab assignments can be an enjoyably funny experience.

Lab assignments are turned in over the campus network and graded electronically. It takes a while to adjust to grading in this way, but there are some real advantages. My grading practices with Maple could be used with Mathamatica in much the same way. I would open a blank Notebook to use as my comment file, and then open student lab assignments, one at a time. (It helps to change the font setting on this comment-notebook to produce text that looks very different from the fonts used on student lab assignments.) I write all of my comments on this comment-notebook and then paste comments onto student lab assignments. Because of the similarity of student mistakes, it takes little time before the same comments are being copied and pasted into several different student work files, and this speeds up the grading process considerably. This practice also encourages detailed comments, since I know they only have to be typed once.

I hope this manual becomes a useful addition to your calculus program. I have included my e-mail address in the Preface. Your comments would be appreciated.

Chapter 1

Some Preliminaries

The Mathematica application consists of two parts, the "Front End," and the "Kernel," each of which is loaded separately. The "Front End" is our visible contact with the program. The "Kernel" is the invisible background engine that drives all of the computations. When Mathematica is opened, the top of your computer screen will be filled with menu items. Some will be discussed later in this chapter, but they are too numerous to discuss fully, and for the most part they will become a part of your routine as you spend time with Mathematica and experiment with the menu accessories.

This chapter contains critical initial details about using Mathematica that are important to become aware of at an early stage of development. Try to pick up as many of the key points as possible. Our first objective is to learn how to move around in Mathematica's environment and how to enter and process information. Another critical matter involves the way information is processed on a computer screen and the way previously saved files are loaded and processed. Along the way, some of the menu bar items will be discussed, and we will finish the chapter with an introduction to Mathematica's on-line help. The **Help Menu will be a fundamental tool in our discovery of Mathematica**.

1.1 Mathematica Notebooks

A Mathematica **work session** is defined to be the activity that takes place between the time that the Mathematica application is opened up and the very next time it is terminated. It is sometimes useful to quit Mathematica, and then immediately open it up again to continue working on the same document. Occasionally, we are forced to do this. Thus, one could sit at a computer terminal and pass through several work sessions, while working on the same document.

Double clicking on the Mathematica icon or on a previously saved Mathematica file, will load the "Front End," producing a new, blank work sheet or the contents of the old file. These work sheets or files are called **notebooks**.

A notebook has the appearance and essential functionality of a typical word processor (and much more). If Mathematica is activated by opening up a previously saved file, then all of the information from that file will appear on the computer screen after the front end is loaded. Actually, this information appears on the computer screen, but it has **not yet been entered into Mathematica's computational environment—the "Kernel"**. (More on this important point later.) A notebook consists of three types of cells, **input cells, output cells and text cells**. There are obvious visible differences between the three types of cells that will be apparent as soon as information is entered onto the notebook.

At this point, open up the Mathematica application, by double clicking on the Mathematica icon, to create a new, blank notebook. The purpose of this exercise is not to do any mathematics, but rather to experience, first hand, some of the idiosyncrasies of Mathematica that may affect your work rather quickly. Move the pointer around. It's **"horizontal"** appearance indicates that it is **in a region where a cell type has not yet been declared.** Press any key on the key board (even the space bar) and a cell type will automatically be selected. Notice that the pointer **becomes a vertical blinking cursor**. Unless other arrangements were already in place, **the new cell will be an input cell by default.** As soon as a cell type is declared, a **right handed square bracket will appear at the far right side of the window**. These brackets serve as markers for individual cells (of any type). As work on a notebook evolves, a hierarchy of these brackets will appear to group cells together in various ways. Brackets can be used to organize a notebook into sections, subsections, subsubsections, etc. The word "notebook," in fact, comes from this feature.

An input cell was created by default, so type something clever, like (**2+3**) and hit the **enter key, (not the return key)**. Mathematica **loads its Kernel** when the **Enter Key** is pressed for the **first time** and so this first computation **takes a long time**. Eventually, Mathematica computes and displays the answer in a new, different type of cell, called an **output cell**. Something else happens, as well. Mathematica labels the two cells as **In[1]** and **Out[1]**. These labels can be used later in the work session to refer to particular cells, but be careful! Labels can be somewhat short-lived (More on this later). Finally, Mathematica runs a separator line across the notebook at the end of the current output cell. We're ready for the next input statement. Type (**b=17**) and hit the enter key again. Follow this by entering, in the same way, (**b=8*A**). With this, we have enough on our computer screen to discuss Mathematica's behavior.

In the current Mathematica work session, the letter b **now has a value of** $8A$. To see this, type (**b**) and press the enter key. The **new statement is automatically entered at the location of the separator line that runs across the page**. Unless we click the pointer somewhere else on the notebook, the separator line will remain at the bottom of our work. Mathematica will respond with "$8A$." Our computer screen should look as follows, except that the **Right hand square brackets** appearing on your monitor are not shown below. **They will be omitted throughout this manual**.

```
In[1]:= 2+3
Out[1]= 5

In[2]:= b=17
Out[2]= 17

In[3]:= b=8*A
Out[3]= 8 A

In[4]:= b
Out[4]= 8 A
```

The letter A has no value—Mathematica treats it as an unknown (**possibly complex valued**) number. The letter b initially had a value of 17, but that was simply erased and replaced by $8A$ after its last assignment. It is important to realize that Mathematica is **case sensitive**. The letters A and a are treated as absolutely different objects by Mathematica.

Move the pointer without clicking around the notebook. Notice how the **pointer switches** between its **vertical** and **horizontal** orientation as we **move into cells** and **move between cells**. If we click, while the pointer is **vertically** oriented, we **move into that cell**, and the pointer becomes a vertical blinking cursor. If we click, while the pointer is **horizontally** oriented, the **separator line moves to that location**, and Mathematica is prepared to **enter a new cell** at that location.

Pay attention to what we do next. Move the mouse pointer back to the input statement In[2] and **click** when the pointer becomes vertical. Press the enter key. Notice that the corresponding output statement disappears and then reappears. The labels In[2], Out[2] also change to In[5], Out[5], but nothing else below this point changes. What is the value of b now? To find out, move the pointer down to the input cell In[4], click and again press the enter key. The last time we checked the value of b it turned out to be $8 * A$. Now we see that its value is 17, and our computer screen looks as follows.

```
In[1]:= 2+3
Out[1]= 5


In[5]:= b=17
Out[5]= 17


In[3]:= b=8*A
Out[3]= 8 A


In[6]:= b
Out[6]= 17
```

This simple example gives us a way to view an important facet of Mathematica's behavior. As you can see, we can move back and forth on a Mathematica worksheet, entering input statements in one order and then another (with or without changes in the input statements). This is a common and effective way of working with Mathematica. When we go back and reactivate an input statement in this way, the corresponding output statement disappears and then reappears (with an appropriate change if the input statement was changed), but nothing below this point appears to change on the computer screen. Of course, Mathematica knows, internally, if changes have been made, and subsequent work will reflect these changes.

In summary, **Mathematica pays no attention to the location of an input statement on a worksheet**. All that matters, is the **sequence in which input statements are entered relative to the passing of time**.

This means that a Mathematica worksheet may not make sense if it is read from top to bottom as a paper document would be read. It would help somewhat to pay attention to the numbers appearing in the labels $In[j]$ and $Out[j]$, but more is needed. If a Mathematica file is being prepared for another reader, it will eventually have to be presented in the order in which statements are entered before it is given to the reader. The usual cut and paste activity can be used to prepare a document in this way. After this, each of the input statements can be entered again from top to bottom to create the finished document.

There is one more type of notebook cell that warrants our attention. In addition to input cells and output cells, we can also **create text cells** in a notebook. Text cells are ignored by Mathematica. Their purpose is to supply the human reader with information that might be useful. Text statements **should be used extensively** in order to motivate, explain what one is doing, and otherwise make a document more readable, interesting, and enjoyable to a reader. It is worth mentioning that Mathematica is, in fact, a very good scientific word processor. We will explore some of Mathematica's tools for creating more involved mathematical text in the exercise sets, but not until later, after we have had time to become familiar with more critical issues.

Move the pointer around the notebook. Notice again how the **pointer switches between its vertical and horizontal orientation as it slips into cells and between cells. Move the pointer between the cells** *Out*[1] and *In*[5] and **click the pointer**. The separator line appears between cells, where the pointer is located. The default cell type is an input cell. If we enter anything from the keyboard, we will immediately create an input cell, so **do not touch the keyboard**. Pull down the **Format Menu**, then pull down the **Styles Menu**, then **click on "Text"**. Now, **be careful**! If you click the mouse pointer, the default input type will return, so **do not click the pointer**. Instead, begin typing the text statement. A text cell is immediately created. (Notice the appearance of a right hand bracket enclosing the cell. A text cell can be created in this way, between any two cells, regardless of their types. After creating a text cell, in a second place, our notebook looks as follows:

```
In[1]:= 2+3
Out[1]= 5
```

This is where we entered the input $b = 17$ a second time.

```
In[5]:= b=17
Out[5]= 17
```

```
In[3]:= b=8*A
```

Text cells can be placed between any two cells, even between an input cell and its corresponding output cell.

```
Out[3]= 8 A
```

```
In[6]:= b
Out[6]= 17
```

The usual menu driven and key board commands can be used to edit, delete, copy or paste statements in input, output, and text cells. The **delete key**, however, **will not eliminate the cell itself**. To **eliminate a cell, use the pointer to click and select the right hand bracket which bounds a cell**. Then pull down the **Edit Menu and click on Clear**. This will delete, not only the contents of a cell, but delete the cell as well. The same procedure can be used to delete a group of cells.

1.2 Opening Old Notebooks

When you work with Mathematica, you will want to save your work, quit and then return at some later date to do additional work on the same file. This can be a puzzling experience unless you understand how Mathematica files are loaded.

In order to experience this first hand, let us save the simple Mathematica file we just created. Files are saved in the same way for all applications. Pull down the File Menu and click on Save. You will be asked to name your file (just call it "junk") and to select a place to put it.

Now quit Mathematica. Don't just close the worksheet, but actually quit the application as well. Then double click on the file we just named "junk," and wait for Mathematica to open. When it opens, notice, to begin with that the input output labels *In*[j] and *Out*[j] are no longer displayed. Mathematica can be configured to display the old input, output labels, but that is not the default configuration. It is probably a good idea that they are not displayed. It should suggest that our notebook is not the same as it was before we quit.

Several other files could be opened at the same time. Without doing so, let us imagine, for the sake of argument, that we have, indeed, opened several files or notebooks. All of our previous work will show on these notebooks. One of these notebooks might say that $b = 17$ and another might say that $b = \sqrt{1 - x^2}$. Which worksheet carries the current value of b, or are both values correct, each on its own worksheet?

The answer is that b has no current value. Test this by placing the pointer somewhere on the last input cell of the file we called "junk," and press enter (or create a new input cell, type (b) and press enter. Mathematica will respond with an answer of b, which simply means that b has no current value.

When we first open a Mathematica file, all of our previous work will show on the computer screen, but **nothing is entered into Mathematica's computational environment—its Kernel**. If there are values on the old worksheet that we still need, in order to continue our work, **all of the appropriate input statements must be reactivated**, by allowing the pointer to pass through each of these input statements (in the appropriate order, of course), and pressing the enter key for each one.

Furthermore, **Mathematica offers us only one computational environment**. We can open up several notebooks, but they must all coexist in the same computational environment. They all use the same Kernel. **There can only be one current value for** b (or any other name, for that matter). If we move the mouse pointer back to the input statement ($b = 8 * A$), click and enter, then the **current value of b is $8A$ on all of the notebooks, regardless of what appears on the notebooks.**

1.3 Frequent Saving

One of the first suggestions made in virtually any computer application is to **save your work frequently**—at least every 10 to 15 minutes. Such a practice will prevent much of your work from being lost if a problem is encountered. Some input mistakes can cause Mathematica to quit unexpectedly. If this happens only a few minutes after saving a file, then very little work will be lost. On the other hand, it is easy to get so absorbed in your work, that an hour or more passes without saving your work. If Mathematica then quits unexpectedly, or you have a computer problem, or a power failure, much of your work will be lost.

1.4 Mathematica's On-Line-Help

This introductory chapter could hardly end without a word about the on-line help which is available from the menu bar or from the keyboard. The help apparatus is excellent! There is little we can offer in the way of guidance, except to say, ..., **use it**! If you are puzzled about how Mathematica is behaving, about how to use a command, if you are searching for a command to use in your work, help is only a click away. Mathematica's help is good enough and easy enough to access, that when help is needed, it might serve as your preferred first line of help, over and above human help. The main purpose of this short section is to entice you to exploring the help file and to offer just a few tips.

Pull down the **Help Menu**, then click on **Help ...** (the first entry in the help menu). When the Help Window appears, click one of the buttons and then click an item in the list. Explore a variety of items in this way. Click a different button and explore again. All of the buttons are worth looking at with the exception of the "Add-ons" button, which will be discussed at a more appropriate time.

Suppose you are in the process of using a Mathematica command, and for what ever reason, you are puzzled by Mathematica's response. You could seek help as we described above, but **there is a more efficient way**. Place the **blinking cursor at the beginning, the end, or somewhere inside the command name**, then Pull down the **Help Menu** and click on **Find in Help ...** (the second item in the help menu). This immediately brings up a help file on the selected word. Notice how the command name (where the cursor is located) is highlighted. Occasionally, Mathematica fails to highlight the whole command name, and this may bring up help on the wrong topic. A more reliable procedure is to **highlight (select) the command name rather than just put the cursor somewhere in the word**, before the Help Menu is pulled down.

If you are unsure, how to spell a command name, or if you have only vague notions about unfamiliar tools that might possibly be useful in solving a problem, it is frequently beneficial to simply **browse through a list of command names** to see if such tools exist and to find their names and spellings. Like browsing in any library, you may or may not find what you want, but you will surely find all sorts of interesting items along the way.

There is another more **systematic way to hunt for ideas, when only vague notions** are available. **The asterisk * is a kind of"wild card"**. It represents **zero or more arbitrary characters**. To look up everything in the help file that **begins** with the letters "Startletters," type

?Startletters*

in a new input cell. To look up everything in the help file that **ends** with the letters "endletters," type

?*endletters

in a new input cell. **Using two question marks (??)** instead of one in such an input statement, brings up a **more detailed help file** on the topic.

All Mathematica names are **capitalized**, which is why we capitalized the letter grouping "Startletters." Furthermore, most, but not all, Mathematica command names are **fully spelled English words**. If two words are involved, like "double," and "name," for example, the Mathematica name is probably going to be "DoubleName," capitalized just like this, with no space between the two words. This makes typing input statements tedious at times, but the advantage is that we already know the spelling of a large number of Mathematica commands.

If you make a **mistake in an input statement**, Mathematica will frequently **beep**. To find out why, pull down the **Help Menu**, and click on **Why the Beep? ...**. In particular, **Mathematica will beep, if you use too many closing parentheses "},],)"**. This can be extremely useful, since even the most experienced Mathematica user can easily make this kind of mistake. Pay attention when Mathematica beeps.

Take some time, especially during your early Mathematica experiences to browse through the help file. Many treasures can be discovered by continuing this practice throughout your work with Mathematica.

The Beginning

There are many other introductory, general features of Mathematica, that we have not discussed. **Sectioning Mathematica notebooks** is not essential, but it can be a real convenience. Pull down the **Cell Menu** and experiment with the **Cell Grouping** submenu if you are interested is sectioning. Mathematica is capable of doing **high level mathematical type setting**, within (unprocessed) **text cells**. This feature can be used to create highly readable notebooks. If you are interested in how to place mathematical symbols within text cells, pull down the **File Menu** and experiment with the **Palettes** submenu.

There are many chapters ahead of us to cover these and other issues, and we are anxious to begin our mathematical adventure. At this point, we can discard the file named Junk that we created earlier. We are ready to experience the power of Mathematica.

Chapter 2

Basic Mathematica

This introductory material may be more readily understood, if you open up the Mathematica application (by double clicking the Mathematica icon) and repeat the same calculations on your own notebook. Mathematica opens up to a blank notebook, ready to accept the first input statement. The default cell is an input cell, so as soon as you begin to type, an input cell is automatically created. Remember that Mathematica's Kernel does not load until the enter key is pressed for the first time. After a lengthy wait for the loading process, the first calculation is performed. With that introduction, we begin our first Mathematica computations.

2.1 Mathematica as a Calculator

As you can see, Mathematica can be an ordinary calculator. Multiplication is denoted by (*) and exponentiation by (^). A blank space can also be used to denote multiplication.

Enter a calculation from the keyboard. As soon as you begin, a blinking cursor appears and an input cell is created. Hit the enter key (**not the return key**), and the calculation is performed and displayed in a new cell—an output cell. Mathematica prepares itself automatically for more input. Type anything from the keyboard and a new input cell is created at the separator line.

```
In[1]:= 7*(8 - 5)^2
Out[1]= 63
```

```
In[2]:= 2*7 + 6/2
Out[2]= 17
```

The **blinking cursor can be anywhere within an input cell**, when the enter key is pressed, activating a calculation. It **does not have to be at the end** of the cell. The pointer, however, **must be a blinking cursor** (or, it's not in the cell). Any input cell can be selected—it does not have to be the last input cell. Just move the pointer to an input cell, click to go inside the cell and hit the enter key.

The return key is only a carriage return key. This is useful. It allows us to break long input lines where ever we choose to make a break. Try using the return key while the blinking cursor is within a cell.

Mathematica follows the usual rules for the order of computations, and parentheses have the same meaning in Mathematica as they do on paper. Negative numbers usually do not have to be enclosed in parentheses, but it is a good practice to use parentheses whenever there is any doubt about whether they are necessary. On the other hand, it is easy to get lost in a calculation when parentheses are used excessively.

Try entering an extra right—)—parenthesis. Mathematica will immediately beep. Pay attention to this. If you don't know why it beeped, pull down the **Help Menu** and click on **Why the beep? ...**. If you enter an extra left—(—parenthesis, Mathematica won't know you made a mistake until after you hit the enter key. You will find out then. With that, we perform a few more calculations, paying particular attention to parentheses.

```
In[3]:= 80*2^-3
Out[3]= 10
```

```
In[4]:= 80*2^(-30)
Out[4]= 10
```

```
In[5]:= 5/(4*6)
```
Out[5]= $\frac{5}{24}$

```
In[6]:= 5/4/6 - 5/4*6
```
Out[6]= $-\frac{175}{24}$

```
In[7]:= 5/(4/6)
```
Out[7]= $\frac{15}{2}$

Notice that the output is in the form of a fraction which has been reduced to lowest terms. The answers are not expressed as decimals. This is characteristic of a symbol processor, and it affects all Mathematica output in a very significant way. **Mathematica will always be exact unless you allow it to approximate answers with a decimal**. In the following, this is demonstrated in several ways.

If we continued to enter statements as we did above, this manual would fill an enormous number of pages. As you can see, **very little horizontal space is being filled**. While that may or may not be acceptable in your own notebook, it is not acceptable in this manual, and so we are about to take measures to **fill up more of the page**.

Curly brackets are used to create ordered lists in Mathematica. **Lists are central objects** that we will devote more time to later. For now we use them only as a devise to fill up the page from left to write.

```
In[8]:= {5/Sqrt[2], 2/Sqrt[2]}
```
Out[8]= $\left\{\frac{5}{\sqrt{2}}, \sqrt{2}\right\}$

```
In[9]:= {Cos[Pi/3], Cos[Pi/4], Cos[Pi/5], Cos[Pi/6], Cos[Pi/7]}
```
Out[9]= $\left\{\frac{1}{2}, \frac{1}{\sqrt{2}}, \frac{1}{4}\left(1+\sqrt{5}\right), \frac{\sqrt{3}}{2}, \cos\left[\frac{\pi}{7}\right]\right\}$

```
In[10]:= Cos[Pi/12]
```
Out[10]= $\frac{1+\sqrt{3}}{2\sqrt{2}}$

```
In[11]:= {Cos[3], 2/Sqrt[2]}
```
Out[11]= $\{\cos[3], \sqrt{2}\}$

Notice that the mathematical constant π **is spelled Pi. The letter P must be capitalized.**

Almost everything in Mathematica is **capitalized**. In particular the square root function is capitalized, and the trigonometric functions are capitalized. Something else, characteristic of Mathematica, is exhibited in the work space above. The **arguments of all commands and all functions** in Mathematica are enclosed in **square brackets**. Finally, notice that the **arguments of the trigonometric functions** are assumed to be in **radian measure**.

As you can see above, Mathematica will carry out as much of the computation as it can under the constraint that the output must represent an exact answer. The value of $\cos(\pi/4)$ is irrational, but algebraic, and so its value is represented algebraically. Actually, it turns out, that the value of $\cos(\pi/n)$ is algebraic for every integer n, but these algebraic expressions can be quite complicated. Notice that Mathematica is not able to express the answer to $\cos(\pi/7)$ algebraically. The value of $\cos(3)$, on the other hand, is transcendental. It is not only irrational, but nonalgebraic as well, and so Mathematica has no other way of expressing its value exactly, except as $\cos(3)$. Initially, this realization may be disconcerting, but eventually, this will be realized as a powerful feature of Mathematica. Decimal numbers and approximations play an important role in mathematics, and as we will see, in Mathematica as well, but mathematics is also a study of symbols and the exact relationships among its symbols.

We will return in a moment to discuss the mechanism for triggering decimal approximations, but first, let us take a mathematical break and discuss some of the more basic software issues that were raised in the preliminary chapter.

At some point in time you will want to save your work and quit Mathematica. Saving a Mathematica file is the same as saving any other file in the Macintosh operating system. If you need help, consult the manual that came with your computer. **When you save a file, be sure to give it a name that is peculiar to just you, so that your file does not have the same name as someone else's file**. Even if you do not plan to quit, **your work should be saved frequently**, so that you have a copy, if something should go wrong with the software or hardware. This is a common practice that is used with all software **to avoid losing work**. After you have done some work, save your file, and quit Mathematica by pulling down the **File Menu** and clicking on **Quit**.

If you have not yet read Chapter 1 carefully, it would be a good idea to study that chapter, before you resume your work with Mathematica. Working with old Mathematica files can be a confusing experience unless you understand how Mathematica files are activated. Pay particular attention to Section 1.2. When you are ready to resume your work, open your previously saved file, by double clicking on its icon. The next topic is decimal approximations.

There are two ways to get Mathematica to approximate answers as decimals. Most of the time we will use a command based approach. If a is a real (or complex) number, then **N[*a*]** is the decimal value of a computed with Mathematica's machine precision. If j is a positive integer, then **N[*a*, *j*]** computes the decimal expansion of a with j digits of accuracy.

The other way to generate decimal output is to have a decimal number appear somewhere in the input statement. We show this approach first. Mathematica makes the reasonable assumption that even <u>one</u> decimal number in an expression compromises the integrity of a pure symbolic answer, and so there is no significance represented by the answer other than as a decimal approximation. Compare the answers above with the following.

```
In[12]:= {Cos[3.], 2/Sqrt[2.]}
Out[12]= {-0.989992, 1.41421}
```

```
In[13]:= N[2/Sqrt[2]]
```

```
Out[13]= 1.41421
```

```
In[14]:= N[Cos[3], 30]
Out[14]= -0.989992496600445457271572794 73
```

Notice above, that Mathematica will **compute with as many digits of accuracy** as you wish. Machine precision is typically between 16 and 19 digits, and there are commands which give the user flexible control over the accuracy of a calculation. If you are interested in this matter, look up the commands **SetPrecision** and **SetAccuracy** in the Help File. The number of significant digits in a decimal calculation, however, is a subtle and complicated matter, and we shall not discuss it at this time.

The **symbol N**, can also be used as **a postfix symbol** to initiate decimal computations. It is absolutely equivalent to the more standard command based form, N[ ], but sometimes it looks nicer or is more convenient to use.

```
In[15]:= 5Sqrt[3]//N
Out[15]= 8.66025
```

Something else in this computation is worth discussing. Notice that Mathematica interprets the **operation between the number** 5 **and** $\sqrt{3}$ **as multiplication**, even though there is **no asterisk (*), nor blank space between the numbers**. Multiplication can be entered in this way **as long as the the term on the left is a number and the term on the right is a letter or name**, so that the expression looks acceptable in a kind of "pencil and paper" way. If x and y are variables, Mathematica interprets xy as just another name, having no connection to either x or y. Likewise, the name $x3$ would be regarded as another name with no relationship to x. The expression $3x$, on the other hand, is interpreted as the product, $3 * x$, of 3 and x.

The Mathematica name **Degree evaluates to $\pi/180$**. This is the multiplier which converts degree measure into radian measure. We compare the decimal values of "Degree" and $\pi/180$ below. The expression $90Degree$ is simply the product of 90 and "Degree." It provides a straight forward way of converting degree measure to radian measure.

```
In[16]:= {N[Pi/180],N[Degree]}
Out[16]= {0.0174533, 0.0174533}
```

```
In[17]:= Sin[90Degree]
Out[17]= 1
```

We continue our discussion about Mathematica's role as a calculator by displaying a few more calculations. Some of them may surprise you. Remember that Mathematica will always be exact unless you allow it to approximate.

```
In[18]:= 2^50
Out[18]= 1125899906842624
```

```
In[19]:= 80!
Out[19]= 715694570462638022948115337231865321655846573423657525
           771094450582270392554801488426689448672808140800000000
```

These are not decimal approximations. We asked for exact calculations, not decimals.

```
In[20]:= Abs[-7]
Out[20]= 7
```

```
In[21]:= Abs[Sin[4]]
Out[21]= - sin[4]
```

```
In[22]:= (17 + 2*5)^(1/3)/(32^(1/5) + 1)^2
```
Out[22]= $\frac{1}{3}$

Mathematica will compute and display complex valued answers just as it does real valued answers. The **symbol "I"**(expressed as a capital letter only) is reserved by Mathematica to represent the **imaginary number** $\sqrt{-1}$, and it should not be used for any other purpose. **A capital letter I must be used in an input statement, but Mathematica displays a small case letter *i* in the output statement.**

```
In[23]:= Sqrt[-2]
```
Out[23]= $i\sqrt{2}$

The commands **Simplify[ ]** and **Expand[ ]**can be used in a variety of situations. To find out more about them, **consult Mathematica's on-line Help File**. If the cursor is at the beginning, the end, or somewhere inside the the word "Simplify," or "Expand," just pull down the Help Menu and click on **Find in Help ....** Try this now and explore the consequences.

We use the simplify and expand commands in the next calculation. The results are unsuccessful, but they seem like such reasonable commands to use. Actually, this next calculation demonstrates **really surprising behavior on Mathematica's part**. It turns out that Mathematica has good reason for its behavior, and this matter definitely **deserves our attention**.

```
In[24]:= a = (-27)^(1/3)
```
Out[24]= $3\ (-1)^{1/3}$

```
In[25]:= {Expand[a], Simplify[a]}
```
Out[25]= $\{3\ (-1)^{1/3}, 3\ (-1)^{1/3}\}$

```
In[26]:= N[a]
Out[26]= 1.5 + 2.59808 i
```

```
In[27]:= b = ComplexExpand[a]
```
Out[27]= $\frac{3}{2} + \frac{3\ i\ \sqrt{3}}{2}$

Instead of the obvious real valued answer of $b = -3$, Mathematica has given us a complex number. Is this answer even correct? In response to this question, we ask Mathematica to compute b^3. If the above answer is correct, then we should get $b^3 = -27$. This time we are more successful with the command Expand[ ].

```
In[28]:= b^3
```
Out[28]= $\left(\frac{3}{2} + \frac{3\ i\ \sqrt{3}}{2}\right)^3$

The symbol % represents the last output.

```
In[29]:= Expand[%]
Out[29]= -27
```

Evidently, the above answer is correct, but it is certainly a disappointment. As long as we plan to live in a real valued world we should still insist that -3 is the only acceptable answer for the cubed root of -27. Mathematica, on the other hand, is quite inflexible about this matter and will refuse to grant us our request. It turns out that Mathematica has good reason for its behavior, although it is rather hard to explain why until after the completion of a more advanced course in complex variables. It turns out that if a formula was used to create nice clean real valued roots of negative numbers, that same formula would cause major mistakes to be made in a broad range of other mathematical computations. All things considered, it is better to accept Mathematica's awkward evaluation for roots of negative numbers.

Mathematica treats all **roots of negative numbers** in a similar fashion. **They will always be complex valued.** An even root of a negative number should be complex valued, but surely we want an odd root of a negative number to be real (and negative), even though there are complex valued roots as well. One way to get Mathematica to cooperate is to make it a practice of using $-(-x)^{1/n}$ instead of $x^{1/n}$ in our Mathematica calculations whenever x is negative and n is an odd integer. This simple solution, however, will not always be enough. We will discuss other ways of extracting real roots when the need arises.

This feature of Mathematica is not altogether bad. After all, we know that there are quite a few symbolic complications to working with roots of negative numbers. The usual rules of manipulating exponents frequently fail for negative bases, even with pencil and paper techniques. Mathematica's behavior here should serve as an warning to us, that we need to exert more caution in dealing with roots of negative numbers.

A new symbol in Mathematica's library was introduced above. **The last output statement is represented by the symbol %. Additionally, %% evaluates to the second last output statement and %*n* evaluates to the output cell *Out*[*n*].** This is a very useful way to refer to previous output statements, but it must be used with caution. If we decide to move around a notebook and re-enter old input statements, then the value of $\%n$ may change. There are other more permanent ways of naming cells. In the above work, we let $a = (-27)^{1/3}$. This is a slightly more permanent way of naming a cell. We will talk more about this matter shortly.

2.2 Assigned and Unassigned Names

A **name** or **string** in Mathematica is any letter followed by zero or more letters, digits, and underscores (lower and upper case letters are distinct). Spaces are not allowed; an underscore should be used instead. The maximum length of a name is big enough that we do not have to be concerned about it. If names are unassigned, they are simply treated as variables or unknown real (or complex) numbers.

To assign a value to a name we use the **assignment operator (=)**. Absolutely **anything in Mathematica can be represented by a name**. Using names is a real convenience, since complicated statements can be referred to again by simply entering their names. Equally important, carefully chosen names, carry (human) meaning, and this can help to motivate what we do, making our work easier to create and easier to explain. As a consequence **names should be used frequently**. All of the names reserved by Mathematica are capitalized. In order to avoid any conflict with names reserved by Mathematica, it is a good idea (although it is not required) to begin user-defined names with lower case letters. If you try to use a name reserved by Mathematica, your assignment will be disallowed, and a warning message will offer an explanation.

There is more to the assignment operator (=) then may be evident at first glance. Suppose a and b are assigned or unassigned names, or expressions involving other assigned or unassigned names. When Mathematica reads the input statement $a = b$, it fully evaluates the right hand side and leaves the left hand side unevaluated. The fully evaluated right hand side is then assigned to the unevaluated name a, replacing any previous value of a, if there was one, in the process. In particular, **a name does not have to be unassigned, before a new value is assigned to it**. The name b **can even include the name** a, although this **must be done cautiously**. If a **already has a value**, then an assignment such as

$a = a + 1$

is quite acceptable. If the previous value of a was 10, then the right hand side evaluates to 11, and this becomes the new value of a. **If** a **were unassigned**, however, then in its attempt to fully evaluate the right hand side, Mathematica would look for, and find, a value for a back in the same statement. This forces Mathematica into **an infinite loop**, something to be avoided.

```
In[1]:= (p12q + 2*x + 7*george)^2
Out[1]= (7 george+p12q+2 x)^2


In[2]:= poly = x^2 + 5*x - 9
Out[2]= -9 + 5 x + x^2


In[3]:= a = (6 + 8)^2/4
Out[3]= 49
```

Since the **equality symbol (=)** is used for the **assignment operator**, something else must be used for the **equals symbol** that we are accustomed to seeing in **equations**. To denote the **equals** symbol in an **equation**, Mathematica uses a **double equality (==)** symbol. The distinction between (=) and (==) is very significant. **Using (=) in an equation is never appropriate**.

```
In[4]:= eq = x == 4
Out[4]= x == 4
```

To see the significance of these assignments, notice the following.

```
In[5]:= eq
Out[5]= x == 4
```

```
In[6]:= Sqrt[a*poly]/25
```

Out[6]= $\frac{7}{25}\sqrt{-9+5\ x+x^2}$

```
In[7]:= x
Out[7]= x
```

Here is an interesting point. Notice that x **is unassigned**. It does not have the value of 4. The equation $x == 4$ and the assignment $x = 4$, are treated very differently by Mathematica.

As you proceed with a Mathematica work session, you are likely to encounter situations where you **attempt to use a letter or name as a variable (an unassigned name) only to discover that you assigned some value to that letter or name in some**

previous problem. The most direct way to **return a name to its unassigned status** is to enter the input statement **name=.** (Notice the period). For example, the name "poly" was assigned to a quadratic above. To return this to its unassigned status, enter the following. This is followed by an evaluation of "poly," just to demonstrate that it no longer has a value.

```
In[8]:= poly =.
```

```
In[9]:= poly
Out[9]= poly
```

To return **several user defined names** to their **unassigned status** at the same time we use the **Clear[ ] command**. The **next** input statement **clears the Kernel of all user defined names**.

```
In[10]:= Clear["Global`*"]
```

This puzzling sequence of symbols needs to be entered in exactly this way. To explain why, so that it is easy to remember, we must introduce the notion of "context." We stray, briefly, from the main theme of this section, to discuss this new idea.

Everything in Mathematica has a "context" associated with it. Context is a way of grouping together names which belong in the same "family." There may be several people named Ralph in a neighborhood, but by including a family name along with Ralph, we identify a unique person. In much the same way, there may be several objects in Mathematica with the same name, but as long as we include the "context" of a name, we identify a unique object.

All of the built-in objects in Mathematica (available when the Kernel is loaded) **have the context**, called **System`**. Notice the **back quote.** Every context **ends with a back quote**. Thus, for example, the **full name** of the Simplify[ ] command is **System`Simplify**. All **user defined symbols and names** carry the context **Global`**. Before we unassigned the name "poly" in the above work space, its full name was Global`poly.

We mentioned on page 6, that the asterisk (*) represents a "wild card," or more specifically, zero or more unspecified characters. It follows that **Global`*** represents any user defined name. Finally, **Double quotes are used to enclose text, which forms a "string"**. Notice, that double quotes were used in the input statement Clear["Global`*"] above to first turn Global`* into a string.

With that background, the above input statement should make sense, and the sequence of symbols used in the statement should make sense as well. The input statement **Clear["Global`*"] should be used frequently to clear away all user defined names.**

This combination of the Global` context and the asterisk (*) can be used in another very interesting way. We have already cleared our Kernel of all user defined names, so it is too late to show the next output statement, but it is worth point out, that the input statement

?"Global`*"

will return a list of all current user defined names. See page 6 on the use of the question mark (?) to engage the Help File.

Another way to return all user defined names to their unassigned status is to **quit Mathematica and then reopen the same notebook**. If memory is in short supply, this has the added advantage of recovering spent memory set aside for computations. This is a drastic step to take, however, since it takes a long time to restart. Once a file is opened up again, its contents will be visible on the screen, but nothing will be entered into Mathematica's computational environment. This **important issue** was raised in Chapter

1. It may help to read carefully the appropriate material in Chapter 1 at this time. The critical input statements will have to be reactivated before they can be used again.

When we use the **assignment operator (=)** to assign a value to a name, let's say A, any **previous value A** may have had is **simply dropped and replaced by the new value**. Actually, this **doesn't always happen**. Everything in Mathematica has a **"Tag"** associated with it. Tags are used to classify objects in Mathematica into different types of objects. While that is hardly an adequate explanation, it is sufficient for our purpose. We bring this matter up only to mention that occasionally, when assignments are made, there is a **clash of Tags**, and as a consequence, Mathematica **prevents the assignment**. We will be warned when this happens, and it is no cause for alarm. If Mathematica prevents an assignment for the name A, because its Tag is protected, simply use the input statement **($A = .$) to first return A to its unassigned status**. The name A will then be free for further assignment.

2.3 Functions and Expressions

Before we start this section, an unrelated Mathematica-input idea is introduced. We have been using curly brackets as a means of packing several input statements in one input cell and displaying all of the results in one output cell.

Frequently, we wish to perform a **sequence of computations** in a certain order. We could enter them in separate input cells, in which case we would see all of the intermediate results. If we **only want to see the final result**, we can put all of the input statements **into one input cell, by separating them with semicolons (;)**. **Only the last computation is displayed**. A **semicolon (;)** can also be used at the end of **any input cell** to perform a computation and **suppress the output display**.

```
In[1]:= y=3x;p=y/5;q=p+2
             3 x
Out[1]= 2 + ---
              5
```

The calculations are performed sequentially, from left to right, but only the last result from above is displayed.

```
In[2]:= r=p^2;
```

Because of the semicolon, $r = p^2$ is computed, but not displayed. To finish the example, we unassign the names.

```
In[3]:= y=.;p=.;q=.;r=.;
```

In the pencil and paper world of mathematics, functions, or expressions are described in two different ways. We write equations like $y = 5x^2 + 1$, for example, and say that y is a function of x, or we define f as a function of x, by a statement like $f(x) = 5x^2 + 1$. Both of these methods are available in Mathematica. The first approach is straightforward enough.

```
In[4]:= y=5x^2+1
                2
Out[4]= 1 + 5 x
```

Actually, this is slightly different from what we encounter in a pencil and paper world, where we are accustomed to viewing $y = 5x^2 + 1$ as an **equation**. This is not what Mathematica gives us. We enter y in an input cell, and $5x^2 + 1$ appears in an output cell. Consequently, we **cannot**, for example, **solve for x in terms of y**, with this form. Still, this is a common and convenient way to describe the above correspondence. Actually, as we shall see in a moment, this is **not a "function" in the strict Mathematica sense** of the word. What we have above is simply an **expression in x**.

An equation is often a awkward way to access the idea of a correspondence. Unless there is a specific reason for wanting an equation, this approach is not advised. There will be times, however, when we may want to enter this correspondence as an equation, rather than as an expression—we just mentioned one reason above. Remember that the "equals" symbol in an equation is denoted by (==).

```
In[5]:= y=.;eq=y==5x^2+1
Out[5]= y == 1 + 5 x^2
```

Notice that we had to first unassign $y = 5x^2+1$, otherwise, Mathematica would have simply responded with the output statement, *true*, implying that $5x^2+1$ is indeed equal to $5x^2+1$.

Finally, we must always remember, that if x ever picks up a value, intentionally, or perhaps, from a previous problem, the idea of a general correspondence is lost. We should **unassign names when we are done with them**, or make it a practice to either start or end each problem with the command:

Clear["Global`*"]

A **function, in a strict Mathematica sense**, is one of the most basic and important structures. The following examples speak for themselves, but there are many important issues to bring up, so pay attention to all of the details. We begin by letting $x = 3$. The idea that **$x = 3$ is not a variable will influence our work in an interesting way**.

```
In[6]:= x=3;
```

```
In[7]:= f[x_] := 5x^2 + 1
```

```
In[8]:= {f[x], f[t], f[t^2], f[p + q], f[1], f[1]^2}
Out[8]= {46, 1 + 5 t^2, 1 + 5 t^4, 1 + 5 (p + q)^2, 6, 36}
```

The symbol $x\_$, is like an "empty box," that is identified by a "flag" denoted by x. Put something in the empty box, and it goes into every location identified by the flag x. **The lower bar symbol in $x\_$ is an essential part of the notation for the argument of a function.**

One would expect $5x^2 + 1$ to immediately evaluate to 46, since $x = 3$. The reason why this does not happen is due to the **delayed assignment symbol (:=)**. This is an important new assignment operator, similar to the **immediate assignment symbol (=)** except that it **delays full evaluation of the right hand side until after a call to evaluate the function is made**. This matter is critical. Notice what happens when we **define the next function g using (=) instead of (:=)**.

```
In[9]:= g[x_] = 5x^2 + 1
Out[9]= 46
```

```
In[10]:= {g[x],g[t],g[7]}
Out[10]= {46, 46, 46}
```

We mentioned before that the arguments of all functions, and commands in Mathematica are enclosed in brackets. **Parentheses will not work!**

Notice the first item in the list of values of f above. The symbol $f[x]$ now calls for a full evaluation, and since $x = 3$, we get $f[x] = f[3] = 46$. As you can see, the remaining items in this list are what we would expect from the function f.

We move on to another example, where we **introduce the command If[ ]**. If $x \leq 2$, then $g[x] = x^2$, otherwise $g[x] = 1 - 3x$. Notice that **(<=) is used to denote the inequality symbol ($\leq$)**

```
In[11]:= g[x_] := If[x <= 2, x^2, 1 - 3x]
```

```
In[12]:= {g[-5], g[2], g[6]}
Out[12]= {25, 4, -17}
```

The **delayed assignment operator** (:=) can also be a source of **confusion**. We discuss one more example, after we unassign x.

```
In[13]:= x=.;p = x^2 + 5x - 8;q[x_]:=p
```

```
In[14]:= q[x]
```
Out[14]= $-8 + 5\ x + x^2$

Our function appears to be well constructed, but this is an illusion. Notice what happens if we evaluate $q[t]$ for any $t \neq x$.

```
In[15]:= q[2]
```
Out[15]= $-8 + 5\ x + x^2$

This baffling result can be explained by looking at the delayed assignment operator (:=). To evaluate $q[2]$, Mathematica **delays the evaluation** of the right hand side of $q[x\_] := p$, until **after x is replaced by 2**. Since p is unevaluated, there is **no "visible" x in p to replace by 2**, and so the first step ends with a value of the unevaluated letter p. Then, p is evaluated to produce $p = -8 + 5x + x^2$.

It is appropriate to include in this section, a comment about the elementary special functions of mathematics, which are already defined by Mathematica. In ordinary mathematical text, we are accustomed to seeing an expression like $\sin^2(x)$. It may be tempting to enter this in a Mathematica expression as sin ^2(x), but **this will not work**. Look at the following evaluations. None of them make sense, except the last one. In pencil and paper mathematics, an expression like $\sin(x)^2$ is well defined, but it is easily misunderstood. Does it mean $\left(\sin(x)\right)^2$ or $\sin(x^2)$? In Mathematica, because square brackets are used to enclose arguments of functions, there can be no doubt that $Sin[x]$^2 means $\left(\sin(x)\right)^2$. One can use an extra set of parentheses and enter $(Sin[x])$^2 instead, into a Mathematica expression, but the extra set of parentheses are clearly unnecessary.

```
In[16]:= {Sin^2[Pi/4],N[Sin^2[Pi/4]],(Sin^2)[Pi/4],
         N[(Sin^2)[Pi/4]], Sin[Pi/4]^2}
```

Out[16]= $\left\{\sin^2\left[\frac{\pi}{4}\right], \sin^{2.}[0.785398], \sin^2\left[\frac{\pi}{4}\right], \sin^2[0.785398], \frac{1}{2}\right\}$

We end this section with an important summary. In Mathematica, a symbol f can represent a **function** or an **expression**, but they are very different structures. Students of mathematics are reminded frequently to appreciate the difference between the (function) symbol f, which denotes the "whole idea" of the correspondence and the symbol $f(x)$ which denotes the value of the function f at x. Mathematica's understanding of these symbols is very similar, but , it does not have the human ability to translate one into the other when a symbolic mistake is made. **If these symbols are misused, Mathematica will give answers which are inappropriate and look strange.** If f represents an expression, then it makes no sense to write $f[x]$. If f represents a function, then we must use $f[x]$ and not f when we are referring to the value of f at x— the expression. The words expression and function will be used carefully in this manual, and they should be read carefully as well.

2.4 Algebra

Mathematica has a large supply of commands which can be used to manipulate algebraic expressions and equations. There are too many commands to cover all of them in this chapter, and so only a sample of basic commands will be introduced. Others will be discussed in future chapters. Use your experience in algebra to predict the nature of other commands. They are all capitalized and most are unabbreviated English words, so we already know the names of many Mathematica commands. **Explore the Help File**. Exhibited below, are a few brief examples using the commands **Factor[ ], Expand[ ], ReplaceAll[ ], Together, Numerator[ ], Denominator[ ], Solve[ ], NSolve[ ], FindRoot[ ].**

There are several different "input-types" for commands (operators) in Mathematica. Let f represent a Mathematica or user defined command or operator, and let x and y be expressions. The following table describes (very roughly speaking) four different input types for operators.

$f[x, y]$	*Traditional form*
$f@x$	*prefix form for* $f[x]$
x~f~y	*infix form for* $f[x, y]$
$x//f$	*postfix form for* $f[x]$

Some commands are available in several of these forms, some in all four forms. For some operators, there is a notational change from one form to another, so the table must be interpreted rather loosely. The added flexibility this gives us for writing input statements can be a real convenience. As we shall see, Mathematica output is frequently set up so that we can use in effectively in postfix form.

```
In[1]:= (x - 1)*(x + 4)^3*(x^2 + 8)
Out[1]= (-1 + x) (4 + x)^3 (8 + x^2)


In[2]:= Expand[%]
Out[2]= -512 + 128 x + 224 x^2 + 104 x^3 + 44 x^4 + 11 x^5 + x^6


In[3]:= Factor[%]
Out[3]= (-1 + x) (4 + x)^3 (8 + x^2)
```

One way to replace x, in the above expression, by, let's say $\cos(t)$, is to enter the assignment $x = Cos[t]$ in an input cell. This permanently and globally changes the value of x, and while it might be desirable to do this, frequently it is not. As an important alternative, we can **replace x by $\cos(t)$** in just **one expression** (the one above) and do it **without changing the value of x**. The commands used to accomplish this replacement are Replace[ ], ReplaceAll[ ], to name two of the most common. We discuss the second of these two, which is the one used most often.

Many commands require us to specify what **"Rules"** are to be used in a computation. Actually, the **whole of Mathematica** can be thought of as an **elaborate system of rules**, many of which are applied automatically. When we have to specify a "Rule," we **use**

the notation ($->$) (a hyphen followed by a greater than sign). A **"Rule"** is a statement of the form ***lhs*$->$ *rhs***. Mathematica tries to **transform *lhs* into *rhs***.

In order to **replace x by $\cos(t)$, we use the rule $x->$ *Cos*[*t*]**, inside the **ReplaceAll[ ]** command. This command has a useful **postfix form denoted by the symbol (/.)** (a division or slash bar followed by a period—no space in between). This **postfix form** is essentially the **only way this command is ever used**.

```
In[4]:= % /. x -> Cos[t]
Out[4]= (-1 + cos[t]) (4 + cos[t])^3 (8 + cos[t]^2)
```

```
In[5]:= expr = (2x*a + 3y^2*b)^3
Out[5]= (2 a x + 3 b y^2)^3
```

Compare carefully, the next two input/output statements. In both cases, we are replacing x by y and y by 2. The two forms, however, are significantly different. Notice the differences in the input statements. In the **first case** the two substitutions are **performed simultaneously**.

```
In[6]:= expr /. {x -> y, y -> 2}
Out[6]= (12 b + 2 a y)^3
```

In the **next case**, they are **performed sequentially**. First x is replaced by y, and then, after this is finished, y is replaced by 2.

```
In[7]:= (expr /. x -> y ) /. y -> 2
Out[7]= (4 a + 12 b)^3
```

Let's take one final look at the above two substitution statements. When there is **only one substitution or rule** being applied, the **curly brackets are optional**. Actually, the two individual substitutions in the second input cell would be examples, where the curly brackets are not used. Finally, the parentheses used in the second substitution input cell are optional, but one can easily see the logic for their use, and it should be our practice to use them, if there is any doubt.

```
In[8]:= 5 + 3x + (2x^3 + 1)/(x^2 - 4)
```

$$\text{Out[8]= } 5 + 3\ x + \frac{1 + 2\ x^3}{-4 + x^2}$$

```
In[9]:= Together[%]
```

$$\text{Out[9]= } \frac{-19 - 12\ x + 5\ x^2 + 5\ x^3}{-4 + x^2}$$

```
In[10]:= {Numerator[%], Denominator[%]}
Out[10]= {- 19 - 12 x + 5 x^2 + 5 x^3, -4 + x^2}
```

```
In[11]:= eq = x^3 - 4*x^2 - 25*x + 28 == 0
Out[11]= 28 - 25 x - 4 x^2 + x^3 == 0
```

```
In[12]:= s = Solve[eq, x]
```

```
Out[12]= {{x → -4}, {x → 1}, {x → 7}}
```

Look at the above output cell. First of all, it is a **list of lists**, and ultimately, the **solutions are given as rules**. Mathematica output from the solve command is given in this way so that **solutions can be substituted directly into other expressions.** Notice, how this is used in the next two input statements. In the first, we ask Mathematica to check to see if the solutions are correct.

```
In[13]:= eq /. s
Out[13]= {True, True, True}
```

```
In[14]:= a = x /. s[[3]]
Out[14]= 7
```

A thorough understanding of the above input cell is important. We applied the **substitution rule $\{x->7\}$** to the **expression x**, and then **assigned** this value to a, so that **a now has a value of 7**. The rule **$s[[3]]$ is the third item in the list s.**

This is the first time we have used a list in this way, and we should make note of this new Mathematica notation. **If L is a list, and j is a positive integer, then $L[[j]]$ is the jth entry in the list L.** The use of the **double brackets is essential**.

```
In[15]:= Solve[x^3 + 7x^2 - 3x - 6 == 0, x];
```

Output is omitted because it is so complicated. We must remember that, if possible, **Mathematica will always give us exact answers, unless we allow it to approximate with decimal answers**. The solutions to a cubic equation are always available, but frequently they are algebraically complicated and/or complex valued. Even more striking, there is no formula for solving (exactly) a general polynomial of degree greater than 4. (See page 72 for more information on this interesting topic.)

The Solve[ ] command can be used to find exact solutions to equations, which are more general than polynomials. They must be at least algebraic, however and not too complicated.

```
In[16]:= eq=Sqrt[x+5]-Sqrt[x]==3/2
```

Out[16]= $-\sqrt{x}+\sqrt{5+x}==\frac{3}{2}$

```
In[17]:= Solve[eq,x]
```

Out[17]= $\{\{x\to\frac{121}{144}\}\}$

The equation above is algebraic. The next equation is transcendental, which means that it is not algebraic.

```
In[18]:= Solve[Cos[x] - x == 0, x];
Solve :: tdep : The equations appear to involve
    transcendental functions of the variables in
    an essentially non - algebraic way."
```

Surely, there is a solution to this equation for some x between 0 and $\pi/2$. A simple plot (the topic of the next section) would reveal this fact, but we can see this without a plot.

The left hand side, $\cos(x) - x$, of the equation is a continuous function which is clearly positive (above the x-axis) at $x = 0$ and clearly negative (below the x-axis) at $x = \pi/2$. Consequently, it must be 0 (cross the x-axis) somewhere between $x = 0$ and $x = \pi/2$.

Some equations—one could say, most equations—are very difficult or impossible to solve exactly. Even when solutions exist, they may be too complicated to express in terms of ordinary mathematical symbols. This is not Mathematica's fault, but the nature of mathematics itself.

The way around this fundamental problem is to drop the severe demands of exactness in favor of numerical approximation. Then, almost any equation can be solved. Mathematica has **several significantly different ways to solve equations numerically**, and it is important to have a good understanding of at least some of the basic methods.

The **NSolve[*eqn*,*x*] command solves a polynomial equation *eqn* for *x*** and gives a **complete list of its solutions as decimals**. It is interesting to note that the equation ***eqn* can contain other unassigned letters besides *x***. This method is reserved, however, for **polynomial equations**.

```
In[19]:= NSolve[x^3 + 7x^2 - 3x - 6 == 0, x]
Out[19]= {{x → -7.29841},
          {x → -0.769686}, {x → 1.06809}}
```

Mathematica uses a fairly straightforward numerical approximation technique to search for all of the decimal roots of a polynomial equation, but the same technique does not work on equations in general. Instead we search for roots, one at a time, using the command FindRoot[ ]. *Newton's Method* lies at the heart of this command. To use it effectively, to understand how to make it work and why it sometimes fails, it helps to have a **good geometric understanding** of *Newton's Method*. This is a standard topic in any first semester calculus course, and it should be studied or reviewed carefully. *Newton's Method* uses a start-up value x_0, to search for an approximate solution to an equation of the form $f(x) = 0$. If x_0 is reasonable close to a solution, the method usually produces an approximate solution to the equation. However, if the approximation scheme bumps into a point x_n along the way, where $f'(x_n) = 0$, then the method fails. This is what happens in our first example.

```
In[20]:= FindRoot[Cos[x] == 2/3, {x, 0}]
FindRoot::jsing : Encountered a singular
    Jacobian at the point x = 0.`.
    Try perturbing the initial point(s).

Out[20]= FindRoot[Cos[x] == 2/3, {x, 0}]
```

The method failed, because $f'(x_0) = 0$ for the start-up value $x_0 = 0$ used in the equation $f(x) = \cos(x) - \frac{2}{3} = 0$. All we have to do to produce a solution is change the start-up value slightly. To find the solution in the interval $[-\pi/2, 0]$, use a start-up value like $x_0 = -1$.

```
In[21]:= FindRoot[Cos[x] == 2/3, {x, -1}]
Out[21]= {x → -0.841069}
```

To find the solution in the interval $[0, \pi/2]$, use a start-up value like $x_0 = +1$.

```
In[22]:= FindRoot[Cos[x] == 2/3, {x, 1}]
Out[22]= {x → 0.841069}
```

The command **FindRoot[*eqn*, {x, x_0, x_{min}, x_{max}}] looks in the interval [x_{min}, x_{max}]** for an approximate solution to the equation *eqn* in the unknown x, using **x_0 as a start-up value**. Look at the output, when a solution does not exist (or can't be found) in the specified interval.

```
In[23]:= FindRoot[Cos[Pi/x]==1,{x,4,4,10}]

FindRoot :: regex :
  Reached the point {13.8953041660043474}
     which is outside the region {{4.,10.}}.

Out[23]= {x → 13.8953}
```

The point $x = 13.8953$ is not a solution.

The equation solving commands Solve[ ] and FindRoot[ ] can also be used to solve systems of equations. The command NSolve[ ] cannot be used in this way.

```
In[24]:= Solve[{3x + 5y == 8, 7x + 4y == 2}, {x, y}]
```

Out[24]= $\{\{x \to -\frac{22}{23}, y \to \frac{50}{23}\}\}$

Notice the use of **set brackets, { }, in the above input cell**. The first argument of the solve command is the place for an equation, and the second argument is the place for an unknown. When we have **two or more objects to place in the same argument** of a command, the usual Mathematica policy is to **group them together with set brackets**. This is an **important issue**, which plays a role in many input statements.

2.5 Plotting Graphs

Certainly one of the distinct advantages of using any Computer Algebra System is the relative ease of graphing functions. We will have ample occasions to take advantage of a large number of exciting graphing tools found in Mathematica. At this time, we merely introduce the most basic graphing command, **Plot**[ ], by graphing below, first one function and then several.

The first argument in the Plot[ ] command must be an **expression in one variable**. In particular, the **first argument cannot be an equation**. which means that the first argument cannot include the equality symbol (==). The second argument in a plot statement is a list containing the plotting variable and a plotting range.

```
In[1]:= Plot[x^2 + x*Sin[5*x], {x,3, 3}];
```

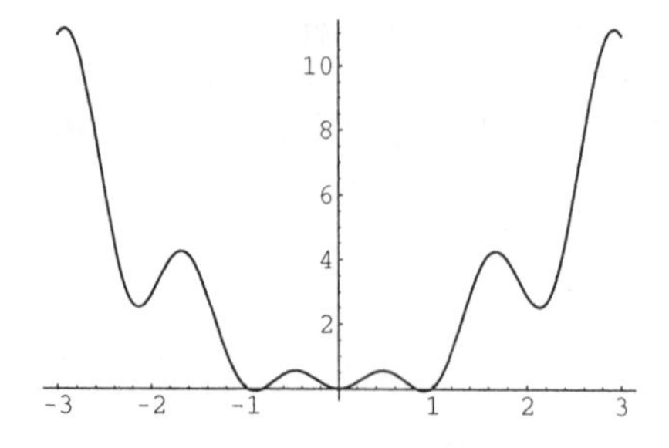

In the next plot, notice how **set (curley) brackets { } are used to group terms together, so that they all belong to the first argument of the plot command.**. Set brackets played a similar role in the last solve command, and **they will be used in this way in most commands.**

```
In[2]:= Plot[{x^2, 80 - 8*x, 80 + 8*x}, {x, -10, 10}];
```

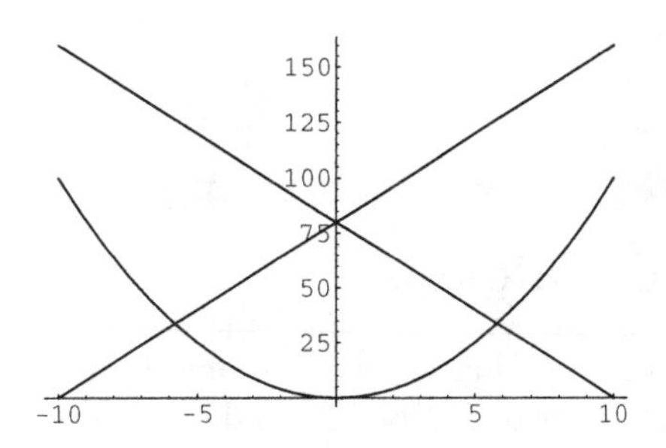

Unless Mathematica is told otherwise, it **chooses its own scale on the vertical axis**, as it did in the last two plots. **A <u>Rule</u> of the form PlotRange− > $\{y_{min}, y_{max}\}$ can be inserted as an optional third argument in the plot command to control the scale on the vertical axis.** This will be demonstrated later in the chapter as our study continues.

Much can be said about the plot command, and these two examples fall far short of giving a complete description of this important command. As you become acquainted with Mathematica over the next several chapters of this manual, take advantage frequently of Mathematica's on-line Help File to learn more about the plot command. Experiment and use the command often. Plotting functions is no longer something to be avoided. It should be your first response to many problems. Using Mathematica, functions are so easy to graph, and the graphs provide us with such a wealth of information that there is no longer a reason to avoid them.

Now that the plot command has been introduced, we can continue our discussion of certain complications regarding fractional powers. We mentioned earlier that Mathematica has decided to accept a complex valued interpretation for $x^{1/n}$ when $x < 0$ and $n \geq 3$ is an odd integer . This may be undesirable, but it is necessary in order to avoid even more serious problems manipulating expressions involving radicals. Some problems, such as evaluating $(-8)^{1/3}$, are easily dealt with using $-8^{1/3}$, instead of $(-8)^{1/3}$ in a Mathematica expression. Other problems are not so easily handled, and plotting $f(x) = x^{1/n}$ over an interval that includes both positive and negative values of x is certainly one of them.

In order to plot $f(x) = x^{1/3}$ on an interval that includes both positive and, negative numbers, the evaluation process mentioned in the last paragraph can be used to first define this function. Recall that the input statement If[$cond, val_1, val_2$] evaluates to val_1, if $cond$ is true and evaluates to val_2 otherwise.

```
In[3]:= f[x_] := If[x < 0, -Abs[x]^(1/3), x^(1/3)]
```

```
In[4]:= Plot[f[x], {x, -10, 10}];
```

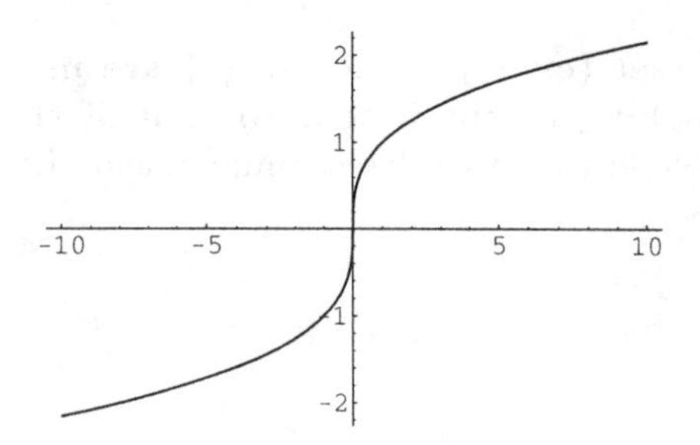

2.6 The Differential Calculus

Calculus is the study of limits, and Mathematica will compute both full limits and one sided limits with the command **Limit[ ]**. Just like "Pi", "I", and a few other names, Mathematica has also reserved the name **Infinity**. Like all other Mathematica assigned names, care should be taken to avoid assigning any other value to this name. Here are some examples.

```
In[1]:= Limit[(x^2 + 5*x - 6)/(2*x^3 + 7), x -> 4]
```
Out[1]= $\frac{2}{9}$

```
In[2]:= Limit[Sin[x]/x, x -> 0]
Out[2]= 1
```

```
In[3]:= Limit[Tan[x], x -> Pi/2]
```
Out[3]= $-\infty$

Notice that **Mathematica has made a mistake**. Only the right hand limit of $\tan(x)$ is $-\infty$. The left hand limit is $+\infty$. This follows immediately from elementary trigonometry, but it is a worthwhile experience to let Mathematica compute these one sided limits as well. Think of the **rule Direction$->-1$ as an arrow pointing in the negative x-direction**, which specifies the **path x follows** in the limiting operation. This makes it easier to remember that the next limit is a **right hand limit**

```
In[4]:= Limit[Tan[x], x -> Pi/2, Direction -> -1]
```
Out[4]= $-\infty$

Similarly, think of the **rule Direction$->1$ as an arrow (the direction for x) pointing in the positive x-direction**. The next limit is a **left hand limit**.

```
In[5]:= Limit[Tan[x], x -> Pi/2, Direction -> 1]
```
Out[5]= ∞

When **no direction is specified**, it **appears** that Mathematica **always computes a right hand limit**. This may be a mistake by Mathematica, or it may have been intentional. Nevertheless, this is an example of why **we must keep a watchful eye** over all of our computations. **A good scientist is a skeptic**. Does an answer seem reasonable? Can you verify it through another means? Computers are not infallible, nor are we.

The next example shows that Mathematica can compute a limit of an expression, which contains unassigned letters. This is followed by an example showing how Mathematica responds to a **limit, which does not exist**. Notice the **interval given in its output cell**.

```
In[6]:= Limit[(a*x + b)/(c*x + d), x -> -Infinity]
Out[6]= a/c
```

```
In[7]:= Limit[Sin[x], x -> Infinity]
Out[7]= Interval[{-1, 1}]
```

There are two basic commands in Mathematica for computing a derivative. To compute the **derivative of an expression**, we use the command **D[ ]**. To compute the **derivative of a function**, we use the command **Derivative[ ]**. The words "expression" and "function," remember, are used very carefully. If f is a Mathematica function, then $f[x]$ is an expression, not a function. We start with the command D[ ]. Not surprisingly, the output from such a command will be, itself, an **expression**.

```
In[8]:= f[x_]:=x^2*Cos[x]
```

```
In[9]:= D[f[x], x]
Out[9]= 2 x cos[x] - x^2 sin[x]
```

The first argument in this command must be an expression, not a function, so we entered $f[x]$ rather than f.

```
In[10]:= f = (a*x^2 + b*x + c)/(p*x^2 + q)
Out[10]= (c + b x + a x^2)/(q + p x^2)
```

```
In[11]:= fp = D[f, x]
Out[11]= -(2 p x (c + b x + a x^2))/(q + p x^2)^2 + (b + 2 a x)/(q + p x^2)
```

The name fp is just a name, and any other name would do just as well. It is, however, an appropriate name for a derivative, if we think of it as representing *f-prime*. Appropriate names help to motivate our work, and help to make it more readable to other human beings.

Mathematica differentiates with respect to the variable placed in the second argument of the D[ ] command. Any other unassigned name appearing in the expression is treated as a arbitrary constant. To compute a third derivative, or a higher order derivative we proceed as follows.

```
In[12]:= D[x^3*Sin[x], x, x, x]
Out[12]= 18 x cos[x] - x^3 cos[x] + 6 sin[x] - 9 x^2 sin[x]
```

```
In[13]:= D[x^20, {x, 10}]
Out[13]= 670442572800 x^10
```

As we mentioned above, we can also differentiate a function (as opposed to an expression) using the Derivative[ ] command. Not surprisingly, the output will itself be a function. If f is a function, and n is a positive integer, then **Derivative[*n*][*f*] is the *n th* derivative of *f*, in the form of a function**. As a function, **it can be evaluated by writing Derivative[*n*][*f*][*x*]**. Finally, this command has **another very interesting form**, which makes it look very much like familiar notation. If f is a function, then the symbols f', f'' **in an input cell, are equivalent to Derivative[1][*f*] and Derivative[2][*f*], respectively**.

```
In[14]:= f=.;f[x_]:=x^10;
```

Notice above, that we first unassigned the name f. It had a previous value as an expression and its "expression tag" would have clashed with the new "function tag."

```
In[15]:= {f'[x],f''[x],f''[t-1]}
Out[15]= {10 x^9, 90 x^8, 90 (-1 + t)^8}
```

This primed notation can be used as often as you wish for higher order derivatives, but, as in ordinary mathematics, it just becomes awkward after more than three derivatives. The nth derivative of a function can be computed using Derivative[n][ ].

```
In[16]:= Derivative[5][f][x]
Out[16]= 30240 x^5
```

This notation can also be awkward, if a higher order derivative is to be used in subsequent work. In this case, it would be convenient to assign a name to a particular derivative. Notice the output, when this is done in the next input cell.

```
In[17]:= g=Derivative[5][f]
Out[17]= 30240 #1^5&
```

```
In[18]:= {g[x],g[t-1]}
Out[18]= {30240 x^5, 30240 (-1 + t)^5}
```

There is no need to be overly concerned with the rather abstract notation above, but since we are likely to see it again, we should offer an explanation. The symbol & is used in Mathematica to represent a pure, unnamed function. The arguments of a pure function can be referred to without naming them by using the symbols #1, #2, ..., #n. In the case above, #1 is the only argument. Compare the structure of 30240 #1^5& to the values of $g[x]$ and $g[t-1]$ above to gain further insight into this combination of symbols. If you are interested, consult the Help File for more information, but as we mentioned, this is not a critical issue. If such abstract notion makes your heart skip a beat, **avoid looking at the output, but using a semicolon (;) at the end of the input cell**. Recall that this suppresses the output.

Finally, it should be mentioned that when we used the Derivative[ ] command, we did not have to specify the variable as a second argument. This is entirely appropriate if we simply consider the mathematical significance of the choice of letter used to define a function. We used the letter "x," but using any other letter would have defined the same function f. The argument of a function f is like an empty box. It becomes letter dependent, only when it is evaluated at a letter x to form $f(x)$.

Mathematica can handle differentiation on a much more abstract level. As we mentioned earlier, unassigned names in the command D[ ] are treated as constants except for the name (variable) used in the differentiation process. In view of this, it is not surprisingly that the answer to the first computation in the next input cell is zero. In the second computation, when y is replaced by $y[x]$, it denotes an arbitrary expression that depends on x.

```
In[19]:= {D[y, x], D[y[x], x]}
Out[19]= {0, y'[x]}
```

Because of this ability to interpret abstract symbols, Mathematica can perform implicit differentiation.

Example 2.1 *Suppose that* $y = y(x)$ *is defined implicitly by the equation*

$$xy^2 + 5x^2y + y^3 = 4x^3 + 7.$$

Find the derivative $y'(x)$ *in terms of* x *and* y.

Imagine how you would solve this problem with pencil and paper. Mathematica can be used to solve the problem in basically the same way. This is a good way to review the basic principle of implicit differentiation.

```
In[20]:= eq = x*y^2 + 5*x^2*y + y^3 == 4*x^3 + 7;
```

Before we differentiate, we must make a statement that y depends on x.

```
In[21]:= eqx = eq /. y -> y[x]
```

$$\text{Out[21]= } 5\ x^2\ y[x] + x\ y[x]^2 + y[x]^3 == 7 + 4\ x^3$$

```
In[22]:= deq = D[eqx, x]
```

$$\text{Out[22]= } 10\ x\ y[x] + y[x]^2 + 5\ x^2\ y'[x] + 2\ x\ y[x]\ y'[x] + 3\ y[x]^2\ y'[x] == 12\ x^2$$

```
In[23]:= Solve[deq, y'[x]]
```

$$\text{Out[23]= } \{\{y'[x] \to -\frac{-12\ x^2 + 10\ x\ y[x] + y[x]^2}{5\ x^2 + 2\ x\ y[x] + 3\ y[x]^2}\}\}$$

```
In[24]:= % /. y[x] -> y
```

$$\text{Out[24]= } \{\{y'[x] \to -\frac{-12\ x^2 + 10\ x\ y + y^2}{5\ x^2 + 2\ x\ y + 3\ y^2}\}\}$$

```
In[25]:= yp = y'[x] /. %[[1]]
```

$$\text{Out[25]= } -\frac{-12\ x^2 + 10\ x\ y + y^2}{5\ x^2 + 2\ x\ y + 3\ y^2}$$

2.7 Solving Problems in Mathematica

Much remains to be done in subsequent chapters, but with this brief introduction, we already have a sufficient knowledge of Mathematica to be able to solve a large number of interesting problems.

Example 2.2 *Plot the function*

$$f(x) = x^6 - 189x^5 - 2192x^4 - 1736x^3 + 27207x^2 - 1547x + 29300.$$

Supply evidence that you have the complete graph (that there are no hidden turns, which are not displayed).

```
In[1]:= f[x_] := x^6 - 189x^5 - 2192x^4 - 1736x^3 + 27207x^2 - 1547x + 29300
```

```
In[2]:= Plot[f[x], {x, -10, 10}];
```

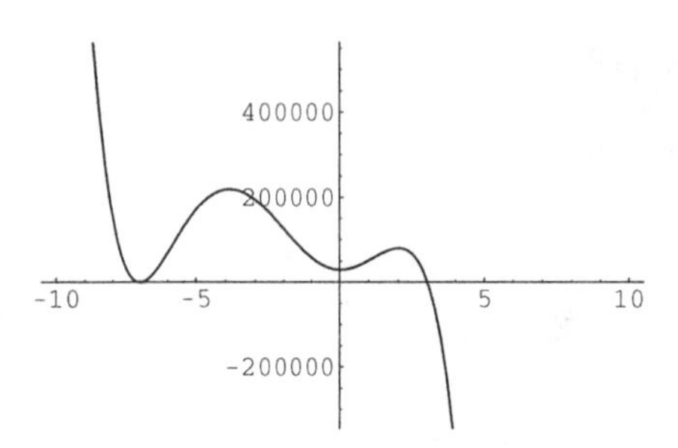

The values of this polynomial get large very quickly, and Mathematica responds to this by placing a very large scale on the vertical axis. It has done a good job of showing the twists and turns in the graph of this polynomial, but there is always a possibility of hidden features in any graph, especially when the scale is so large. **The vertical axis can be scaled as desired by inserting the rule PlotRange− > $\{y_{min}, y_{max}\}$ as an optional third argument in the plot command.** For example, to scale the vertical axis between -100 and 100 we would write.

Plot[$f[x]$, $\{x. - 10, 10\}$,PlotRange− > $\{-1000, 1000\}$];

in an input cell. We won't actually do this, because Mathematica has given us a better vertical scale, or at least that would be our guess.

Do we have a complete graph? How do we know? **Make it a habit to be a skeptic.** One way to answer such questions is to use the differential calculus. There are other ways as well.

```
In[3]:= fp = D[f[x], x]
Out[3]= -1547 + 54414 x - 5208 x^2 - 8768 x^3 - 945 x^4 + 6 x^5
```

```
In[4]:= s = NSolve[fp == 0, x]
Out[4]= {{x → -7.}, {x → -3.85785},
         {x → 0.0285117}, {x → 2.01344},
         {x → 166.316}}
```

Pay attention to the next input statement. It speaks for itself, but it is an interesting way to make all of the appropriate assignments at once.

```
In[5]:= {a, b, c, d, e} = x /. s
Out[5]= {-7., -3.85785, 0.0285117,
         2.01344, 166.316}
```

The derivative fp can only change signs at these five points. By looking at the last plot, we see clear evidence for sign changes occurring at the first four of these points, but s also contains an unexpected large positive value of x where a sign change could occur. We should test the sign of fp for some value of x larger than e.

```
In[6]:= fp /. x -> 200
Out[6]= 337658561253
```

This positive value implies that the graph is increasing for x larger than $x = e$, and so there is **one last hidden turn in the graph which is not shown** in the above plot. In response, we plot on a larger interval

```
In[7]:= Plot[f[x], {x, -200, 200}];
```

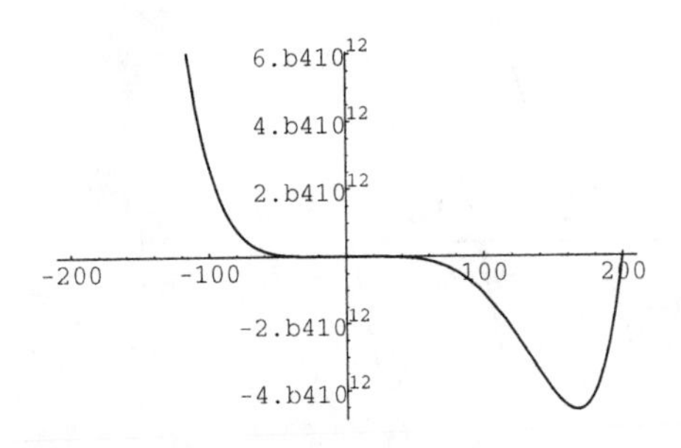

Not surprisingly, the scale is now so large that the twists and turns closer to the origin are now lost in the huge scale—a not uncommon problem. Any further graphing using the plot command it probably not going to be very effective. Perhaps the best way to finish this graph is to just evaluate the relative maximums and relative minimums. Surely this is easy to do. Our results are expressed in the form $\{x_0, y_0\}$. This is an appropriate way to express an ordered pair, since a list is an ordered set.

```
In[8]:= RelMin1 = {a, f[a]}
Out[8]= {-7., -100.}
```

```
In[9]:= RelMax1 = {b, f[b]}
Out[9]= {-3.85785, 219131.}
```

```
In[10]:= RelMin2 = {c, f[c]}
Out[10]= {0.0285117, 29278.}
```

```
In[11]:= RelMax2 = {d , f[d]}
Out[11]= {2.01344,  80099.3}
```

```
In[12]:= RelMin3 = {e, f[e]}
Out[12]= {166.316, -4.57102 10^12}
```

Example 2.3 *Find all solutions (as decimal numbers) to the system of equations*

$$5\cos(2x) - y - 2 = 0, \; x^2 - 3x - y - 1 = 0.$$

We could jump right into the use of the FindRoot[ ] command, but what would we use for start-up value? Does a solution exist, and how many are there? **Questions such as these can only be answered by graphing the equations.** Remember, we can **only plot expressions (not equations) with the Plot[ ] command.** We could use a new command for plotting equations, but the equations in this example can easily be solved for y in terms of x by simple observation.

```
In[13]:= {y1 = 5*Cos[2*x] - 2, y2 = x^2 - 3*x - 1};
```

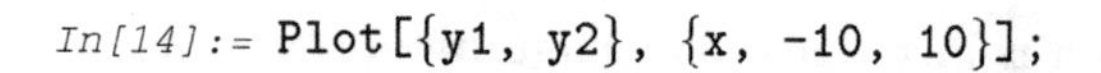

```
In[14]:= Plot[{y1, y2}, {x, -10, 10}];
```

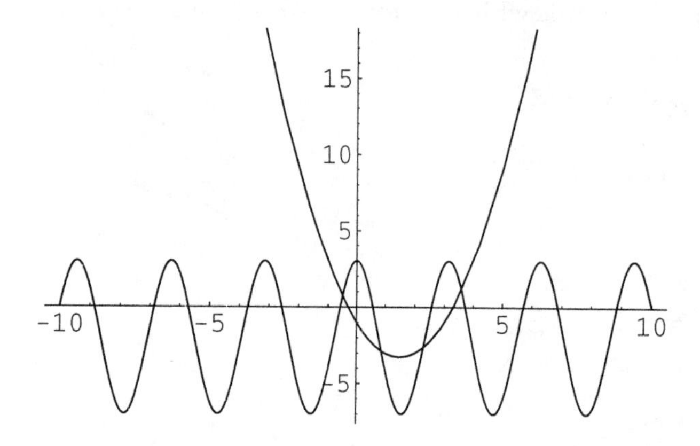


```
In[15]:= Plot[{y1, y2}, {x, -2, 5}];
```

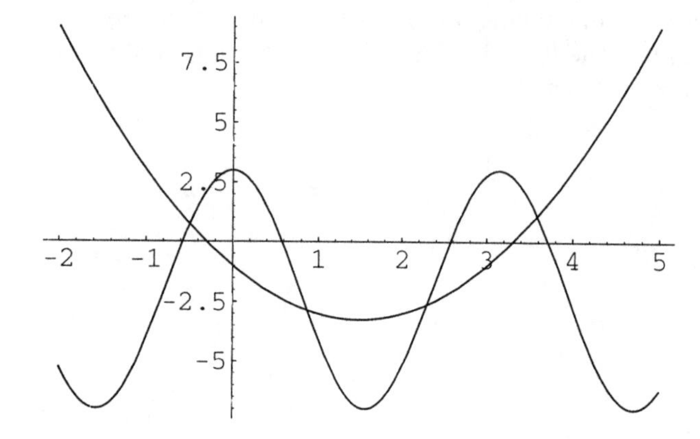


```
In[16]:= FindRoot[y1 == y2, {x,-1}]
Out[16]= {x → -0.496086}
```

```
In[17]:= FindRoot[y1 == y2, {x,1}]
Out[17]= {x → 0.871293}
```

```
In[18]:= FindRoot[y1 == y2, {x,2}]
Out[18]= {x → 2.29409}
```

```
In[19]:= FindRoot[y1 == y2, {x,3.5}]
Out[19]= {x → 3.5902}
```

Imagine how difficult this solution would have been without the graphical information. Plots can be very helpful, and they should be used frequently

Example 2.4 *Compute the following limit, and verify the limit through alternate means.*

$$\lim_{x\to 1} \frac{6\sin x - 1 + 5 - 3x - 3x^2 + x^3}{(x-1)^5}$$

```
In[20];= h[x_]:=(6Sin[x-1]+5-3x-3x^2+x^3)/(x-1)^5
```

```
In[21]:= L=Limit[h[x],x->1]
          1
Out[21]= --
         20
```

Is the answer correct? Some limits have obvious values, but this is not one of them. The numerator and denominator of $h(x)$ both go to 0 as $x \to 1$, and so the limit cannot be evaluated purely by observation. The best way to answer the question is to use *L'Hopital's Rule*, a rule, which may or may not be presented in a first calculus course. When it is presented later, in Section 5.6 of this manual, we will show how limits such as this can be verified using this very powerful rule. For the time being, however, we will have to resort to other methods in order to confirm the limit value. The most obvious way is to simply evaluate $h(x)$ for a range of x values which are close to $x = 1$. A more effective way of doing basically the same thing is to plot $h(x)$ over a small interval centered at $x = 1$. As x approaches $x = 1$, are the points $(x, h(x))$ on the graph close to the point $(1, 1/20) = (1, .05)$?

```
In[22]:= N[{h[1.1],h[1.01],h[1.001],h[1.0001]}]
Out[22]= {0.0499881, 0.0499978,
          0.222045, 22204.5}
```

Something is **clearly wrong here**! The values of $h(x)$ are not getting close to 0.05 as x gets close to $x = 1$. Also, notice above that the **decimal command N[ ] acts on a list of values and produces a list of decimals**. Further comments about this **important matter follow this example**. Can we confirm our limit value with a plot? We try that next.

```
In[23]:= Plot[h[x],{x,0.8,1.2}];
```

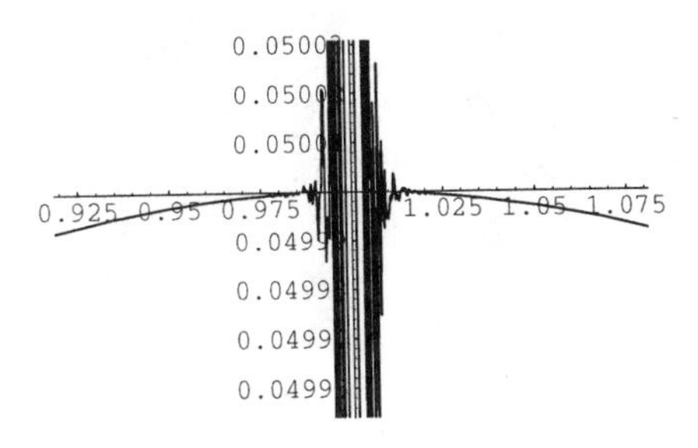

Something is clearly wrong with the plot as well! It is not at all obvious, but the limit value $L = 1/20 = .05$ happens to be correct. The failure of our attempt to confirm this cannot be blamed on Mathematica, but on unavoidable problems dealing with decimal approximations. When decimal arithmetic is performed on a fraction of the form A/B, where both A and B are small, round off errors become significant. They are unavoidable! We can ask Mathematica for more decimal accuracy (by using N[$expr, n$] for a large integer n), but regardless of how large n is, all we have to do is choose x close enough to $x = 1$, and decimal arithmetic on the value of $h(x)$ will fail to be accurate. We will not be able to confirm the value of this limit until Section 5.6.

Our work here should serve as another example of why **we must maintain a constant vigil over the accuracy of our results**.

In spite of this problem, the above method frequently will confirm the value of a limit. Because of the ever present possibility of round off errors, however, it is just not a very reliable way to confirm a limit value. Other methods sometimes work. Certainly, all of the methods that were used to compute limits in a pencil and paper environment can be used, as well, in a Mathematica notebook, to confirm the value of a limit. Until *L'Hopital's Rule* is established, however, our ability to confirm Mathematica's limit calculations is somewhat restricted.

Let us return to the above decimal calculations one more time. The command **N[ ]** normally operates on a real number, but, as we see above, it can also **act on a list** of real numbers, and when it does, it also **returns a list** of decimals. We say that the command **N[ ] is listable. Most (but not all) commands in Mathematica are listable**.

Our next and final example in this chapter is admittedly a bit involved. Think of it as an example, whose purpose is to demonstrate Mathematica's problem solving power. If we keep our minds on the ideas and let Mathematica take care of the calculations, then, perhaps, we can avoid getting bogged down in the computational details. **Mathematica lets us focus all of our energy on ideas**. Now, we should not turn a blind eye to the calculations. We can at least glance at the calculations to see if they look reasonable.

Example 2.5 *An Air Traffic Controller is monitoring the positions of two airplanes on a radar screen. Both are flying at the same altitude. At a certain moment the passenger jet is 75.4 miles South of the airport and traveling North at 547 miles per hour. At the same moment the air force fighter jet is 126 miles Northeast of the airport and traveling Southwest at 714 miles per hour. Neither plane will be landing at the airport, and both will be flying over the airport with no change in speed or direction. How close do the planes get to each other?*

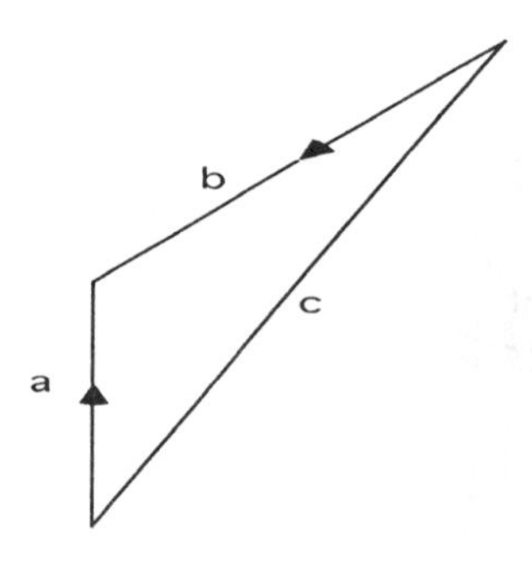

The Law of Cosines gives us the equation $c^2 = a^2 + b^2 - 2ab\cos(\alpha)$. The angle α is 135°. We let t be time in hours with $t = 0$ corresponding to that "initial moment," and we wish to minimize c as a function of t.

```
In[24]:= {a = 75.4 - 547*t, b = 126 - 714*t};
```

```
In[25]:= c = Sqrt[a^2 + b^2 - 2*a*b*Cos[135*Pi/180]]
```

```
Out[25]= √((126 - 714 t)^2 +
          √2 (126 - 714 t) (75.4 - 547 t) +
          (75.4 - 547 t)^2)
```

As the planes pass over the airport the values of a and b will turn from positive to negative values. This situation needs to be considered because the minimum distance between the planes could occur after one of the planes passes the airport. At first glance it is not clear that the *Law of Cosines* stated above is still valid for negative values of a and/or b. However, a moments thought will convince you that as exactly one of the variables becomes negative, and then as they both become negative, the angle α changes from 135° to 45° and then back to 135°, introducing, in the process, just the right combination of negative signs to make the formula valid for all values of a and b. The key idea here is that $\cos(135^o) = -\cos(45^o)$.

```
In[26]:= cp = D[c, t]
Out[26]= (- 1428 (126 - 714 t) - 547 √2 (126 - 714 t)-
          1094 (75.4 - 547 t)-
          714 √2 (75.4 - 547 t))/
        (2 √((126 - 714 t)^2 +
            √2 (126 - 714 t) (75.4 - 547 t) +
            (75.4 - 547 t)^2))
```

```
In[27]:= Plot[c, {t, 0, 1}];
```

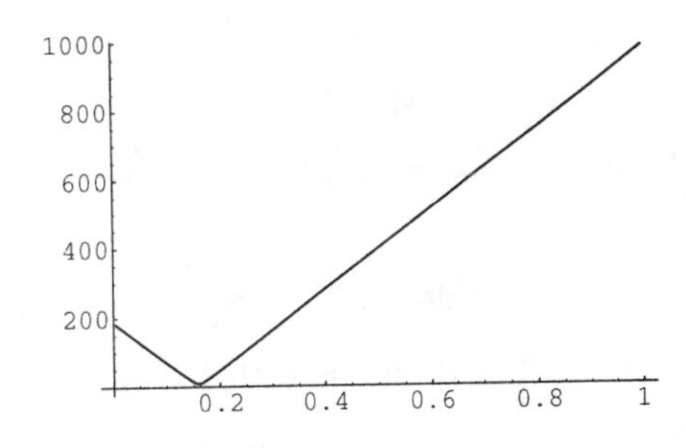


```
In[28]:= s = FindRoot[cp == 0, {t, {0, 0.4}}]
Out[28]= {t → 0.160144}
```

The minimum distance between the planes, in miles is

```
In[29]:= distance = c /. s
Out[29]= 9.14298
```

To see where the planes will be at this moment, we evaluate the triangle lengths A. and B.

```
In[30]:= A = a /. s
Out[30]= -12.199
```

```
In[31]:= B = b /. s
```

```
Out[31]= 11.6569
```

These answers indicate that the passenger jet will have passed the airport (A is negative), and the fighter jet will still be approaching the airport (B is positive) at the moment of closest approach.

2.8 Exercise Set

Some Mathematica features and Commands from this Chapter

- assignment $a = expr$
- delayed assignment: $a := b$
- Expand[ ]
- function format $f[x\_] := expr$
- input seq separator ;
- Plot[ ]
- N[,j]
- Simplify[ ]
- suppressing output ;
- wild card (*)
- Clear[ ]
- Denominator[ ]
- Factor[ ]
- "Global`*"
- Limit[ ]
- PlotRange -> $\{y_{min}, y_{max}\}$
- NSolve[ ]
- Solve[ ]
- Together[ ]
- D[ ]
- Derivative[n][ ]
- FindRoot[ ]
- If[ ]
- list $\{a_1, a_2, \ldots, a_n\}$
- N[ ]
- Numerator[ ]
- substitution: $expr/.\ x->a$
- unassignment $a = .$

1.Evaluate the following expressions exactly.

$a) 3 \times \frac{6\times 7^2-8}{5} - 2 + \frac{14}{6\times 5+8}$ $\quad b) \frac{8}{5} \times 3 - \frac{8}{5\times 3}$ $\quad c) (-2^4 + \frac{1}{5^2})^{-3}$

$d) \frac{\frac{5}{3}}{14} - \frac{6}{\frac{2}{3}}$ $\quad e) \frac{\sqrt[3]{-8}}{\sqrt{25}} + 81^{1/4}$ $\quad f) 7\cos^3(\pi/6) + \frac{\tan(\pi/12)}{5}$

2. Evaluate the following expressions as decimals with 10 digits of accuracy.

$a) \frac{\sqrt{43.7}}{5^2}$ $\quad b) 8\sqrt{\cos(13^\circ)}$

$c) \frac{(73)^{2/3}+17}{(-3)^{1/5}}$ $\quad d) \sin^2(-3\pi/7)$

3. Evaluate the expressions in problem 2 as decimals with 3 digits of accuracy.

4. Evaluate the expressions in problem 2 as decimals with 15 digits of accuracy.

5. Evaluate the following expressions, and express the answers in simplest form. (Do not express them as decimals)

$a) (2 + \sin(\pi/3))^6$ $\quad b) \frac{(27)^{2/3}+5}{6\times(8+\frac{1}{3})}$ $\quad c) (\tan(\pi/8) + 2\cos(3))^2$

6. Graph the function $f(x) = \frac{3x+7}{x^2+5}$. Verify that you have a complete graph. Find the x-intercepts and the coordinates of the relative maximums and minimums.

7. Graph the function $f(x) = x^4 - 50x^3 - 28x^2 - 280x - 642$. Verify that you have a complete graph. Find the x-intercepts and the coordinates of the relative maximums and minimums.

8. Simplify the following rational function. How do you know that the answer Mathematica gives you is in simplest terms?

$$f(x) = \frac{x^5 + 14x^4 + 30x^3 - 235x^2 - 506x + 1456}{2x^5 - 5x^4 - 85x^3 + 336x^2 - 72x - 504}$$

9. Compute the following limits if they exist in a finite or infinite sense. Verify each answer by some means using an algebraic approach whenever appropriate.

$a)\ \lim_{x\to 4}(\frac{3x^2-17x+20}{5x^2-19x-4})$ $b)\ \lim_{x\to 3^+}(2x^2\cot(\pi x))$ $c)\ \lim_{x\to 4}(\frac{2-\sqrt{x}}{x-4})$
$d)\ \lim_{x\to 0}(\frac{5x-\sin(5x)}{x^3})$ $e)\ \lim_{x\to\infty}(\frac{7x^2+3x-9}{2x^2+8x+5})$

10. Use the D[ ] command to compute $f'(4)$ for $f(x) = x^2\sin(\pi x)$. Repeat the computation using the Derivative[ ]) command. What are the advantages and disadvantages of each? Compute $f'(4)$ by using the definition of a derivative as the limit of a certain difference quotient.

11. Use Mathematica to generate the rule $f'(x) = 10x^9$ for $f(x) = x^{10}$ and then verify the rule using the definition of a derivative as the limit of a certain difference quotient. Also verify the value of the limit algebraically.

12. Compute the derivatives of the following functions, where a denotes an arbitrary constant. Use the D[ ] command, then repeat the computation using the Derivative[ ] command. What are the advantages and disadvantages of each?

$a) f(x) = x^2\sqrt{a^2-x^2}$ $b) g(w) = \sqrt{1+\sqrt{1+\sqrt{1+w^2}}}$
$c) H(t) = \sec^5(\frac{t^2}{t^2+a^2})$ $d) S(y) = \cos(2y)\sec^2(y) - \cos(2y)\tan^2(y)$

In part d) think about what this answer should be and force Mathematica to give you that answer.

13. Compute $\frac{d^8y}{dx^8}$ for $y = \tan(x)$. Express the answer as a sum of powers of $\tan(x)$.

14.(a) Use implicit differentiation and the D[ ] command to find the slope of the tangent line to the graph of the equation $x^3y^2 + 4xy = 85 - (2x^4 - 8y)^2$ at the point on the curve corresponding to $x = -5$. Use the point with the smallest y-coordinate. How do you know there isn't a smaller one that Mathematica simply failed to give you? It would be nice (and reassuring) if we had a graph of the equation to look at, but we have not yet discussed the command for plotting equations. This problem can be done, just as well, without a graph. (b) compute the second derivative of y with respect to x at the same point.

15. Use Mathematica to generate a Product Rule for differentiating the product of three arbitrary functions f, g, and h. Recall how to set up an unassigned name which depends on another variable.

16. Use Mathematica to generate a Chain Rule for the composite function $f \circ g \circ h$. Recall how to set up an unassigned name which depends on another variable.

17. Find the absolute maximum and absolute minimum of $f(x) = x^4+2x^3-76x^2-242x+100$ on the closed interval $[-9, 9]$. Do this problem without graphing. Use a plot to verify your results only after you are finished.

18. At 12:00 noon a passenger ship is 100 miles South of a oil tanker. The passenger ship travels due North at 35 miles per hour, while the oil tanker travels due East at 17 miles per hour. At what rate is the distance between the ships changing at 2:30 pm of the same day?

19. Radar at an airport spots a passenger jet traveling towards the airport at a constant elevation of 25,000 feet. Radar determines that when the line-of-sight distance between the jet and radar is exactly 12 miles, this line-of-sight distance is decreasing at the rate of 370 miles per hour. What is the ground speed of the jet at this moment? (Recall that 1 mile = 5280 feet.)

20. A rectangular pool-deck complex is being built for a hotel. The pool must have a water surface area of 4,000 square feet, and the pool must be enclosed on three sides by a 10 foot wide deck and on the fourth side (the side facing the building) by a 30 foot wide patio. Find the dimensions which minimize the total area of the pool-deck complex.

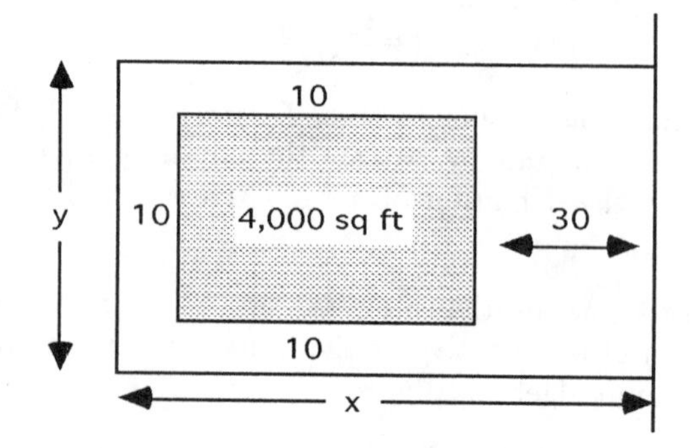

Project: Swimming Coach

Suppose that you are the head coach for a large team of athletes who all plan to compete in a combination swimming/running event. The event begins on an island 2 miles out from a straight shore line. The end of the race is 10 miles down the shoreline, as shown in the picture below. Each competitor in the race will swim to some *"transfer point"* along the beach, and then run from that point to the finish line. The choice of a transfer point is entirely up to each athlete. Simply by choosing the appropriate transfer point an athlete may decide to run the whole 10 miles or decide to completely avoid running, and any transfer point between (and including) these extremes is allowed.

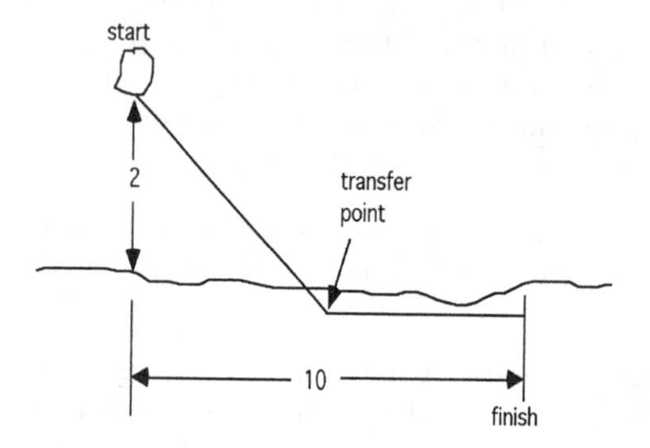

As the head coach, it is your responsibility to chose the best possible transfer point for each member of your team. Your only concern is that each athlete turn in his/her best total time (Typical head coach attitude, right?). You know the swimming and running speeds of each member of your team, but these speeds change from day to day, and so what you need is a means of determining, quickly, and just before the race begins, a transfer point for each member of your team. The Department of Mathematics will graciously loan you one of its portable computers, and Mathematica to take to the race as long as you promise not to drop it in the water. Suppose we let s and r denote the swimming speed and running speed in mph for any one of your athletes. Find a function which depends on r and s, which will quickly determine the transfer point and total race time for that athlete. (This is why they hired a mathematician as head coach.) Good luck!

Project: Road Construction

Towns A and B are separated by a river running basically east-west, and plans are being made to connect the towns by a road and a bridge over the river.

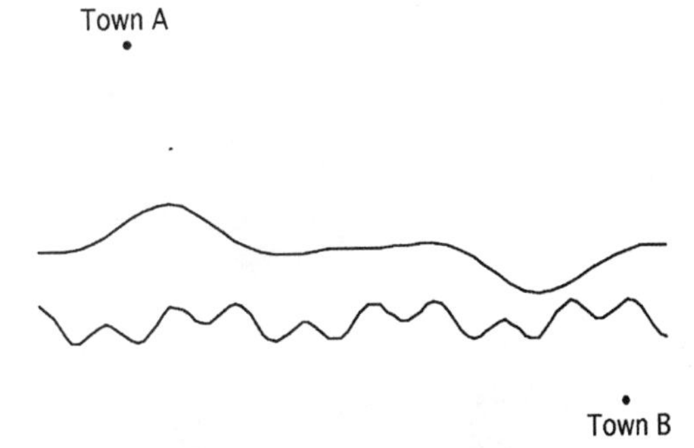

To locate everything precisely, a north-south and an east-west axis are placed over the region with units in kilometers (km). The origin is located at some point in the river as can be seen in the picture below. According to these axes, Town A is located 15 km west and 10 km north of the origin, and Town B is located 17 km east and 6 km south of the origin. The bridge can be built at any point along the river, but it must cross the river due north-south. To save money, it is decided to build the road straight from Town A to the point P on the north side of the bridge and straight again from the point Q on the south side of the bridge to Town B. (To simplify matters, assume that the road/bridge complex intersects the river bank only at P and Q.)

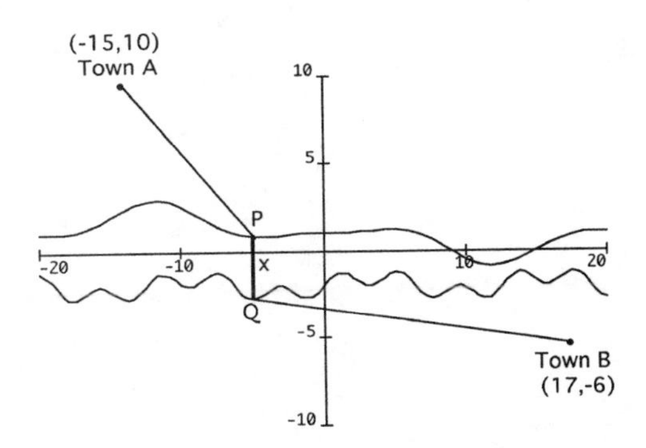

Cost however, is a major complication. The north side of the river is hilly, expensive farm land, and it will cost \$500,000 per km to build the road. The bridge itself will cost \$2,000,000 per km to build. The south side of the river is flat, undeveloped, government land, and this part of the road can be built for only \$190,000 per km. By trial and error it is determined that the curves representing the northern and southern boundaries of the river can be described reasonably well by the equations

$$y = 1 + 2\cos(x/4)\sin(x/7)^3, \text{ and } y = -2 + \cos(x/2)\sin(x),$$

where the northern boundary is listed first.

At what point along the river should the bridge be built in order to minimize the cost of the whole project?

Chapter 3

Basic Integration

3.1 Area and the Definite Integral

The definition of a definite integral as the limit of its Riemann sums is so important to mathematics and its applications that, we begin this chapter by using Mathematica to gain some intuitive insight into this concept. Even if we downplay the many mathematical reasons for understanding this definition thoroughly, we are still left with some very important applied reasons.

As long as it is known that the integral of f on $[a, b]$ exists, its value can be determined by looking at a restricted, more manageable class of Riemann sums. In this case, if

$$a = x_0 < x_1 < \cdots < x_n = b$$

is the partition obtained by dividing $[a, b]$ into n subintervals of equal length

$$\Delta x = \frac{b-a}{n},$$

and if

$$x_0 \le c_1 \le x_1, x_1 \le c_2 \le x_2, \ldots, x_{n-1} \le c_n \le x_n,$$

then,

$$\int_a^b f(x)dx = \lim_{n \to \infty} \sum_{k=1}^{n} f(c_k)\Delta x. \tag{3.1}$$

Typically, the points $c_j (j = 1, 2, \ldots, n)$ are chosen to be left end points, right end points, or mid points of their respective subintervals, in which case, the corresponding Riemann sums are called **left sums, right sums, and middle sums**, respectively. Any other scheme, however, for determining these points is just as acceptable. Furthermore, if f is continuous on $[a, b]$, then it is integrable on $[a, b]$, and so this simplified setting is usually sufficient. Applied problems, in particular, are usually continuous in nature.

Frequently when problems from the real world are formulated into mathematics, they take the form of the right hand side of (3.1). A complicated problem might not be solvable "as a whole," but if we partition the problem up into a large number of parts, we might conclude that the "whole" problem is just the sum of its parts. We might be able to easily estimate each of the small parts, to get approximations $P_1, P_2, \ldots P_n$. Then a good approximation for the whole problem would be

$$\sum_{k=1}^{n} P_k.$$

If the approximations get better as n (the number of subdivisions) gets larger, then the answer to the original problem would be

$$Answer = \lim_{n \to \infty} \sum_{k=1}^{n} P_k. \tag{3.2}$$

Without a knowledge of the definition of a definite integral, we would be left with a formidable problem. Frequently, however, (3.2) can be made to look like (3.1), so by using the definition of an integral, we can identify our problem as an integral. At this point, we have a well understood problem, which can easily be evaluated.

Left, right, and middle Riemann sums can easily be computed with Mathematica's **Sum[ ] command**.

$$Sum[q[j], \{j, 1, n\}] = q[1] + q[2] + \ldots + q[n] = \sum_{1}^{n} q[j].$$

This is **another basic command** that will be used often in a variety of situations besides the computation of Riemann sums. The letter j may have a previous value, without having any effect on the sum command. The letter j inside the sum command is treated as a different variable. We say that the **index** j **is a "local" variable.**. We will discuss this important matter of local variables later in this chapter.

Graphing the rectangles in the approximation process is more challenging. It is not important to understand how to put commands like this together, so the user defined commands we need are simply given.

```
In[32]:= LeftRiemannDisplay[f_,a_,b_,n_]:=Show[Plot[f[x],{x,a,b},
         DisplayFunction->Identity],
         Graphics[Table[Line[{{a+(b-a)/n*j,0},
         {a+(b-a)/n*j,f[a+(b-a)/n*j]},
         {a+(b-a)/n*(j+1),f[a+(b-a)/n*j]},
         {a+(b-a)/n*(j+1),0},{a+(b-a)/n*j,0}}],{j,0,n-1}]],
         DisplayFunction->$DisplayFunction];
```

```
In[33]:= RightRiemannDisplay[f_,a_,b_,n_]:=Show[Plot[f[x],{x,a,b},
         DisplayFunction->Identity],
         Graphics[Table[Line[{{a+(b-a)/n*j,0},
         {a+(b-a)/n*j,f[a+(b-a)/n*(j+1)]},
         {a+(b-a)/n*(j+1),f[a+(b-a)/n*(j+1)]},
         {a+(b-a)/n*(j+1),0},{a+(b-a)/n*j,0}}],{j,0,n-1}]],
         DisplayFunction->$DisplayFunction];
```

```
In[34]:= MidRiemannDisplay[f_,a_,b_,n_]:=Show[Plot[f[x],{x,a,b},
         DisplayFunction->Identity],
         Graphics[Table[Line[{{a+(b-a)/n*j,0},
         {a+(b-a)/n*j,f[a+(b-a)/n*(j+1/2)]},
         {a+(b-a)/n*(j+1),f[a+(b-a)/n*(j+1/2)]},
         {a+(b-a)/n*(j+1),0},{a+(b-a)/n*j,0}}],{j,0,n-1}]],
         DisplayFunction->$DisplayFunction];
```

It is easy enough to see, by example, that these commands draw the appropriate rectangles, so rather than labor excessively over an explanation of the formulas, let's just accept these commands on face value. The first one, unfortunately, will have to be typed, quite carefully, into a Mathematica notebook. Only small changes are necessary to produce the second and third formula—a copy and paste strategy for entering them is recommended. In the second, j is replaced by $j+1$ inside the argument of $f(x)$ in two locations, and in the third, j is replaced by $j+\frac{1}{2}$ in the same two places.

These formulas should be saved on a separate Mathematica notebook—think of it as a special library of commands, and name it "MyLibrary." A few more commands will be added to this library file later as this manual evolves. These commands could be saved quite formally as a special Mathematica package, but this is entirely unnecessary. We have not even discussed Mathematica's standard packages yet, although we will shortly.

With this preparatory work behind us, we are ready to study the meaning of a definite integral such as

$$\int_1^6 (9-x^2)dx.$$

This integral can be approximated by a Riemann sum obtained by dividing the interval $[1,6]$ into, say 20 subintervals of equal length, by using, let's say, left end points to establish $(\pm)$heights of rectangles, and by computing the sum of the $(\pm)$areas of all the rectangles. To draw the corresponding picture, we use our newly created command LeftRiemannDisplay .

```
In[35]:= f[x_]:=9-x^2
```

```
In[36]:= LeftRiemannDisplay[f,1,6,20];
```

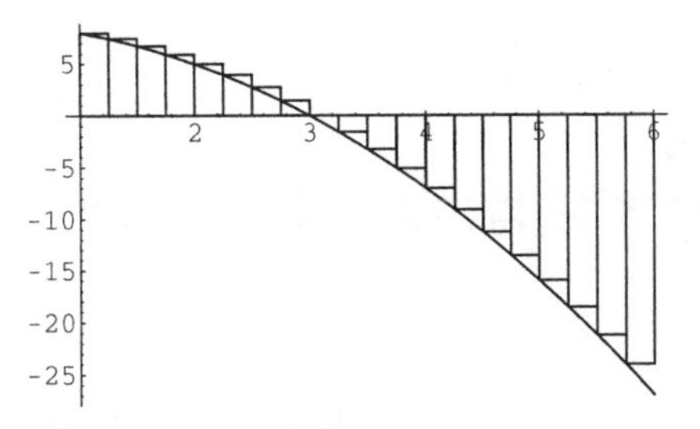

Next, we compute the corresponding left Riemann sum. The areas of rectangles below the x-axis should naturally subtract from the areas of rectangles above the x-axis. From the above picture, we should expect a negative answer, when we are done.

```
In[37]:= a=1; b=6; n=20; delta=(b-a)/n;
```

We divide the interval $[1,6]$ into 20 subintervals of equal length

$$\Delta x = \frac{6-1}{20}.$$

The partition points

$$1 < 1+\Delta x < 1+2\Delta x < 1+3\Delta x < \ldots < 1+20\Delta x = 6$$

are created by the function p we define next. We don't need to display all of the terms, but we do anyway, just to get a good "feel" for the idea of an integral. **To create a list which**

contains more than just a few terms, it helps to use Mathematica's Table[ ] command. This command is worth remembering.

```
In[38]:= p[j_]:=a+j*delta;
```

```
In[39]:= see_points=Table[p[j],{j,0,n}]
```

Out[39]= $\{1, \frac{5}{4}, \frac{3}{2}, \frac{7}{4}, 2, \frac{9}{4}, \frac{5}{2}, \frac{11}{4}, 3, \frac{13}{4}, \frac{7}{2}, \frac{15}{4}, 4, \frac{17}{4}, \frac{9}{2}, \frac{19}{4}, 5, \frac{21}{4}, \frac{11}{2}, \frac{23}{4}, 6\}$

Using left end points to establish ($\pm$)heights we have the following.

```
In[40]:= lrs=Sum[f[p[j]]*delta,{j,0,n-1}]
```

Out[40]= $-\frac{715}{32}$

Notice that lrs, the left Riemann sum, is negative as we anticipated it would be.

```
In[41]:= RightRiemannDisplay[f,1,6,20];
```

Output omitted to save space.

```
In[42]:= rrs=Sum[f[p[j+1]]*delta,{j,0,n-1}]
```

Out[42]= $-\frac{995}{32}$

```
In[43]:= MidRiemannDisplay[f,1,6,20];
```

Output omitted to save space.

```
In[44]:= mrs=Sum[f[(p[j]+p[j+1])/2]*delta,{j,0,n-1}]
```

Out[44]= $-\frac{1705}{64}$

For polynomial functions, we can carry this analysis out to its ultimate conclusion and actually take the limit of the Riemann sums as $n \to \infty$. After unassigning n, divide the interval $[1, 6]$ into n subintervals of equal length $\Delta x = (6-1)/n$. The partition points

$$1 < 1 + \Delta x < 1 + 2\Delta x < \ldots < 1 + n\Delta x = 6.$$

are created by the function p defined next.

```
In[45]:= n=.; p[n_,j_]:=a+j*(b-a)/n
```

Notice that p **is actually a function of two variables**, although n remains fixed for each Riemann sum. The corresponding left Riemann sum is computed next. Mathematica assumes, given the location of n in the argument of the Sum[ ] command, that this unassigned letter represents a positive integer.

```
In[46]:= ns=Sum[f[p[n,j]]*(b-a)/n,{j,0,n-1}]
```

Out[46]= $\frac{5\left(9\ n - \frac{25-105\ n+86\ n^2}{6\ n}\right)}{n}$

These Riemann sum approximations for the integral turn into an exact value when we take a limit as $n \to \infty$.

```
In[47]:= actual=Limit[ns,n->Infinity]

Out[47]= -80/3
```

Is this limit correct? Just to gain some insight, we simplify ns until its limit value is obvious.

```
In[48]:= Simplify[ns]

Out[48]= -(5 (25 - 105 n + 32 n^2))/(6 n^2)

In[49]:= Apart[%]

Out[49]= -80/3 - 125/(6 n^2) + 175/(2 n)
```

Now the limit is obvious.

```
In[50]:= compare=N[{lrs,rrs,mrs,actual}]
Out[50]= {-22.3437, -31.0937, -26.6406,
          -26.6667}
```

We used a **new, basic command** in our work above. The command **Apart[ ] decomposes a rational expression into a sum of simpler terms**. The command Apart[ ] is essentially the **opposite of Together[ ]**.

Experiment with these ideas for **different values of n** (n =number of rectangles) and for **different functions**. This is an excellent opportunity to **gain genuine insight into the meaning of a definite integral**. Evaluate some left, right, and middle Riemann sums. Draw the rectangles used in the approximation process using the commands introduced at the beginning of this section. Notice that for n large, it doesn't visually seem to make much difference whether we use left, right, or mid-points to establish ($\pm$)heights of rectangles. Take the limit as n (the number of rectangles) goes to infinity. (Mathematica cannot always compute the limit, but it can when $f(x)$ is a polynomial.)

3.2 Integration and The Fundamental Theorem of Calculus

According to the *Fundamental theorem of calculus*, if f is continuous on the interval $[a, b]$, then the function F defined by

$$F(x) = \int_a^x f(t)dt, \tag{3.3}$$

is differentiable on $[a, b]$, and $F'(x) = f(x)$ for all x in $[a, b]$. As a consequence of 3.3, if G is any antiderivative of $f(x)$, then

$$\int_a^b f(x)dx = G(b) - G(a) = G(x) \mid_a^b . \tag{3.4}$$

Formula 3.3 is important for reasons other than the fact that it gives us formula 3.4, which is the most basic formula for evaluating definite integrals. Formula 3.3 also implies that every continuous function has an antiderivative, and it gives us a formula for it. This is of considerable consequence in its own right, and so we begin by using Mathematica to study F.

The command for evaluating a definite integral is (no surprise here) **Integrate[ ]**. Recall the function $f(x) = 9 - x^2$ from our Riemann sum example.

```
In[1]:= Integrate[9-x^2,{x,1,6}]
```

Out[1]= $-\frac{80}{3}$

Example 3.1 *Let $f(x) = \sin(x^3)$. Find an antiderivative F of f and verify. Compute the sample of values $F(2), F(3), F(6.4)$ as decimal values.*

Finding an antiderivative, is equivalent to evaluating the indefinite integral of $f(x)$. The same **Integrate[ ] command is used for both the definite and indefinite integral.** Only the **second argument is different** for the two types of integrals. We try the easy approach first, by asking Mathematica for a formula for the antiderivative.

```
In[2]:= Integrate[f[x],x]
```

Out[2]= $\int \sin[\cos[x]]dx$

Mathematica cannot compute the indefinite integral so it just returns the integral symbol. It turns out that the antiderivative is quite complicated and hard to express in any other way except as an integral using the *Fundamental Theorem of Calculus.* We can use any number as the lower limit. Different choices for the lower limit only change the value of an additive constant.

```
In[3]:= f[x_]:=Sin[Cos[x]]
```

```
In[4]:= F[x_]:=Integrate[f[t],{t,0,x}]
```

```
In[5]:= F[x]
```

Out[5]= $\int_0^x \sin[\cos[t]]dt$

Mathematica still can not evaluate the integral, but the function $F(x)$ is, nevertheless, well defined. For a particular x its value is just $A_1 - A_2$, where A_1 and A_2 are just certain areas associated with the graph of $y = f(t)$ which are above and below the t-axis on the interval between $t = 0$ and $t = x$.

```
In[6]:= F[2]
```

Out[6]= $\int_0^2 \sin[\cos[t]]dt$

Mathamatica cannot evaluate the above integral either, and so it just returns the integral symbol. It can, however, compute the derivative of $F(x)$.

```
In[7]:= D[F[x],x]
Out[7]= sin[cos[x]]
```

This shows that $F(x)$ is an antiderivative of $f(x)$ and confirms the *Fundamental Theorem of Calculus* for this example.

Mathematica cannot find the exact value of $F(x)$ for any x (other than $x = 0$), and so we will only seek a decimal approximation for $F(2)$. We can do this without an antiderivative.

The command **NIntegrate[ ] computes a numerical approximation to the value of an integral.**

```
In[8]:= NIntegrate[f[t],{t,0,2}]
Out[8]= 0.803863
```

The function $nF(x)$ defined next is a kind of pseudo-antiderivative of $f(x)$. For each x, $nF(x)$ is a decimal approximation for the exact antiderivative $F(x)$ of $f(x)$. We us it to compute our sample of decimal values of $F(x)$.

```
In[9]:= nF[x_]:=N[F[x]]
```

```
In[10]:= {nF[2],nF[3],nF[6.4]}
Out[10]= {0.803863, 0.11889, 0.0981523}
```

How does Mathematica perform these integral approximations? We could make them ourselves by approximating the integral with an appropriate Riemann sum. This is another reason why this definition is so important. Actually, Mathematica uses a much more sophisticated approach to approximating the value of a definite integral. We will discuss this later in the chapter.

Mathematica is able to compute exact values for a large number of integrals without any help from us (other than the initial command to integrate), but its capacity to integrate is not boundless. It is the nature of mathematics and not a weakness in Mathematica, that frequently forces it to fail at exact evaluation.. It can give exact values to definite integrals only if it can find a formula for an antiderivative. For continuous functions, antiderivatives always exist—that's what formula (3.3) says. "most" of the time, however, there is no formula, other than (3.3), for expressing an antiderivative in finite terms using elementary functions. Even when a formula exists, and can be found by Mathematica, it may look so awkward, or be so computationally expensive to compute, that it is undesirable to attempt an exact value. When an antiderivative for a definite integral is hard or impossible to compute exactly, we should turn instead to a numerical approximation using the command NIntegrate[ ]. The input statement:

NIntigrate[$f(x), \{x, a, b\}$]
immediately activates a genuine numerical approximation technique, without a search for an antiderivative, regardless of whether $f(x)$ has or does not have an elementary antiderivative. By way of contrast, the input statement:

N[Intigrate[$f(x), \{x, a, b\}$]]

first attempts an exact evaluation by way of antidifferentiation, before it gives a decimal approximation, and it uses a genuine numerical approximation technique only if an antiderivative cannot be found.

To end this discussion on a positive note, we should add that Mathematica is pretty good at evaluating integrals.

```
In[11]:= Integrate[2x*Cos[x^2],{x,1,3}]
Out[11]= -sin[1] + sin[9]
```

Mathematica **does not supply the arbitrary constant** for an indefinite integral.

```
In[12]:= Integrate[7*x^5+1/Sqrt[x]+5,x]
```

Out[12]= $2\sqrt{x} + 5\ x + \frac{7\ x^6}{6}$

Example 3.2 *Evaluate the integral* $\int_2^5 x^2 \sin(x)dx$. *How do you know the answer is correct?*

```
In[13]:= Integrate[x^2*Sin[x],{x,2,5}]
```

Out[13]= $2\ \cos[2] - 23\ \cos[5] - 4\ \sin[2] + 10\ \sin[5]$

This should be the answer. Is it?

```
In[14]:= F=Integrate[x^2*Sin[x],x]
```

Out[14]= $2\ \cos[x] - x^2\ \cos[x] + 2\ x\ \sin[x]$

This should be an antiderivative. Is it?

```
In[15]:= D[F,x]
```

Out[15]= $x^2\ \sin[x]$

So F(x) is an antiderivative.

```
In[16]:= (F/.x->5)-(F/.x->2)
```

Out[15]= $2\ \cos[2] - 23\ \cos[5] - 4\ \sin[2] + 10\ \sin[5]$

This verifies the result.

Example 3.3 *Find a function* $y = y(x)$ *which satisfies the following differential equation, and verify that that your answer is correct.*

$$\frac{dy}{dx} = \frac{3x^4 + 7x^3}{x^6}, \text{ and } y(-7) = 13$$

```
In[17]:= yp = (3*x^4 + 7*x^3)/x^6;
```

```
In[18]:= y = Integrate[yp, x] + c
```

Out[18]= $c - \frac{7}{2\ x^2} - \frac{3}{x}$

We form an equation with x replaced by -7, and y replaced by 13. Notice how this is done.

```
In[19]:= eq = (y /. x -> -7) == 13
```

Out[19]= $\frac{5}{14} + c == 13$

```
In[20]:= s = Solve[eq, c]
```

Out[20]= $\{\{c \to \frac{177}{14}\}\}$

```
In[21]:= y = y /. s[[1]]
```

Out[21]= $\frac{177}{14} - \frac{7}{2 x^2} - \frac{3}{x}$

This is the solution.

```
In[22]:= chk = {y /. {x -> -7}, D[y, x]}
```

Out[22]= $\left\{13, \frac{7}{x^3} + \frac{3}{x^2}\right\}$

This verifies the result.

3.3 Integration by Substitution

Mathematica does a great job at finding antiderivatives. If an antiderivative is expressible in terms of elementary functions, Mathematica will probably find it. Admittedly, antiderivatives can be found, without any knowledge of the mathematical techniques used to compute them. Should we just accept Mathematica's answer to an integral, or should we be concerned with how an answer evolves? The potential for error or misinterpretation is always present. Regardless of our level of involvement in mathematics, it is important to have sufficient skills to be able to look at output and decide whether or not it is correct.

On a deeper level, when an integral is evaluated, it is natural to wonder how the answer was determined, especially if the result is unexpected. By differentiating the answer, we can verify that it is correct, but this doesn't solve the mystery.

To gain a deeper understanding of integration, we need to turn off the computer and learn how to integrate with pencil and paper. Some pencil and paper work is probably essential. As a second step in the learning process, we can turn our computer back on and mimic this pencil and paper environment on a Mathematica notebook. With a computer, we can also do more involved pencil and paper manipulations then we could do otherwise. This is where we take up the study.

We plan to **set up an integral and prevent it from being evaluated**. While it is in this **unevaluated state**, we plan to **manipulate the integral until it is in a familiar elementary form**. Three user-defined commands are given below.

The first of these is so surprisingly simple, it needs to be explained. The notation (command) **int[ ] has no Mathematica meaning** what so ever, nor does the symbol (command) d[ ]. They both represent arbitrary functions. We use int[$f[x]$, $d[x]$] to represent the integral $\int f(x)\, dx$ in its unevaluated form. With this notation, we can keep the pair $f(x)$ and dx together but separated. The brackets around the x in $d[x]$ are important. To avoid the temptation to drop the brackets, we incorporate all of these features together to form the first of our three commands. As you can see below, the **InertInt[ ] command** is entered much like the regular Integrate[ ] command. The central tool for our work is the **IntBySubs[ ] command** defined below. Finally, the third command, **BasicInt[ ]**, is simply a call to evaluate an integral. We will use it once we have a recognizable elementary integral. Notice that the argument in this command and the first argument in the IntBySubs[ ] command is an inert integral, not a function.

Earlier in this chapter, we started a separate Mathematica notebook, called "MyLibrary," on which we entered some commands for drawing rectangular partitions over a graph. **Carefully enter the following three commands, exactly as they are given below, onto the same library file**. These commands should be **saved for occasional use in this manual**.

```
In[1]:= InertInt[f_,x_]:=int[f,d[x]]
```

```
In[2]:= IntBySubs[p_,U_,u_]:=Module[{x,g},x=Part[Part[p,2],1];
        g=(Part[p,1]/D[U,x])/.{(U)->u};InertInt[g,u]]
```

```
In[3]:= BasicInt[p_]:=Integrate[Part[p,1],Part[Part[p,2],1]]
```

By way of a partial explanation of the integration by substitution command, recall that when we make the substitution $u = U(x)$ in the integral $\int f(x)dx$, we use the differential $du = U'(x)dx$ to turn dx into du. By solving this for dx, notice that

$$\int f(x)dx, \text{ becomes } \int \frac{f(x)}{U'}du.$$

Of course, the entire integrand must be turned into an expression in u. All this is prominently displayed in the formula for IntBySubs[ ]. No attempt will be made to explain the remaining parts of these formulas. Let's just accept them on face value for now. Explanations will be made later at the appropriate time.

Let's suppose for the sake of the next example, that we are only going to accept the seven basic integration rules that follow immediately from the corresponding differentiation rules. These are the rules for the integrals

$$\int u^n du (n \neq -1), \int \sin(u)du, \int \cos(u)du, \int \sec^2(u)du$$

$$\int \sec(u)\tan(u)du, \int \csc^2(u)du, \int \csc(u)\cot(u)du$$

Example 3.4 *Use Mathematica to evaluate the integral* $\int (5x-3)\sqrt{2-7x}dx$. *Justify the answer by manipulating the integral symbolically until it is in one or more of the above seven basic forms.*

We first compute the integral directly with Mathematica's integration command.

```
In[4]:= f = (5*x - 3)*Sqrt[2 - 7*x];
```

```
In[5]:= a=Integrate[f,x]
```

Out[5]= $\sqrt{2-7\ x}\left(\frac{68}{147}-\frac{46\ x}{21}+2\ x^2\right)$

Why does the answer take this form? To answer this question, we manipulate the original integral, in much the same as we would with pencil and paper. Names like ix and iu are used to suggest integration with respect to x and u, respectively.

```
In[6]:= ix=InertInt[f,x]
```

Out[6]= $\text{int}\left[\sqrt{2-7\ x}\ (-3+5\ x),\ d[x]\right]$

We change the variable of integration to $u = 2 - 7x$. After the substitution, an x remains in the integrand. To complete the transformation over to the new variable u, we express x in terms of u and substitute.

```
In[7]:= iu=IntBySubs[ix,2-7x,u]
```

Out[7]= $\text{int}\left[-\frac{1}{7}\sqrt{u}\ (-3+5\ x),\ d[u]\right]$

```
In[8]:= s=Solve[u==2-7*x,x]
```

Out[8]= $\{\{x \to \frac{2-u}{7}\}\}$

```
In[9]:= iu1=iu/.s[[1]]
```

Out[9]= $\text{int}\left[-\frac{1}{7}\left(-3+\frac{5\ (2-u)}{7}\right)\sqrt{u},\ d[u]\right]$

Once we expand the integrand, each of the resulting terms will be an elementary integral. A **new command** is used to isolate the integrand, so that it can be expanded. If *expr* is an expression and j is a positive integer, then **Part[*expr*, j] is the j *th* term of *expr*.** The first term in int[A, B] is A. A full explanation of this very useful is presented on page 100, but for now, this command will only be used in an obvious way.

```
In[10]:= iu2 =Part[iu1,1]
```

Out[10]= $-\frac{1}{7}\left(-3+\frac{5\ (2-u)}{7}\right)\sqrt{u}$

```
In[11]:= iu3 = InertInt[Expand[iu2],u]
```

Out[11]= $\text{int}\left[\frac{11\ \sqrt{u}}{49}+\frac{5\ u^{3/2}}{49},\ d[u]\right]$

The above expression now has an obvious antiderivative.

```
In[12]:= iu4 = BasicInt[iu3]
```

Out[12]= $\frac{22\ u^{3/2}}{147}+\frac{2\ u^{5/2}}{49}$

To finish, we must return to an expression in x, the original variable.

```
In[13]:= b=iu4/.u->2-7x
```

Out[13]= $\frac{22}{147}\ (2-7\ x)^{3/2}+\frac{2}{49}\ (2-7\ x)^{5/2}$

Notice that our answer b looks different from the answer a that we obtained by a direct Mathematica evaluation of the integral. **Are the two answers the same? An easy way, well worth remembering, to resolve the issue, is to show that $a - b = 0$**

```
In[14]:= Simplify[a-b]
```

Out[14]= 0

We end this example with a more revealing way to show that the two answers are the same.

```
In[15]:= Expand[b]
```

Out[15]= $\frac{68}{147}\ \sqrt{2-7\ x}-\frac{46}{21}\ \sqrt{2-7\ x}\ x+2\ \sqrt{2-7\ x}\ x^2$

```
In[16]:= Expand[a]
```

Out[16]= $\frac{68}{147}\ \sqrt{2-7\ x}-\frac{46}{21}\ \sqrt{2-7\ x}\ x+2\ \sqrt{2-7\ x}\ x^2$

3.4 Numerical Integration

Mathematica uses fairly sophisticated techniques for approximating the value of a definite integral. Rather than discuss these techniques, we look instead at *Simpson's Rule*, one of two numerical integration techniques typically introduced in a standard calculus course. An understanding of this rule will give us some insight into the general idea of numerical integration, and will hopefully give us an appreciation of Mathematica's NIntegrate[ ] command in the process.

We begin by using Mathematica to derive the formula for *Simpson's Rule*. This is a fascinating experience to observe. Recall that in setting up this rule, we divide the interval $[a, b]$ into an even number, $n = 2m$, of subintervals of equal length

$$h = \frac{b-a}{n}.$$

This determines the partition points

$$a = x_0 < x_1 \ldots < x_n = b, \text{ where } x_k = a + kh, (k = 0, 1, \ldots, n).$$

These subintervals are grouped together in adjacent pairs and on each of the pairs of subintervals the function $f(x)$ is approximated by a quadratic function $q(x)$. Rather than integrate $f(x)$ on a subinterval, we integrate its quadratic approximation over the subinterval instead. *Simpson's Rule* is obtained by adding these m $(m = n/2)$ quadratic integrals.

Mathematica is an excellent environment to carry out all of these details. For each k $(k = 1, 2, \ldots, n)$, let $y_k = f(x_k)$. We **determine the function $q(x)$ on the first pair of subintervals and then integrate**. It is an easy matter to generalize the result to the other subinterval pairs. Let $q(x)$ be the quadratic function with $q(x_0) = y_0, q(x_1) = y_1$ and $q(x_2) = y_2$.

```
In[1]:= q[x_]:=a*x^2+b*x+c
```

```
In[2]:= Q=Integrate[q[x],{x,x0,x0+2h}]
```

$$\text{Out[2]=}\ 2\ c\ h + 2\ b\ h^2 + \frac{8\ a\ h^3}{3} + 2\ b\ h\ x0 + 4\ a\ h^2\ x0 + 2\ a\ h\ x0^2$$

```
In[3]:= {eq1=q[x0]==y0,eq2=q[x0+h]==y1,eq3=q[x0+2*h]==y2}
```

$$\text{Out[3]=}\ \{c + b\ x0 + a\ x0^2 == y0,\ c + b\ (h + x0) + a\ (h + x0)^2 == y1,\ c + b\ (2\ h + x0) + a\ (2\ h + x0)^2 == y2\}$$

```
In[4]:= s=Solve[{eq1,eq2,eq3},{a,b,c}]
```

$$\text{Out[4]=}\ \{\{a \to -\frac{-y0 + 2\ y1 - y2}{2\ h^2},\ b \to -\frac{3\ h\ y0 + 2\ x0\ y0 - 4\ h\ y1 - 4\ x0\ y1 + h\ y2 + 2\ x0\ y2}{2\ h^2},\ c \to -\frac{1}{2\ h^2}(-2\ h^2\ y0 - 3\ h\ x0\ y0 - x0^2\ y0 + 4\ h\ x0\ y1 + 2\ x0^2\ y1 - h\ x0\ y2 - x0^2\ y2)\}\}$$

```
In[5]:= {a,b,c}={a,b,c}/.s[[1]];
```

```
In[6]:= Q
```

```
Out[6]= -3 h y0 -2 x0 y0 +4 h y1 +4 x0 y1 - 4/3 h (-y0 +2 y1 -y2) -
    2 x0 (-y0 +2 y1 - y2) - x0^2 (-y0 +2 y1 -y2)/h - h y2 -2 x0 y2 -
    x0 (3 h y0 +2 x0 y0 -4 h y1 -4 x0 y1 +h y2 +2 x0 y2)/h -
    1/h (- 2 h^2 y0 -3 h x0 y0 -x0^2 y0 +4 h x0 y1 +2 x0^2 y1-
      h x0 y2 -x0^2 y2)
```

```
In[7]:= Simplify[Q]
Out[7]= 1/3 h (y0 +4 y1 +y2)
```

An expression identical to this except for the subscripts on the symbols y_j will exist for each of the adjacent pairs of intervals. Simpson's Rule is easily seen to be the consequence of adding all m such expressions together.

Example 3.5 *Use* Simpson's Rule *with $n = 50$ to approximate the value of $\int_1^4 \frac{x^3}{\sqrt{x^2+9}}dx$. Estimate the size of the error term using Simpson's 4th derivative error estimate. Find an exact value for the integral and compare.*

If the 4 *th* derivative $f^{(4)}$ is continuous on the interval $[a, b]$, then the error term

$$E_S = \mid \int_a^b f(x)dx - S_n \mid$$

involved in approximating $\int_a^b f(x)dx$ by its *Simpson's Rule* approximation S_n, satisfies

$$E_S < \frac{b-a}{180} M h^4$$

where,

$$h = \frac{b-a}{n}, \text{ and } M = \max \mid f^{(4)}(x) \mid \text{ for } x \text{ in } [a, b]$$

While the error term can be a chore to compute by hand, notice in the next Mathematica work session how effectively this term can be estimated.

```
In[8]:= f[x_]:=x^3/Sqrt[x^2+9]
```

```
In[9]:= a=1;b=4;n=50;m=n/2;h=(b-a)/n;
```

#To set up the sum for *Simpson's Rule* notice that the multipliers in front of the function values appear as $(1), (4), (2), (4), (2), (4), (2), \ldots, (4), (1)$, ending with a (4) and then a (1). They can be grouped so that they appear as

```
[(1), (4)], [(2), (4)], [(2), (4)], \ldots ,[(2), (4)], [(1)].
```

This may help explain why the pairs with $(2), (4)$ multipliers are grouped together as they are in the sum, leaving us with two exceptional terms in the beginning and one exceptional term at the end. Then notice that the (2) multipliers appear in front of $f(x_j)$ (in other words $f(a + j * h)$) for j **even** and the multipliers of (4) appear in front of $f(x_j)$ for j **odd**. This is why we used indices of the form $(2k)$ and $(2k + 1)$ in the sum instead of simply (j).

```
In[10]:= S=N[h/3*(f[a]+4f[a+h]+Sum[2f[a+2*k*h]+4f[a+(2k+1)*h],
         {k,1,m-1}]+f[b])]

Out[10]= 14.5862
```

This is the Simpson's Rule approximation. The use of the **decimal command N[ ] is critical** in the above computation. Mathematica wants to be exact. **Imagine how complicated such a sum would be if we asked for an exact value.**

```
In[11]:= error=(b-a)/180*h^4*M
```

$$\text{Out[11]}= \frac{27\ M}{125000000}$$

The value of M, which we find next, is the maximum of $|f^4(x)|$ on [1, 4]. Think about what the value of M is going to be used for, and you will realize that it is inappropriate to attempt to find M to any degree of accuracy. All we need is to replace M by any convenient number a little larger than M. To find such a number we plot $f^{(4)}(x)$ and by eye sight read off a convenient number slightly bigger than $|\ f^{(4)}(x)\ |$ on the interval. The choice is in the eye of the beholder.

```
In[12]:= fp4=D[f[x],{x,4}]
```

$$\text{Out[12]}= \frac{105\ x^7}{(9+x^2)^{9/2}} - \frac{270\ x^5}{(9+x^2)^{7/2}} + \frac{225\ x^3}{(9+x^2)^{5/2}} - \frac{60\ x}{(9+x^2)^{3/2}}$$

```
In[13]:= Plot[fp4,{x,1,4}];
```

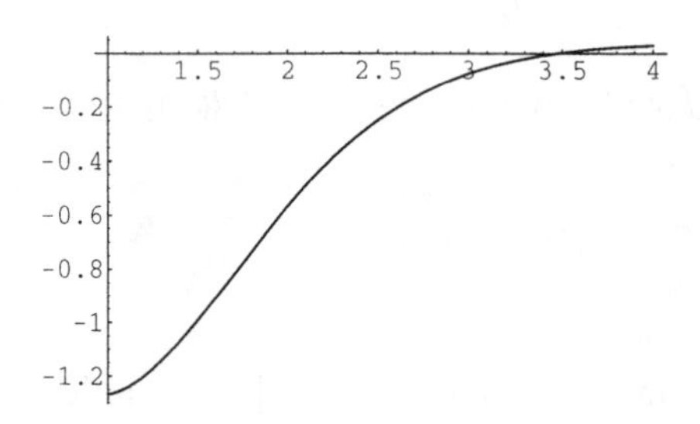


```
In[14]:= M=1.5;error
Out[14]= 3.24 10^-7
```

So the error term satisfies Error < 0.0000004.

The next and last example presents an opportunity to make use of one of the most versatile of Mathematica's commands. We have used the **Module[ ] command** without comment before. This is a good time to take a closer look.

The Module[ ] command has two arguments. The first is a list of local symbols, and the second argument is an ordered sequence of input statements (possibly just one) separated by semicolons (;). We've used the set-up appearing in the second argument before, and it works much the same way in this command. The input statements are performed sequentially from left to right. The semicolons not only separate input statements, but they suppress the display of output. Only the last output statement is displayed (as long as its input statement does not end with a semicolon).

Local symbols are very useful. An assignment such as $a = 3$, is a **global assignment**. The value of a is permanently 3 everywhere in Mathematica, until its value is changed by another assignment, or we quit Mathematica. In contrast, local assignments are very temporary and local. When a command is used, we would not want to inadvertently assign permanent values to unknown names, nor would we want our results to depend in other unknown assignments that were made earlier. If p, q, r are local symbols in a Module command their values are independent of whatever global values these symbols might have, and whatever local values they acquire when the Module command is entered disappear when the output appears.

If no local symbols are needed, then the Module command is not needed. Just enter the sequence of input states as we did before. Here is a simple example that serves no other purpose except to explain the significance of local and global symbols, and the sequential nature of the input statements. Look at the initial global assignments (and unassignments), the **output** of the command, and the **unchanged values of the symbols after the command is used**.

```
In[15]:= Clear[y,h];x=3;k=1;
```

```
In[16]:= p[x_,y_]:=Module[{h,k},h=x+y;k=x*y;h+k]
```

```
In[17]:= p[100,5]
Out[17]= 605
```

Notice the unchanged values of x, y, h, k.

```
In[18]:= {x,y,h,k}
Out[18]= {3, y, h, 1}
```

We will use the module command to formulate *Simpson's Rule* in a much more convenient way, and in the process we hope to demonstrate the power of this very useful command.

Example 3.6 *Create a procedure command for determining the Simpson's Rule approximation of order n for $\int_a^b f(x)\,dx$. Use it to determine the Simpson's Rule approximation of order 20 for $\int_1^8 \sqrt{3x+1}dx$ and of order 50 for $\int_0^\pi \sin(x)dx$*

We define a module such that **simpson[f, a, b, n]**, is the Simpson's Rule approximation of order n for the integral $\int_a^b f(x)dx$. The main formula is essentially the same as the formula we used for the Simpson's Rule approximation in the last example. Just copy, paste and make small adjustments. **The N[ ] command plays an important role**. An exact sum would be much too complicated.

```
In[19]:= simpson[f_,a_,b_,n_]:=Module[{h,m,s1,s2},
         h=(b-a)/n; m=n/2;s1=f[a]+4f[a+h]+f[b];
         s2=Sum[2f[a+2*k*h]+4f[a+(2k+1)*h],{k,1,m-1}];
         N[(s1+s2)*h/3]]
```

We used several local symbols just to demonstrate their function. All of this could have been put together in the form of one input statement, but **notice how the use of local symbols helps to break up the process into a sequence of simple steps**. With this command, look how easy it is to form the formulas. The integrals in this example are easily evaluated by antidifferentiation techniques, so we can compare the approximations with their exact values of 26 and 2.

```
In[20]:= g[x_]=Sqrt[3x+1];simpson[g,1,8,20]
```

```
Out[20]= 26.
```

```
In[21]:= simpson[Sin,0,Pi,50]
Out[21]= 2.
```

3.5 Exercise Set

Some New and Some Old Mathematica features and Commands

Apart[ ]	BasicInt[ ]	Clear[ ]
D[ ]	delayed assignment: $(a := b)$	Derivative[n][ ]
Expand[ ]	FindRoot[ ]	input seq separator: (;)
InertInt[ ]	IntBySubs[ ]	Integrate[,x]
Integrate[,$\{x, a, b\}$]	LeftRiemannDisplay[ ]	MidRiemannDisplay[ ]
Module[ ]	N[ ]	NIntegrate[ ]
Plot[ ]	RightRiemannDisplay[ ]	s[[j]]=j th term of s
Simplify[ ]	Solve[ ]	Sum[ ]
substitution: $(expr/.x->a)$	Table[ ]	Together[ ]

Remember the method ($g[x\_]$:=expression in x) used to create a function g.

1. Set up and evaluate (as decimal numbers) the left, right and midpoint Riemann sums for $n = 30$ corresponding to the integral $\int_{-3}^{2} x^3 dx$. Draw the graphs showing the approximating rectangles. Compute the exact value of the integral by taking the limit of the left end point Riemann sums as n, the number of rectangles, goes to infinity.

2. Let $f(x) = \frac{x}{x+\sin(x)}$

a) Use Mathematica to find a function $F(x)$ such that $F'(x) = f(x)$, for x in the interval $[1, \infty)$.If Mathematica cannot find the antiderivative directly, use the Fundamental Theorem of Calculus to define it.

b) Evaluate $F(18)$ and $F(3)$ as decimal numbers. Compare the value of $F(18) - F(3)$ with a direct evaluation of $\int_3^{18} f(x)dx$ as a decimal. Why are the answers the same?

c) Use one of Mathematica's differentiation commands to compute $F'(x)$. Verify the answer by using the definition of a derivative as a limit of a difference quotient to compute $F'(x)$.

d) Use Mathematica to compute the derivatives of the functions

$$g(x) = F(3x^4 + 7x^2 + 9), h(x) = F(\sin(x)).$$

The answers should be anticipated without using Mathematica. Why?

e) Find the equation of the tangent line to the graph of g at $x = 4$, for the function g in part d).

3. Find exact values for the following integrals. Verify that your answers are correct.

$a)$ $\int (2 + x + x^2)\sqrt{3x - 5}\, dx$ $\quad b)$ $\int \sin^3(2y)\cos^4(2y)\, dy$

$c)$ $\int_0^3 (x^3)\sqrt{25 - x^2}\, dx$

4. Evaluate the integral $\int_2^7 \frac{\sin(x)}{\sqrt{x}} dx$ as a decimal.

5. Find y as a function of x if $\frac{dy}{dx} = 3x^3 + 7x^2 - 4x - 8$, and if $y = -17$ when $x = 6$.

6. Find y as a function of x if $\frac{d^2y}{dx^2} = \frac{7x+5}{(3x+8)^3}$, $y(-1) = 2$, and $y\prime(-1) = 4$. Verify that the answer is correct.

7. Show that $\int_a^b f(x)g(x)\,dx = \int_a^b f(x)\,dx \int_a^b g(x)\,dx$ is *false*, by using an appropriate counterexample.

8. Evaluate the integral $\int \sin(3x)\sqrt{4+\cos(3x)}dx$. Verify Mathematica's answer by manipulating the original integral algebraically, and changing the variable of integration, until it looks like one or more of the seven basic integrals.

9. Evaluate $\int_1^3 \sqrt{x^2+1}\,dx$ as a decimal. Use the Trapezoidal Rule and Simpson's Rule with $n = 20$ to estimate the integral. Approximate the size of the error terms by using formulas for the error terms given in your calculus text.

10. Mathematica can "prove" (if that's what you should call it) the Fundamental Theorem of Calculus. This is an interesting (and easy) exercise. It will make you think about the proof, and it will demonstrate how good Mathematica is at handling really abstract mathematical expressions. Define

$$F(x) = \int_a^x f(t)dt$$

and use Mathematica to compute

$$F'(x) = \lim_{h \to 0}\left(\frac{F(x+h) - F(x)}{h}\right).$$

Incidentally, Mathematica assumes continuity of the function f in this problem.

Project: Cubic Rule

Use the following instructions to derive another numerical integration rule for $\int_a^b f(x)dx$. You might want to call it the Cubic Rule. Divide the interval $[a, b]$ into $n = 3m$ subintervals of equal length

$$h = \frac{b-a}{n}.$$

This establishes the partition points $a = x_0 < x_1 < \ldots < x_n = b$ in the usual way. Let $y_k = f(x_k)$ for each $k = 1, 2, \ldots, x_n$. Group the subintervals into adjacent triplets of subintervals. (This can be done since n is a multiple of 3.) In each triplet of subintervals, construct a cubic polynomial which agrees with $f(x)$ at the four partition points in the interval. Approximate the integral of $f(x)$ on this triplet of subintervals by integrating the cubic instead. Add up all $m = n/3$ of these approximations in order to get the sought after Cubic Rule. (Use the derivation of Simpson's Rule in this manual for motivation. Do all your computations on the first triplet of subintervals, simplify, generalize to the other triplets, and add up the results.)

Express your rule as a Mathematica command (use the Module[ ] command, in the variables f, a, b, and n. Remember that Mathematica wants to be exact so put N[ ] around your expression, or it will lead to undesirable complications.

Approximate the integral $\int_0^\pi \sin(x)dx$ using this new rule with $n = 48$ (a multiple of 3). Compare your answer with the answer obtained by using Simpson's Rule with $n = 48$ and also with the approximation obtained by using Mathematica's NIntegrate[ ] command. Finally, notice that this is an easy integral with an exact value of 2, so you can compare all three of these approximations to the exact value. One would expect the Cubic Rule to be an improvement over Simpson's Rule; after all, we're approximating by cubic polynomials rather than quadratics. Cubics should "fit the curve" better. However, you probably will not experience an improvement in approximating this integral.

Chapter 4

Applications of Integration

We began the last chapter with a remark concerning the importance of the definition of a definite integral as the limit of its Riemann sums. As you can now see in your calculus text, science is rich with examples of situations which have an interpretation as a limit of some Riemann sum and hence as a definite integral. In this chapter, we follow up on the ideas presented in your calculus text, and show how Mathematica can be used to solve some interesting applied problems.

In this chapter, we will need at least one command which is not automatically available when Mathematica opens up. To prepare for that event, we introduce another basic part of the notebook environment.

Mathematica has a large number of specialized **packages** of commands. Some packages have several commands, and others have only one. Mathematica is so big, it could not possibly load everything automatically, and so, when a notebook is opened, it only makes available a list (a very large list) of the most common and frequently used commands. To see what packages are available, pull down the **Help Menu**, and click the **Add-ons button**, then click on **Standard packages**. Additional help on working with packages is available. Just look for it after you click the Add-ons button. **The name of the package is the <u>context</u> for the commands inside of the package.** Recall that **each context name ends with a backquote (`), and this same punctuation mark must follow each package name and each subpackage name as well.** For example, inside of the Graphics` package is the ImplicitPlot` package, so the full name of this package is Graphics`ImplicitPlot`. **The backquotes must appear in both places.**

To **load this package**, so that we can use its commands, we enter

```
<<Graphics`ImplicitPlot`
```

in an input cell. **The post fix symbol << is equivalent to the command Get[ ].**There are subtleties, but only one is likely to be an immediate concern. If you try to **use a command** that resides in a package **before you load the package**, your blunder will not be entirely forgotten. When you realize your mistake and try to load the package, Mathematica will **want to use your "false" version of the command**, and it will **not let you load** the command with the same name in the package. Mathematica will tell you that there is a "conflict of names." The rest of the package, however, will load. When this happens, **you must remove the offending name with the Remove[ ] command,** before you can load the command from the package. After the false command is removed, just enter the package again.

4.1 Sample Problems

We remarked earlier that **Mathematica can plot equations, but not with the Plot[ ] command**. An equation, recall, is simply an expression, which involves the equality symbol (==).

The command we use to plot equations in two variables is **Implicitplot[ ]**. It resides in the package we referred to above.

Example 4.1 *Find the area of the region between the curves defined by the equations* $x^2 + y^2 = 25$ *and* $x = 7y - 3y^2 - 1$.

To begin this problem, a plot is essential, and since the command resides in a package, we must first load the package.

```
In[1]:= <<Graphics`ImplicitPlot`
```

```
In[2]:= eq1=x^2+y^2==25; eq2=x==-1+7*y-3*y^2;
```

```
In[3]:= ImplicitPlot[{eq1,eq2},{x,-10,10}];
```

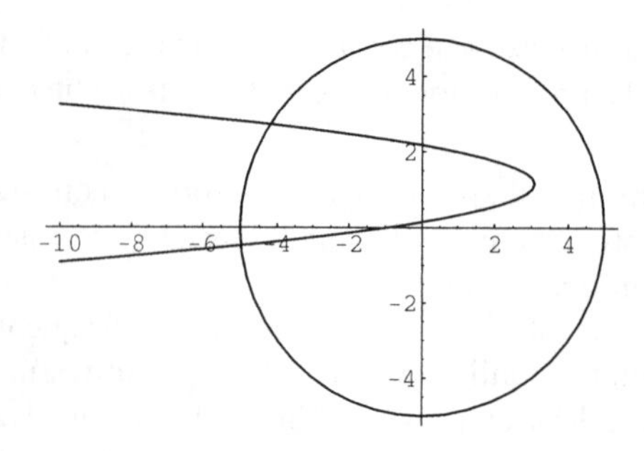

Two equations must be entered into the first argument of the implicitplot command, so they must be **enclosed in set brackets**. This standard practice (for all commands) has been mentioned several times now, so it should become routine. The second argument is obviously a range for x. **If a range for y, in the form $\{y, y_{min}, y_{max}\}$, is entered as a third argument, an entirely different algorithm is used to plot the equations.** Experiment with both forms as you work through problems.

From the above plot, it is clear that it would be very difficult to find the area of the region by dividing it up into "thin" vertical rectangles. The formula for the length of a "thin" rectangle located over (or on) the point x on the x-axis appears to change two times (actually three times) depending on what curves the top and bottom sides of the rectangle intersect. (The third change is quite small.)

On the other hand a "thin" horizontal rectangle always has its right end point on the parabola and its left end point on the left half of the circle. Think of a "thin" horizontal rectangle passing through the point y on the y-axis. Its width (the small dimension) is dy and its length is $x_2 - x_1$, where (x_2, y) is a point on the right boundary ($eq2$), and (x_1, y) is a point on the left boundary ($eq1$).

A horizontal distance is always obtained by taking the x-coordinate on the right minus the x-coordinate on the left. Since we plan to integrate (collect and add rectangles) with respect to y, these x-coordinates must be expressed in terms of y.

```
In[4]:= s=Solve[eq1,x]
```

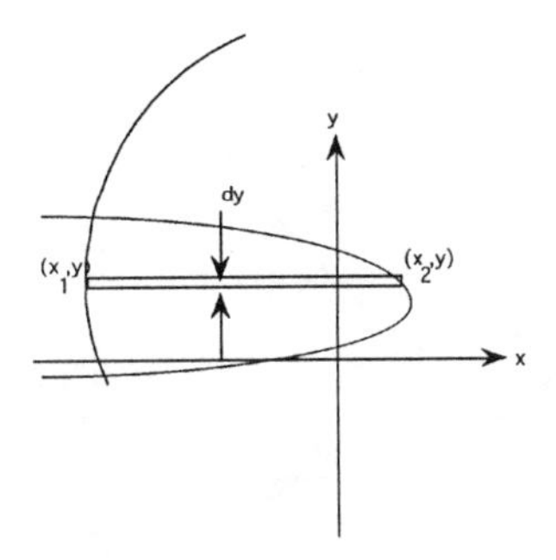


```
Out[4]= {{x → -√(25 - y²)}, {x → √(25 - y²)}}

In[5]:= xl=x/.s[[1]]
Out[5]= -√(25 - y²)

In[6]:= s=Solve[eq2,x]
Out[6]= {{x → -1 + 7 y - 3 y²}}

In[7]:= xr=x/.s[[1]]
Out[7]= -1 + 7 y - 3 y²
```

The area of this "thin" horizontal rectangle is just Ldy. The dy is an assumed part of the integration command. If you write dy inside the integral, Mathematica will not understand. Collecting and adding, in the sense of integration, all of the "little" rectangular areas, means integrating L with respect to y.

```
In[8]:= L=xr-xl
Out[8]= -1 + 7 y - 3 y² + √(25 - y²)

In[9]:= a=y/.FindRoot[L==0,{y,-1}]
Out[9]= -0.472536
```

Notice the efficient way in which a was assigned a value above. The output form, $y- > y_0$, for the FindRoot[ ] command allows for very efficient combinations of commands.

```
In[10]:= b=y/.FindRoot[L==0,{y,2}]
Out[10]= 2.72403

In[11]:= area=NIntegrate[L,{y,a,b}]
Out[11]= 16.9462
```

Example 4.2 *Find the volume of the solid of revolution formed by revolving about the line* $x = 10$, *the region bounded by the curves* $y = x^2 - 8x + 26$ *and* $y = 4 + 12x - x^2$.

```
In[12]:= f[x_]:=4-x^2+12*x; g[x_]:=x^2-8*x+26;
```

```
In[13]:= Plot[{f[x],g[x]},{x,0,10}];
```

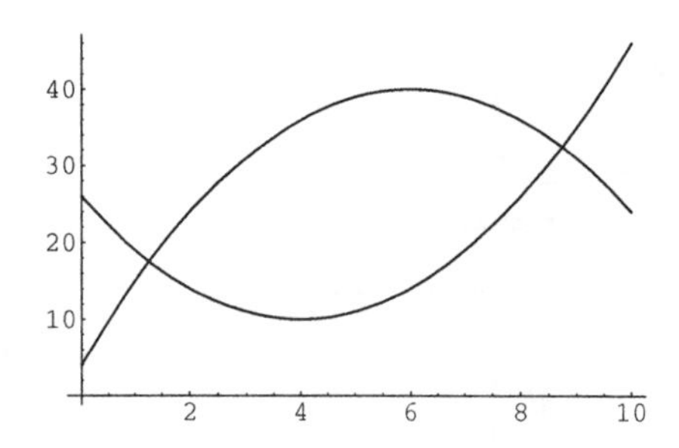

A "thin" vertical rectangle located at the point x on the x-axis inside this region and revolved about the line $x = 10$ would generate a shell. Collecting shells with respect to x, means integrating with respect to x, and so all of the variables must be expressed in terms of x. The volume of a shell is

$$v = 2\pi RLT,$$

where R is the radius, L is the length of the shell, and T is its thickness.

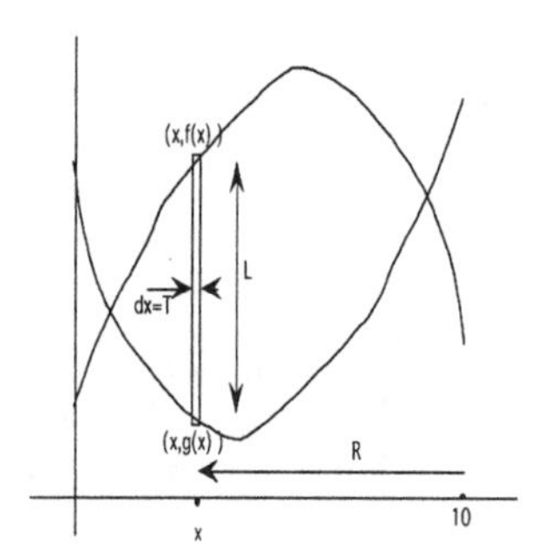

The thickness, $T = dx$, is an assumed part of Mathematica's integration command, and so it is not a part of our expression for v. Collecting and adding up, in the sense of integration, all of the volumes of the "little" shells means integrating $v = 2\pi RL$ once we have R and L in terms of x.#

```
In[14]:= L=f[x]-g[x];
```

This vertical length is the y-coordinate above minus the y-coordinate below.

```
In[15]:= R=10-x;
```

This horizontal length is the x-coordinate on the right minus the x-coordinate on the left.

```
In[16]:= v=2*Pi*R*L
```

Out[16]= $2\ \pi\ (10 - x)\ \left(-22 + 20\ x - 2\ x^2\right)$

```
In[17]:= s=Solve[f[x]==g[x],x]
```

Out[17]= $\left\{\left\{x \to 5 - \sqrt{14}\right\}, \left\{x \to 5 + \sqrt{14}\right\}\right\}$

```
In[18]:= {a,b}=x/.s
```

```
Out[18]= {5 - √14, 5 + √14}

In[19]:= NIntegrate[v,{x,a,b}]
Out[19]= 4388.44
```

Example 4.3 *A swimming pool has an elliptical shape as shown in the picture. The water is 3 feet deep at one end, 12 feet deep at the other end, and the depth changes linearly from one end to the other. How much water (in cubic feet) is in the pool, and what is its weight in pounds?*

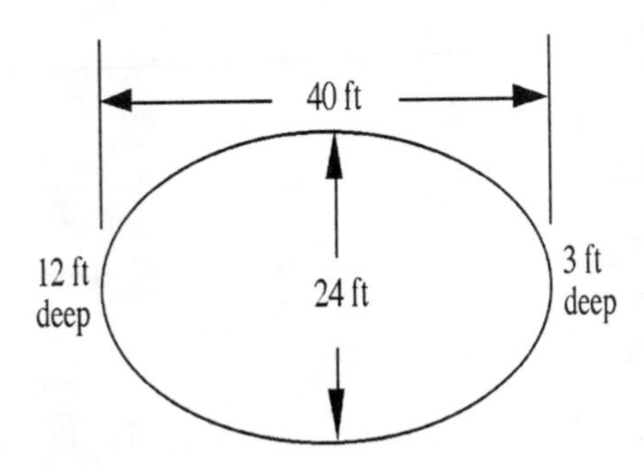

The equation of an ellipse centered at the origin with a horizontal major axis of length 40 and a vertical minor axis of length 24 is

$$\frac{x^2}{20^2} + \frac{y^2}{12^2} = 1$$

It is easy to see that the ellipse crosses the x-axis at $(\pm 20, 0)$ and crosses the y-axis at $(0, \pm 12)$. Using cuts of thickness dx, perpendicular to the x-axis, the (3-dimensional) pool can be partitioned into thin, essentially rectangular boxes, or "slabs" of water. The volume of this rectangular "slab" of water is easily expressed as the product of its three dimensions.

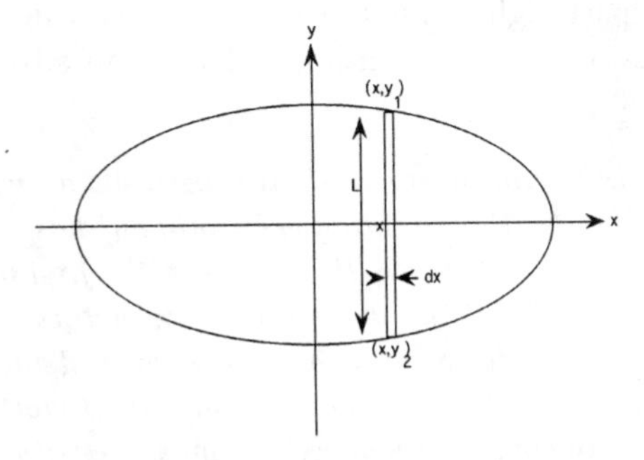

The three dimensions of the slab are L, dx, and the dimension, d (not visible in the diagram), which is perpendicular to the plane of the paper and represents the water depth of the pool at x. The dx is an assumed part of the integration command, and so the volume of the thin slab takes the form $v = Ld$. We add (in the sense of integration) the volumes of all of the thin slabs to form the total volume of the water. The answer, naturally, is expressed in cubic feet, and the total weight of the water is obtained by multiplying its volume by water's density of 62.5 pounds per cubic foot.

Both L and d must be expressed in terms of x in order to do the integration. The relationship between d and x is linear. We simply write the equation of the line through the points $(x, d), (20, 3), (-20, 12)$.

```
In[20]:= eq=x^2/20^2+y^2/12^2==1;

In[21]:= d=3+(12-3)/(-20-20)*(x-20);

In[22]:= s=Solve[eq,y]
```

Out[22]= $\{\{y \to -\frac{3}{5}\sqrt{400-x^2}\}, \{y \to \frac{3\sqrt{400-x^2}}{5}\}\}$

```
In[23]:= {y1,y2}=y/.s
```

Out[23]= $\{-\frac{3}{5}\sqrt{400-x^2}, \frac{3\sqrt{400-x^2}}{5}\}$

```
In[24]:= L=y2-y1;
```

This vertical length is the y-coordinate above minus the y-coordinate below.

```
In[25]:= v=L*d
```

Out[25]= $\frac{6}{5}\left(3-\frac{9}{40}(-20+x)\right)\sqrt{400-x^2}$

```
In[26]:= volume=NIntegrate[v,{x,-20,20}]
Out[26]= 5654.87

In[27]:= weight=62.5*volume
Out[27]= 353429.
```

Captain Ralph is an outer-space fighter pilot, with whom we will share an occasional adventure in the future. Our first encounter with him is hardly an adventure, but it would be if he fails in his mission.

Example 4.4 *Captain Ralph is piloting a small shuttle between a space station orbiting Mars and the planet's surface. The shuttle (standing vertically up) has the shape of the solid of revolution formed by revolving about the y-axis the region in the first quadrant bounded by the coordinate axes and $y = 36 - (9x^4)/4$ (The units are all in meters.) As you can see the ship is quite narrow, being 36 meters tall and only 4 meters wide at the base. (If you wish to get a "true" picture of the ship by plotting this curve, you should plot with the same scale on both axes.) The ship is also quite nose heavy, with a mass density of $\delta = 219 + 50\sqrt{y}$ kilograms per cubic meter at level y. Density here should be thought of as the mass of a "thin horizontal slice" at level y divided by its volume with the mass of a slice centered on the y-axis.*

Captain Ralph has been ordered to transport a 2,000 pound laser gun down to the planet's surface, and Ralph has decided that the only safe way to do this is to place the gun as close to the ship's center of mass as possible. Please help Captain Ralph avoid an adventure by finding this point.

To help Captain Ralph we must find the total mass M of the ship and its moment M_x with respect to the x-axis. Then the $y-$coordinate of the center of mass would be $\overline{y} = M_x/M$. Since the center of mass is on the y-axis, $\overline{y}$ is all we need. To generate these terms we partition the ship into "thin" horizontal disk shaped slices. The thickness of a slice is dy, which is omitted from all of our calculations for the same reason as in previous examples. Except for the omission of dy, we have that the volume v of a slice is $v = \pi r^2$, its mass is $m = vd(d = \delta)$, and its moment with respect to the x-axis is $m_x = my$.

```
In[28]:= f[x_]=36-9/4*x^4;
```

```
In[29]:= Plot[f[x],{x, -20, 20},PlotRange->{0, 40},AspectRatio->1];
```

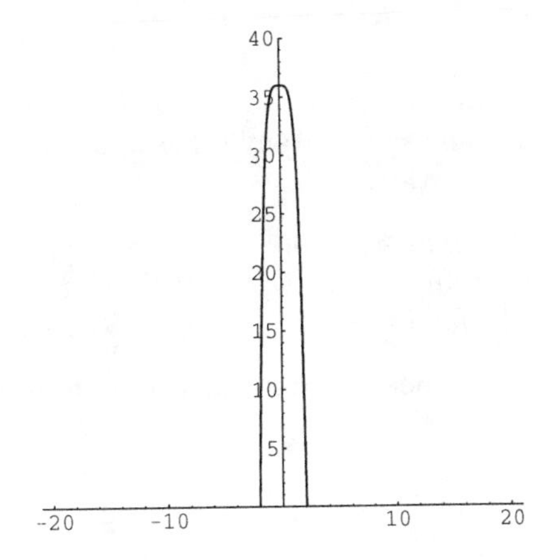

Mathematica normally takes artistic liberty to fill a plot region with a "best fit" graph. Notice the options we included in this plot command to get a true scale picture of the ship. The rule **AspectRatio– >1 (a ratio of horizontal to vertical axis length) creates a plot with the same lengths on the horizontal and vertical axes. the x and y axes the same.**

```
In[30]:= s=Solve[f[x]==y,x]
```

Out[30]= $\left\{\left\{x \to -\sqrt{\frac{2}{3}}\,(36-y)^{1/4}\right\}, \left\{x \to -i\,\sqrt{\frac{2}{3}}\,(36-y)^{1/4}\right\}, \left\{x \to i\,\sqrt{\frac{2}{3}}\,(36-y)^{1/4}\right\}, \left\{x \to \sqrt{\frac{2}{3}}\,(36-y)^{1/4}\right\}\right\}$

```
In[31]:= r=x/.s[[4]]
```

Out[31]= $\sqrt{\frac{2}{3}}\,(36-y)^{1/4}$

```
In[32]:= v=Pi*r^2
```

Out[32]= $\frac{2}{3}\,\pi\,\sqrt{36-y}$

```
In[33]:= d=219+50*Sqrt[y];
```

```
In[34]:= {m=d*v,mx=m*y}
```

Out[34]= $\left\{\frac{2}{3}\ \pi\ \left(219+50\ \sqrt{y}\right)\ \sqrt{36-y},\ \frac{2}{3}\ \pi\ \left(219+50\ \sqrt{y}\right)\ \sqrt{36-y}\ y\right\}$

```
In[35]:= M=NIntegrate[m,{y,0,36}]
Out[35]= 119345.
```

```
In[36]:= Mx=NIntegrate[mx,{y,0,36}]
```

Out[36]= $1.91043\ 10^{6}$

```
In[37]:= ybar=Mx/M
Out[37]= 16.0077
```

So far, all of the applications we have considered were a consequence of the definition of a definite integral as a limit of its Riemann sums. Many other applications are a consequence of interpreting an integral as an antiderivative.

Example 4.5 *A new mutual fund is growing at the rate of* $.83 - 1/\sqrt{(t+5)}$ *in millions of dollars per year, where* t *is the age of the fund in years. If the initial value of the fund was \$94,000, how long will it take for the fund to reach a value of \$7,000,000 ?*

If r denotes the above rate, and if v denotes value at time t in millions of dollars, then $r = \frac{dv}{dt}$.

```
In[38]:= r=0.83-1/Sqrt[t+5];
```

```
In[39]:= v=Integrate[r,t]+c
```

Out[39]= $c + 0.83\ t + \frac{0.}{5+t} - 2.\ \sqrt{5+t}$

Notice that\$94,000 is 0.094 in millions of dollars.

```
In[40]:= Solve[(v/.{t->0})==0.094]
Out[40]= {{c → 4.56614}}
```

```
In[41]:= c=c/.%[[1]]
Out[41]= 4.56614
```

4.2 Exercise Set

Some New and Some Old Mathematica features and Commands

Apart[ ]	Clear[ ]	delayed assignment $a := b$
Expand[ ]	FindRoot[ ]	Get[ ] or <<
ImplicitPlot[ ]	Integrate[,x]	Integrate[,$\{x, a, b\}$]
N[ ]	NIntegrate[ ]	Plot[ ]
Remove[ ]	Simplify[ ]	Solve[ ]
Sum[ ]	substitution: $expr/.x- > a$	Table[ ]
Together[ ]	<<Graphics`ImplicitPlot`	

Remember the method ($g[x\_]$:=expression in x) used to create a function g.

1. Find the area of the region in the first quadrant, bounded by

$$y = \frac{20}{x^2+1}, \quad x = 2, \text{ and } y = 1.$$

Verify your answer by an alternate calculation.

2. Find the area between the curves

$$y = 3x^4 + 4x^3 - 54x^2 - 108x + 1, \text{ and } y = -75 - 36x.$$

3. Find the area between the curves

$$x + 2y^2 - 28y + 96 = 0, \text{ and } x^2 - 8x - 3y + 7 = 0.$$

4. Find the volume of the solid of revolution formed if the region in the first quadrant, bounded by the curves $y = \frac{20}{x^2+1}$, $x = 2$, and $y = 1$ is revolved about

a) the x-axis b) the y-axis
c) the line $x = -4$ d) the line $y = 6$

5. Verify each of your answers in Problem 4 by an alternate calculation.

6. Use Mathematica and exact integration to derive a formula for the volume of a cone of height H and radius (of the base) R.

7. Use Mathematica and exact integration to derive a formula for the volume of a sphere of radius R.

8. Use exact integration to derive a formula for the surface area of a sphere of radius R. Depending on how the problem is set up, the integrand might have an infinite discontinuity at one of the end points of the integration interval. Set the problem up in this way. Mathematica will handle the problem if you simply ignore it, but how do you know Mathematica is correct? Evaluate this integral a second time by trying to avoid these end point problems. Do you get the same answer?

9. Consider the torus generated by revolving, about the x-axis, the circle of radius r centered at the point $(0, R)$ on the y-axis, where r and R are fixed positive numbers with $R > r$. Use Mathematica and exact integration to find a formula for the volume of the torus. Verify your answer with an alternate calculation.

10. The region bounded by the curves $y = x^2 - 8x + 8$ and $y = 34x - x^2 - 45$ has a mass density of $\delta = 7 + \sqrt{x}$ (in grams per square centimeter) at each of its points on a vertical line passing through x on the x-axis. Find the mass and the center of mass of the region.

11. Find the length of a thin wire defined by the curve $y = \sqrt{x} + x\sin(x)$ for x in the interval $[1, 10]$. Assuming constant mass density δ, find the center of mass.

12. A flat, circular, glass port hole with a radius of 9 inches has been built into the side of a pool so that under water activity can be monitored. The top edge of the window is 7 feet below the water surface. What is the total force acting on the glass.

13. A diving bell having a vertical, flat, semicircular, glass window with a radius of 4 feet is used to take tourists down into the ocean to view the wonders of nature. The straight side of the semicircle is the horizontal bottom edge, and the window is built to safely withstand a force of 87,000 pounds. How deep (measure from the water surface to the bottom of the window) can the diving bell safely dive ? If the diving bell is deep enough, one would expect,

intuitively, that pressure is almost constant across the entire window, in which case we could get the force acting on the window simply by multiplying pressure times area. Discuss this matter. Is this a reasonable approximation, at the maximum safe depth?

14. Use Mathematica and exact integration to show that the center of mass of a triangle in the (x, y)-plane having uniform (constant) mass density δ lies at the intersection of the medians, one third up from the midpoint of each side towards the opposite vertex.

Surely, there is no loss in generality, if we place one vertex at the origin and one side along the x-axis. Consequently, we can view the triangle in the form as it appears in the accompanying figure.

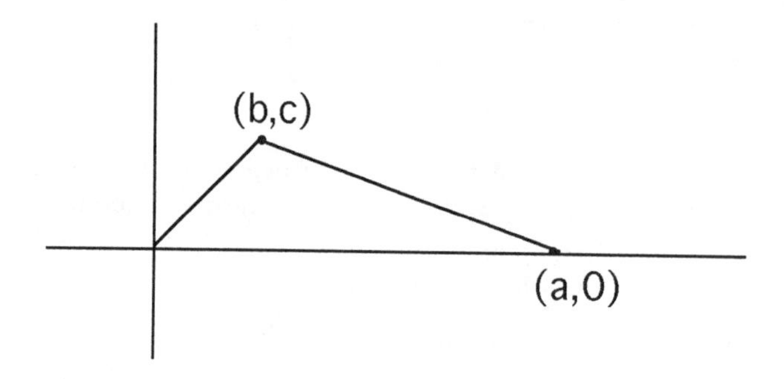

15. Executives for a logging company are considering the construction of a railroad line from a point where raw lumber is collected to its processing mill, and a map of the region involved is being studied to decide where to build the rail line. A rectangular coordinate system has been placed on the map with the origin at the center of the region. The units are in miles. The raw lumber is collected at the point (0,10), and the processing mill is at the point (20,0). Management must choose between two possible routes labeled C_1 and C_2 below which connect these two points. For each of the two rail lines, company engineers have determined a formula for the approximate force that would have to be exerted at the point (x, y) on the road in order to pull a "standard" train at a constant 30 mph pace throughout the course. These approximate force formulas are labeled f_1 and f_2 below. This information will be used to determine the total work energy required to move a standard train over each of the two rail lines under consideration. Management has decided to choose the rail line which requires less total energy. The road C_1 and the force f_1 (in thousands of pounds) at the point (x, y) on this road are defined by

$$C_1 = \{(x,y)|y = \frac{20-x}{2x+2}, 0 \leq x \leq 20\}, f_1 = 317 - x^2 + 10x - 0.4y^2 + 0.8y,$$

The Road C_2 and the force f_2 (in thousands of pounds) at the point (x, y) on this road are defined by

$$C_2 = \{(x,y)|y = \frac{2000 - x^3 + 30x^2 - 300x}{200}, 0 \leq x \leq 20\},$$

$$f_2 = 132 - 2x^2 + 40x - 3y^2 + 30y.$$

Notice that both curves begin and end at the points (0,10) and (20,0). It is now your job to supply management with these total energy numbers.

16.A model rocket is fired straight up from the ground. The rocket's engine fires for 10 seconds producing an acceleration $a(t) = 100 + 58\sqrt[3]{t-5}, (0 \leq t \leq 10)$. At $t = 10$ seconds, the engine turns off and the motion of the rocket is subject to just gravitational acceleration. How high does the rocket go, and how long does it take to reach the high point? What is its maximum speed? How long does it take to fall back to earth, and at what velocity does it strike the earth? Would this be a good toy for a 10 year old child?

17. A mineral survey of private farm land, suggests that the land contains approximately 500,000 barrels of oil. The owner of the land plans to install a low capacity oil well, and the survey crew has estimated that it would be capable of producing oil at the rate of $(\frac{813}{\sqrt{t+7420}} - 1)$ barrels per hour, where t is the age of the well, in hours. The machinery will require very little attention to operate, and so it will be allowed to continue pumping until the production rate drops below 1 barrel per hour, at which time the well will be abandoned. How long, in years, will the well produce oil? How much oil will be recovered, and how much will remain underground when the well is abandoned?

Chapter 5

The Transcendental Functions

According to the dictionary, the word transcendental means something that is beyond common thought or experience, something that is mystical or supernatural. In mathematics, a transcendental expression, is an expression, which is not algebraic. It is an expression, which is somehow beyond algebra. An expression $f(x)$ is algebraic if $y = f(x)$ satisfies an equation of the form

$$a_n(x)y^n + a_{n-1}(x)y^{n-1} + \ldots a_2(x)y^2 + a_1(x)y + a_0(x) = 0,$$

where the expressions $a_0(x)$, $a_1(x), \ldots,$ $a_n(x)$ are all polynomials in x.

The trigonometric functions are all transcendental (beyond algebra), although this is not entirely obvious. In addition to the trigonometric functions, there are several other special transcendental functions which are widely used in mathematics. They will be introduced in this chapter, and their properties relative to calculus will be studied.

Several of these special functions are inverses of other special functions, and so we begin with a section on inverse functions. The chapter ends with a study of the long anticipated *L'Hopital's Rule*. This important rule gives us "the final word" on how to evaluate limits of expressions, when the limits are not obvious. *L'Hopital's Rule* could have been discussed much earlier in the development of calculus, but it is placed in this chapter because so many really mysterious limit examples involve transcendental functions.

5.1 Inverse Functions

Finding the inverse of a function f involves nothing more than solving the equation $y = f(x)$ for x in terms of y. Since a symbolic solution is called for here, the **Solve[ ] command must be used rather than the FindRoot[ ]**, command. Actually, the FindRoot[ ] command could be used to find the inverse of a function at a particular point, but it cannot be used to find a formula for the inverse. As we have seen, Mathematica's Solve[ ] command is a powerful tool for solving equations, but its ability to solve more complicated equations is severely limited by mathematics itself. While mathematicians speak confidently about the existence of solutions, actually finding solutions is another matter. Frequently, solutions cannot be found, even though they clearly exist, and this happens to be the destiny of mathematics and not a reflection of the person doing the work.

Surprisingly, even the solution to an equation as straightforward as a polynomial may be impossible to find. A formula for solving a second degree polynomial equation is well known to every student of algebra, and formulas exist, as well, for solving third and fourth degree polynomial equations, although they are less well known. The Norwegian mathematician Niels Abel proved in 1824, however, that there is no formula for solving a general fifth degree polynomial equation, and later the French mathematician Evariste Galois proved that there is no formula for solving a general polynomial equation of degree n for any $n \geq 5$. By a formula here, we mean a formula based on root taking and the operations of $+, -, \times, \div$ which would be applied to the coefficients of the polynomial. Perhaps even more striking, it can be shown, for example, that not even one root of the equation

$$x^5 - 9x + 3 = 0$$

can be expressed by root taking and applying the operations of $+, -, \times, \div$ to integers (or rational numbers).

Give Mathematica a polynomial equation of degree 5 or more, and Mathematica will not be able to solve it, unless it just happens to be an equation which easily factors. This is unlikely to happen, unless the equation was set up to factor. We will now use Mathematica's powerful solve command to find a few inverses, but at the same time let us appreciate how very limited we are in this practice.

Example 5.1 *Find the intervals where*

$$f(x) = \frac{x}{x^2 - 2x + 4}$$

is one-to-one. In each of these intervals : a) find the corresponding inverse function $f^{-1}(x)$, b) verify that the identities $f(f^{-1}(x)) = x$ and $f^{-1}(f(x)) = x$ hold, c) plot f and its inverse in a way that shows the symmetry between the graphs of f and f^{-1} about the line $y = x$.

A great deal of insight can be gained by first turning our attention to a graph of $f(x)$.

```
In[1]:= f[x_]:=x/(x^2-2*x+4)
```

```
In[2]:= Plot[f[x],{x,-10,10}];
```

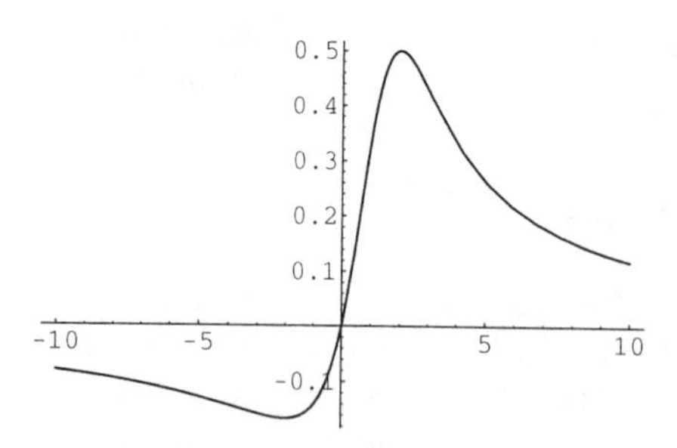

This plot supplies a wealth of information. The function f appears to be one-to-one on three separate intervals, and so there should be three separate inverses. How do we know, however, that this plot does not have hidden features too small to see (or features beyond this window) which would make these conclusions false? To be on the safe side we first look at the sign of $f'(x)$.

```
In[3]:= fp=D[f[x],x]
```

Out[3]= $-\frac{x\ (-2 + 2\ x)}{(4 - 2\ x + x^2)^2} + \frac{1}{4 - 2\ x + x^2}$

```
In[4]:= fp=Together[fp]
```

Out[4]= $\frac{4 - x^2}{(4 - 2\ x + x^2)^2}$

Now it is clear that f is a one-to-one increasing function on the interval $[-2, 2]$, because fp is positive on this interval, and f is a one-to-one decreasing function on the intervals $(-\infty, -2]$ and $[2, \infty)$, because fp is negative there.

The intervals for the inverses will be determined by the high and low points on the graph. There is no need to ask Mathematica to solve $fp = 0$, since the above formula for fp makes the solution obvious.

```
In[5]:= {a=-2,b=2,A=f[a],B=f[b]}
```

Out[5]= $\left\{-2, 2, -\frac{1}{6}, \frac{1}{2}\right\}$

From the graph of $f(x)$, we can see the domains and ranges of all three of the inverses, and so we let g_1, g_2, and g_3 be the inverses to f on these various intervals with inverses and intervals paired as follows.

$$(-\infty, a] \underset{g_1}{\overset{f}{\rightleftarrows}} [A, 0) \qquad [a, b] \underset{g_2}{\overset{f}{\rightleftarrows}} [A, B] \qquad [b, \infty) \underset{g_3}{\overset{f}{\rightleftarrows}} (0, B]$$

The next step is to find formulas for all three of these inverses.

```
In[6]:= s=Solve[f[x]==y,x]
```

Out[6]= $\left\{\left\{x \to \frac{1+2y-\sqrt{1+4y-12y^2}}{2y}\right\}, \left\{x \to \frac{1+2y+\sqrt{1+4y-12y^2}}{2y}\right\}\right\}$

```
In[7]:= {s1[y_]=x/.s[[1]],s2[y_]=x/.s[[2]]}
```

Out[7]= $\left\{\frac{1+2y-\sqrt{1+4y-12y^2}}{2y}, \frac{1+2y+\sqrt{1+4y-12y^2}}{2y}\right\}$

It is not obvious, quite yet, how to specify the formulas for g_1, g_2, and g_3. Notice that neither $s1$, nor $s2$ are defined at $y = 0$. This is not surprising considering the sense in which 0 is excluded from the domains of g_1 and g_3. To help settle this issue we compute the following limits.

```
In[8]:= {Limit[s1[y],y->0],Limit[s2[y],y->0]}
Out[8]= {0, ∞}
```

```
In[9]:= {Limit[s2[y],y->0,Direction->1],Limit[s2[y],y->0, Direction->-1]}
Out[9]= {-∞, ∞}
```

It now appears that $g_2(y) = s1(y)$, and that $g_1(y)$ and $g_3(y)$ both have the same formula $s2(y)$ restricted to the appropriate interval. To establish this beyond any doubt, we show next that for x in the appropriate interval

$$f(g_j(x)) = x, g_j(f(x)) = x \text{ for } j = 1, 2, \text{ and } 3.$$

Actually, we will only show this for $j = 1$. The other two cases are similar.

The contrast between these two identities is interesting. The first is fairly easy to prove without any reference to what interval x is in. The second identity, on the other hand, is not so easy to prove. A hard look at the structure of f and g_1 would explain why there is such a contrast between these two identities, but suffice it to say, that Mathematica fails to simplify $g_1(f(x))$. This identity is only meant to hold for $x \leq -2$, and it turns out that this information is needed in order to proceed.

```
In[10]:= {g1=s2,g2=s1,g3=s2};
```

```
In[11]:= f[g1[x]]
```

$$Out[11]= \frac{1+2\,x+\sqrt{1+4\,x-12\,x^2}}{2\,x\left(4-\frac{1+2\,x+\sqrt{1+4\,x-12\,x^2}}{x}+\frac{\left(1+2\,x+\sqrt{1+4\,x-12\,x^2}\right)^2}{4\,x^2}\right)}$$

```
In[12]:= Simplify[%]
Out[12]= x
```

```
In[13]:= g1[f[x]]
```

$$Out[13]= \frac{(4-2\,x+x^2)\left(1+\frac{2\,x}{4-2\,x+x^2}+\sqrt{1-\frac{12\,x^2}{\left(4-2\,x+x^2\right)^2}+\frac{4\,x}{4-2\,x+x^2}}\right)}{2\,x}$$

```
In[14]:= Simplify[%];
```

Undesirable output is omitted. Mathematica needs to know that $x < -2$. **To force this issue, we resort to some trickery**. The **square of a number is always nonnegative**, and this frequently can be used to our advantage. **We let $x = -2 - t^2$**, where t is an arbitrary real number. **Mathematica has no choice, but to accept that $x \leq -2$**. The Simplify[ ] command works on the resulting expression, but not very well, so we **pull out the heavy artillery** and use the **new command FullSimplify[ ]**. This command **should be used with discretion**, because it takes a fair amount of time to process. **Another new command, called PowerExpand[ ]**, is used after that. Get acquainted with these new commands, by **looking them up in the Help File**. Finally, we return to the original variable x.

```
In[15]:= FullSimplify[%%/.{x->-2-t^2}]
```

$$Out[15]= -\frac{8+4\,t^2+t^4+\sqrt{\frac{t^4\,\left(4+t^2\right)^2}{\left(12+6\,t^2+t^4\right)^2}}\,(12+6\,t^2+t^4)}{2\,(2+t^2)}$$

```
In[16]:= PowerExpand[%]
```

$$Out[16]= -\frac{8+4\,t^2+t^4+t^2\,(4+t^2)}{2\,(2+t^2)}$$

The **PowerExpand[ ] command makes assumptions on variables** that are **not always valid**. Be **careful** when you use it.

```
In[17]:= Simplify[%]/.{-2-t^2->x}
Out[17]= x
```

This establishes that g_1 is the inverse of f on this interval.

To finish this example, we choose $j = 2$ (the cases $j = 1, 3$ are similar) and graph the pair $\{f(x), g_2(x)\}$ along with the line $y = x$, on the same plot to demonstrate that the graph of a function and its inverse are symmetric images of each other with respect to the line

$y = x$. In order to create the **appropriate visual image to show this symmetry, we must tell Mathematica to choose the same scale on the x and y axes**. We have used the rule PlotRange– > $\{y_{min}, y_{max}\}$ before.

The functions f and g_2 have **different domains**, and this **complicates the plotting process** considerably. We **plot them separately, and bring them together with an interesting new command called Show[ ]**. This command is worth knowing. Also, notice how the equation $y = x$ ($y == x$ in Mathematica) is entered in the first plot. **The Plot[ ] command cannot be use to plot equations.**

The rule **DisplayFunction– >Identity** suppresses the initial display of of the individual plots. It may not be critical, but we are interested in saving space in this manual.

```
In[18]:= p1=Plot[{f[x],x},{x,-2,2},PlotRange->{-2,2},
         DisplayFunction->Identity];
```

```
In[19]:= p2=Plot[g2[x],{x,-6^(-1),1/2},PlotRange->{-2,2},
         DisplayFunction->Identity];
```

```
In[20]:= Show[{p1,p2},DisplayFunction->$DisplayFunction, AspectRatio->1];
```

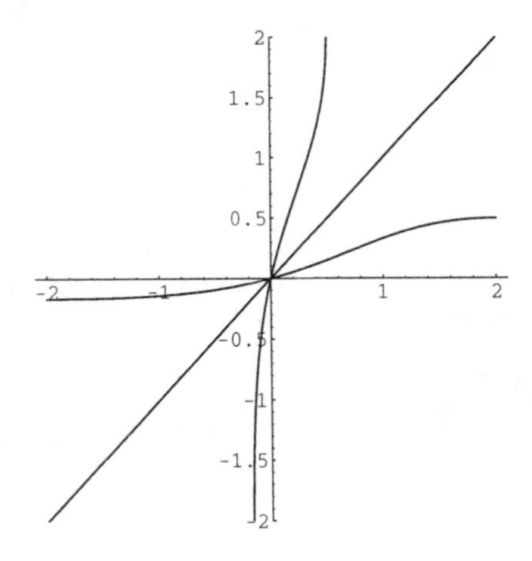

Recall that Aspect ratio is a ratio of the lengths of the x and y axes. This option is included to enhance the display of symmetry between the graphs.

In order to save time and space, we will not discuss a second example, but it would be a valuable experience to look for the inverse of a function f which is a cubic polynomial. Such a function will either be globally one-to-one, or it will be one-to-one on three separate intervals and have three separate inverses. Solving the cubic equation $f(x) = y$ for x in terms of y always leads to three solutions, and **Mathematica will always give you all three**. If f is globally one-to-one, then two of these solutions will involve the imaginary symbol I, and these two parts can simply be discarded. If f has three separate inverses then there will be three real solutions to the equation $f(x) = y$. Even with a function as simple as a cubic, the inverse is quite complicated.

As a final topic in this section on inverse functions, we draw your attention to a theorem in your calculus text, which allows one to evaluate the derivative of an inverse function at a point, even if a formula for the inverse cannot be found. If $g = f^{-1}$ and if $b = f(a)$, then, according to this result,

$$g'(b) = \frac{1}{f'(a)}.$$

It is usually a straightforward matter to show that an inverse exists, but even with Mathematica, it is usually difficult or impossible to actually find a formula for the inverse. This result, then, can be very useful.

Example 5.2 *Show that the function $f(x) = 2x + \sin(x)$ is one-to-one on the whole real line. Compute, approximately, the derivative of $f^{-1}(x)$ at the point $x = 17$.*

```
In[21]:= f[x_]:=2*x+Sin[x]
```

```
In[22]:= fp=D[f[x],x]
Out[22]= 2 + Cos[x]
```

The derivative is always positive, and so f is an increasing function on the interval $(-\infty, \infty)$. This means that f^{-1} exists.

```
In[23]:= Solve[f[x]==y,x]
Solve::tdep : The equations appear to involve
     transcendental functions of the variables in
     an essentially non-algebraic way.

Out[23]= Solve[2 x + Sin[x] == y, x]
```

Not surprisingly, the inverse cannot be found, even though it exists. To find the derivative of the inverse at 17, we use the theorem we mentioned above.

```
In[24]:= FindRoot[f[x]==17,{x,5}]
Out[24]= {x -> 8.00575}
```

```
In[25]:= answer=1/fp/.%
Out[25]= 0.540886
```

5.2 Logarithmic and Exponential Functions

The search for a function whose derivative is $1/x$ ($x > 0$) leads us immediately to the function ln(x) defined by

$$\ln(x) = \int_1^x \frac{1}{t} dt.$$

Mathematica denotes this function by **Log**, and, of course, its value at x is the expression **Log[x]**. Remember that parentheses are used in basically the same way as they are with pencil and paper. The expression **Log[x]^2** in Mathematica **means $\ln(x)^2 = (\ln(x))^2$ and not $\ln(x^2)$**. Writing (Log[x])^2 for $\ln(x)^2 = (\ln(x))^2$ is acceptable, but the extra set of parentheses is unnecessary. As usual, remember that Mathematica will only give exact answers unless you allow it to approximate.

```
In[1]:= {Log[5],N[Log[5]]}
```

```
Out[1]= {log[5], 1.60944}
```

This activity, however, hardly gives us an appreciation of the definition of this function as the area under the graph of the function $y = 1/t$ from $t = 1$, to $t = x$. In order to gain some insight into this definition of $\ln(x)$, we will approximate some of its values directly from its area definition. In the process, perhaps we will demonstrate the importance of the Fundamental Theorem of Calculus. This central theorem in mathematics provides a fail safe means of always getting an antiderivative to a continuous function. We use the name MyLog[x] for our approximation of $\ln(x)$. The value of MyLog[x] is the approximate area under the curve of $y = 1/t$ from $t = 1$ to $t = x$, obtained by computing a Riemann sum corresponding to $n = 50$ and using right end points of subintervals to establish heights of the thin rectangles.

```
In[2]:= f[t_]:=1/t
```

```
In[3]:= MyLog[x_]:=Module[{h},h:=(x-1)/50;N[Sum[f[1+h*j]*h, {j,1,50}]]]
```

```
In[4]:= MyLog[5]

Out[4]= 1.57795
```

Compare this value with N[Log[5]] above. This number represents the sum of the rectangular areas shown in the next graph. Recall, the we defined the graphic tool used to create this figure back in Chapter 3. we saved it then on a special library file, so now we can just copy and paste it onto our current notebook. Actually, all we have to do is open the library file, place the pointer in the appropriate input cell and press the enter key. The command will not be pasted into our current notebook, but it will still be entered into the active (current) Mathematica Kernel. It doesn't matter what notebook it appears in, as long as it is entered.

```
In[5]:= RightRiemannDisplay[f_,a_,b_,n_]:=Show[Plot[f[x],{x,a,b},
        DisplayFunction->Identity],
        Graphics[Table[Line[{{a+(b-a)/n*j,0},
        {a+(b-a)/n*j,f[a+(b-a)/n*(j+1)]},
        {a+(b-a)/n*(j+1),f[a+(b-a)/n*(j+1)]},
        {a+(b-a)/n*(j+1),0},{a+(b-a)/n*j,0}}],{j,0,n-1}]],
        DisplayFunction->$DisplayFunction];
```

```
In[6]:= RightRiemannDisplay[f,1,5,50];
```

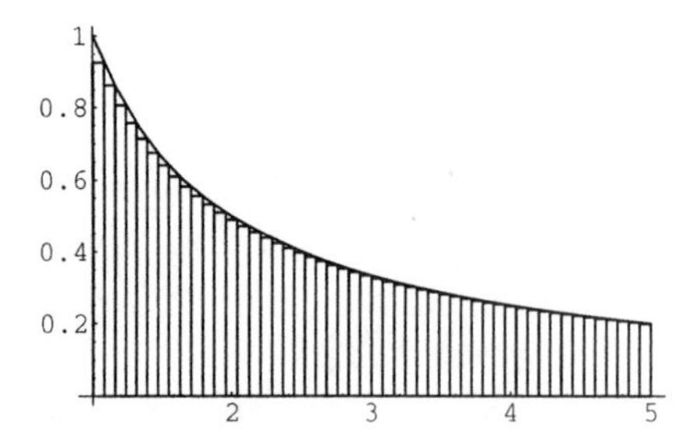


```
In[7]:= MyLog[0.2]
Out[7]= -1.64195
```

```
In[8]:= RightRiemannDisplay[f,0.2,1,50];
```

Output omitted.

Notice in this last command how the range between 1 and $x = 0.2$ had to be reversed so that it went from a smaller to a larger number. Integrating from 1 to the smaller $x = 0.2$ introduces a negative sign in the area interpretation of $\ln(0.2)$ and this is also reflected in our value of LN(0.2). Just how this negative sign entered into this formula can be seen in the definition of LN as a module. Immediately, in that definition, we see from $h := (x - 1)/50$, that h, the "width" of a rectangle, will be negative when $0 < x < 1$. More generally, when we integrate $\int_a^b f(x)dx$ from a larger number a to a smaller number b, we can interpret the symbol dx, the "width" of a thin rectangle, as being negative in the same way as h above.

The definition of $\ln(x)$ gives us the important formula

$$\int \frac{1}{x} dx = \ln(x) + C \ (x > 0),$$

and this is easily extended to the negative real axis by the formula

$$\int \frac{1}{x} dx = \ln(|x|) + C \ (x \neq 0).$$

Notice, however, how Mathematica evaluates this integral.

```
In[9]:= Integrate[1/x,x]
Out[9]= log[x]
```

Since the expression $1/x$ is well defined on the negative real axis, Mathematica is certainly not assuming that x is a positive variable. Curiously enough, Mathematica has not made a mistake, but the way it has chosen to express the antiderivative could cause complications. To fully understand Mathematica's answer requires a more advanced course in complex valued function theory. It turns out that the natural logarithm of a negative number can be defined as the complex number

$$\ln(x) = \ln(|x|) + \pi I, \ (x < 0).$$

Since $\ln(x)$ and $\ln(|x|)$ differ only by the constant πI, it follows that they are both legitimate antiderivatives of $1/x$.

```
In[10]:= Integrate[1/t,{t,-11,-7}]
```

Out[10]= log[7] − log[11]

```
In[11]:= Integrate[1/t,{t,-15,a}]
```

Out[11]= $-i\ \pi - \log[15] + \log[a]$

At this point, we appear to have an unacceptable form for the answer, but notice that the complex values disappear as soon as a is replaced by a negative number. Replacing a by a positive number, on the other hand, would not be appropriate. Do you see why?

```
In[12]:= %/.a->-5
```

Out[12]= log[5] − log[15]

We have the expected answer. Expressions involving $\ln(x)$ will usually be real valued, but we should be prepared for complex valued complications. Complex valued logarithms play an important role in mathematics, but they are never appropriate in our real valued world of calculus. If they are encountered, they should be dealt with accordingly. The command **ComplexExpand[ ]** might help in this case.

On another matter, we mention that Mathematica recognizes all of the algebraic properties of $\ln(x)$.

```
In[13]:= Log[Sqrt[(8+2*x)^3/(a^5*9*b)]]
```

Out[13]= $\log\left[\frac{1}{3}\sqrt{\frac{(8+2\ x)^3}{a^5\ b}}\right]$

The commands Simplify[ ], FullSimplify[ ], Expand[ ] fail to simplify this expression, because what we probably want to get, is only valid when all the numbers involved are **real** and **positive**. The **PowerExpand[ ] command makes these assumptions. It must be used with care**.

```
In[14]:= PowerExpand[%]
```

Out[14]= $-\log[3] + \frac{1}{2}\ (-5\ \log[a] - \log[b] + 3\ \log[8+2\ x])$

Look at the consequences of the commands Simplify[ ] and Expand[ ] on the above logarithmic expression. Both of these commands have useful options. **They should be looked up in the Help File.** Recall how easy it is to do this. With the cursor on "Simplify" or "Expand," pull down the **Help Menu** and click on **Find in Help ....**

The inverse of the natural logarithm is represented in Mathematica by the function **Exp**. Its value at x, naturally, is the expression denoted by **Exp[x]**. In mathematics this expression also has the form $\exp(x) = e^x$, where e is the exponential constant, whose decimal value is roughly 2.71828. **Mathematica denotes this universal constant by the capitalized letter $E = 2.71828\ldots$. The function Exp[x] can also be expressed in its equivalent power form Exp[x]= E^x.**

The use of a capitalized letter E is certainly consistent with general Mathematica policy, but we are so accustomed to using the small case letter e to denote this constant, that it is easy to make a mistake and use e^x instead of E^x in input statements. Before we leave this topic, **let us be perfectly clear**! Mathematica **does not recognize e as the exponential constant**. If you use e^x instead of E^x or $Exp[x]$ in an input statement, your output will look perfectly acceptable, but **e will simply be regarded as an unassigned name** by Mathematica.

```
In[15]:= {Exp[0],Exp[4],Exp[-4],N[Exp[4]],E^0,E^(-4),N[E^4]}
```

Out[15]= $\{1, E^4, \frac{1}{E^4}, 54.5982, 1, \frac{1}{E^4}, 54.5982\}$

```
In[16]:= Solve[Log[y]==x,y]
```

Out[16]= $\{\{y \to E^x\}\}$

When the exponential function is entered in the form e^x, it is not the exponential function, in spite of its appearance. You can see that something is **not quite right** as soon as it is, for example, differentiated. **Certainly Mathematica knows that ln(e) = 1, and it would never leave such an obvious simplification unattended.**

```
In[17]:= {g=e^x,D[g,x]}
```

Out[17]= $\{e^x, e^x \log[e]\}$

Just like the names *Sin*, *Cos*, *Tan*, *Log*, etc., ***Exp* is the name of a function, not an expression**. Consequently, it would be a mistake to enter *Exp*^4 (or *f*^4 for any function *f*) in an input cell.

Example 5.3 *Find the $x-$intercept of the line which is tangent to the graph of $y = \ln(x^3+5)$ at the point on the curve having a $y-$coordinate of 10.*

```
In[18]:= Clear["Global`*"]
```

```
In[19]:= f[x_]:=Log[x^3+5]
```

```
In[20]:= s=Solve[f[x]==10,x]
```

Out[20]= $\{\{x \to -(5 - e^{10})^{1/3}\}, \{x \to (-5 + e^{10})^{1/3}\}, \{x \to (-1)^{2/3} (-5 + e^{10})^{1/3}\}\}$

```
In[21]:= a=x/.s[[2]]
```

Out[21]= $(-5 + e^{10})^{1/3}$

With this, the point of tangency is $(a, 10)$.

```
In[22]:= fp=D[f[x],x]
```

Out[22]= $\frac{3\ x^2}{5 + x^3}$

```
In[23]:= m=fp/.x->a
```

Out[23]= $\frac{3\ (-5 + e^{10})^{2/3}}{e^{10}}$

This is the slope of the tangent line.

```
In[24]:= eq=y-10==m*(x-a)
```

Out[24]= $-10 + y == \frac{3\ (-5 + e^{10})^{2/3}\ \left(-\ (-5 + e^{10})^{1/3} + x\right)}{e^{10}}$

```
In[25]:= Solve[(eq/.y->0),x]
```

Out[25]= $\{\{x \to -\frac{15 + 7\ e^{10}}{3\ (-5 + e^{10})^{2/3}}\}\}$

This is the x-intercept.

```
In[26]:= X=N[x/.%[[1]]]
Out[26]= -65.4234
```

Example 5.4 *Find the area of the region bounded by the graphs of $y = \exp(2x - 3)$, $y = 4$, and $x = -1$. Verify with an alternate calculation.*

```
In[1]:= f[x_]:=Exp[2*x-3]
```

```
In[2]:= Plot[{f[x],4},{x,-1,2.5},PlotRange->{0,5}];
```

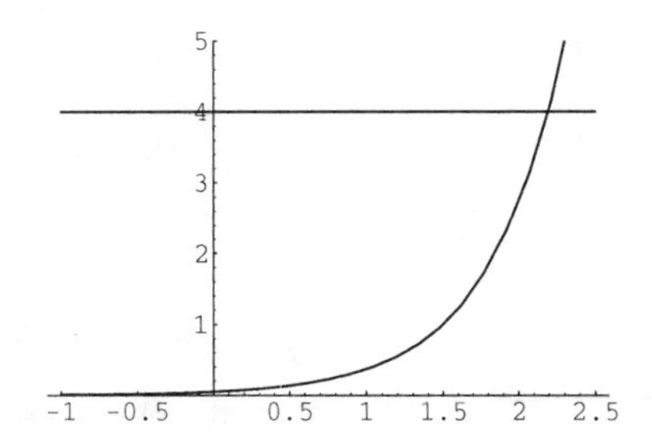


```
In[3]:= FindRoot[f[x]==4,{x,2}]
Out[3]= {x → 2.19315}
```

```
In[4]:= a=x/.%
Out[4]= 2.19315
```

This is area, using "thin" vertical rectangles.

```
In[5]:= A1=NIntegrate[4-f[x],{x,-1,a}]
Out[5]= 10.776
```

```
In[6]:= Solve[f[x]==y,x]
Solve : ifun : Inverse functions are being used
    by Solve, so some solutions may not be found.
```

Out[6]= $\{\{x \to \frac{1}{2}\ (3 + \log[y])\}\}$

```
In[7]:= xr=x/.%[[1]]
```

Out[7]= $\frac{1}{2}(3 + \log[y])$

```
In[8]:= b=N[f[-1]]
Out[8]= 0.00673795
```

The y-coordinate of the lower left-hand corner is small but not zero.

```
In[9]:= A2=NIntegrate[xr-(-1),{y,b,4}]
Out[9]= 10.776
```

This is the area, using "thin" horizontal rectangles.

In order to evaluate $\log_b(x)$ on a calculator, we use the well known formula

$$\log_b(x) = \frac{ln(x)}{ln(b)}.$$

Mathematica uses the same formula and denotes $\mathbf{log}_b(x)$ by **log[b, x]**. The b comes first, as it does when you say "log base b of x."

```
In[10]:= {Log[2,8],Log[5,x]}
```

Out[10]= $\{3, \frac{\log[x]}{\log[5]}\}$

```
In[11]:= eq1=y==Log[3,x];
```

```
In[12]:= Solve[eq1,x]
```

Out[12]= $\{\{x \to 3^y\}\}$

```
In[13]:= eq2=Log[3,(x^2+9)^2]==5*(y^2+1)
```

Out[13]= $\frac{\log[(9+x^2)^2]}{\log[3]} == 5\,(1+y^2)$

```
In[14]:= s=Solve[eq2,x]
```

Out[14]= $\{\{x \to -\sqrt{-9-\sqrt{e^{5\,(\log[3]+y^2\,\log[3])}}}\},$

$\{x \to \sqrt{-9-\sqrt{e^{5\,(\log[3]+y^2\,\log[3])}}}\},$

$\{x \to -\sqrt{-9+\sqrt{e^{5\,(\log[3]+y^2\,\log[3])}}}\},$

$\{x \to \sqrt{-9+\sqrt{e^{5\,(\log[3]+y^2\,\log[3])}}}\}\}$

```
In[15]:= s1=Simplify[s]
```

Out[15]= $\{\{x \to -\sqrt{-9 - \sqrt{243^{1+y^2}}}\}, \{x \to \sqrt{-9 - \sqrt{243^{1+y^2}}}\},$
$\{x \to -\sqrt{-9 + \sqrt{243^{1+y^2}}}\}, \{x \to \sqrt{-9 + \sqrt{243^{1+y^2}}}\}\}$

Only the last two evaluations are real numbers. The first two involve square roots of negative numbers. The real solutions are as follows:

```
In[16]:= s1[[{3,4}]]
```

Out[16]= $\{\{x \to -\sqrt{-9 + \sqrt{243^{1+y^2}}}\}, \{x \to \sqrt{-9 + \sqrt{243^{1+y^2}}}\}\}$

By now we are accustomed to using $s[[j]]$ to extract the $j\,th$ element of a list s. **Notice above that $s[[\{j,k\}]]$ can be used to extract—as another list—the $j\,th$ and $k\,th$ elements of the list s.** This same notation can be used in the form $s[[\{j_1, j_2, \ldots, j_n\}]]$, to extract other sublists from s. The **new command, Range[ ], can often be used in a convenient way to specify this "inner" list.**

5.3 The Inverse Trigonometric Functions

All six of these functions are known to Mathematica. Notation, such as $\sin^{-1}(x)$ for the inverse sine function, may well be a familiar sight in print, but, of course, it would not be a suitable way to name these functions in Mathematica, and so the inverse trigonometric functions are known to Mathematica only by the names **ArcSin, ArcCos, ArcTan, ArcCot, ArcSec, ArcCsc.**

These functions are, of course, not globally one-to-one, and so their inverses only exist if their domains are suitably restricted. This complicates the identities

$$f(f^{-1}(x)) = x, f^{-1}(f(x)) = x.$$

If $f(x) = \sin(x)$, for example, the first of these identities holds as long as $f(f^{-1}(x))$ is well defined, namely, for all x in the interval $-1 \leq x \leq 1$. The second identity, on the other hand, only holds for x in the interval $-\pi/2 \leq x \leq \pi/2$, even though $f^{-1}(f(x))$ is well defined for all values of x. Can you see why these two identities are so different? As we show next, Mathematica treats the expressions $f(f^{-1}(x))$ and $f^{-1}(f(x))$ as they should be treated. The graph of $f^{-1}(f(x))$ is shown below. Can you explain why it looks this way?

```
In[1]:= {Sin[ArcSin[x]],ArcSin[Sin[x]]}
Out[1]= {x, arcsin[sin[x]]}
```

```
In[2]:= Simplify[%]
Out[2]= {x, arcsin[sin[x]]}
```

```
In[3]:= Plot[ArcSin[Sin[x]],{x,-10,10}];
```

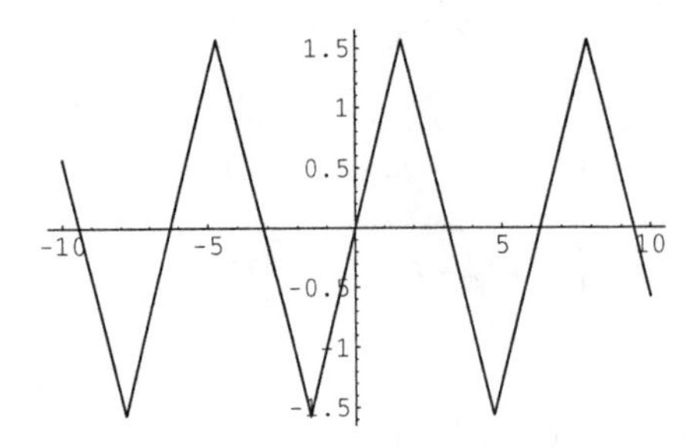

The rules for differentiating the $\operatorname{arcsec}(x)$ and $\operatorname{arccsc}(x)$ may look somewhat unfamiliar, but they are easily seen to be correct.

```
In[4]:= D[ArcSec[x],x]
```

Out[4]= $\dfrac{1}{\sqrt{1-\frac{1}{x^2}}\ x^2}$

```
In[5]:= Plot[ArcSec[x],{x,-10,10}];
```

The output is omitted. Actually, the plot looks correct, but Mathematica complains about plotting over the interval [-1,1], which is not part of the domain of the arcsecant function. To avoid the problem, we plot the left and right hand parts of the graph separately. The "DisplayFunction" option is used to alternately suppress the plot displays, and then, finally, to display the combined plot.

```
In[6]:= p1=Plot[ArcSec[x],{x,-10,-1}, DisplayFunction->Identity];
```

```
In[7]:= p2=Plot[ArcSec[x],{x,1,10}, DisplayFunction->Identity];
```

```
In[8]:= Show[{p1,p2},DisplayFunction->$DisplayFunction];
```

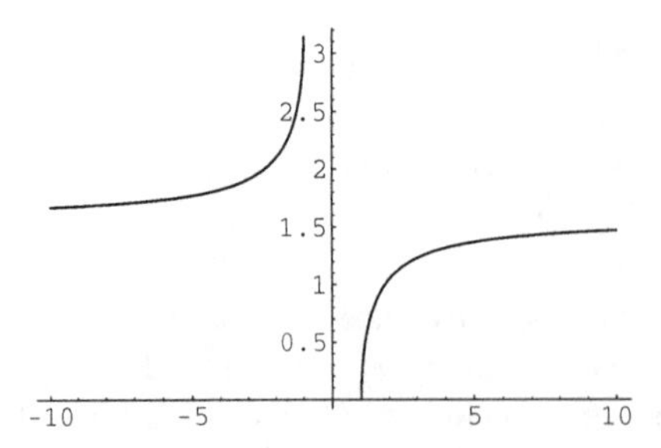

As you can see, the arcsec function maps the domain $|x| \geq 1$ onto $[0, \pi]$ (excluding $\pi/2$) which is the most common way of defining the arcsec function. Mathematica also gives a somewhat unfamiliar answer to the standard integral

$$\int \frac{1}{x\sqrt{x^2-1}}dx = arcsec(|x|) + c$$

```
In[9]:= f[x_]:=1/(x*Sqrt[x^2-1])
```

In[10]:= `Integrate[f[x],x]`

Out[10]= $-\arctan\left[\frac{1}{\sqrt{-1+x^2}}\right]$

Is this answer correct? There are a variety of ways of showing that it is. We show that the derivative of the answer is $f(x)$.

In[11]:= `D[%,x]`

Out[11]= $\frac{x}{(-1+x^2)^{3/2}\left(1+\frac{1}{-1+x^2}\right)}$

In[12]:= `Simplify[%]`

Out[12]= $\frac{1}{x\sqrt{-1+x^2}}$

Example 5.5 *A 300 foot long ship whose bow is 57.2 miles east and 37.6 miles north of an observer is steaming due west at 27 miles per hour, while the observer travels due north at 19 miles per hour. When the ship is within visual range, the observer plans to focus a telescope on the ship, adjusting the viewing angle of the lens so that it always fits exactly the 300 foot length of the ship. When is this viewing angle a maximum, and what is the distance between the observer and the bow of the ship at that time? When is the distance between the observer and the bow of the ship a minimum, and what is this distance?*

We establish the angles with some care, allowing them to be positive or negative, as the ship or the observer or both pass through their common intersection point (the origin). In this way the viewing angle θ is well-defined in all of these situations.

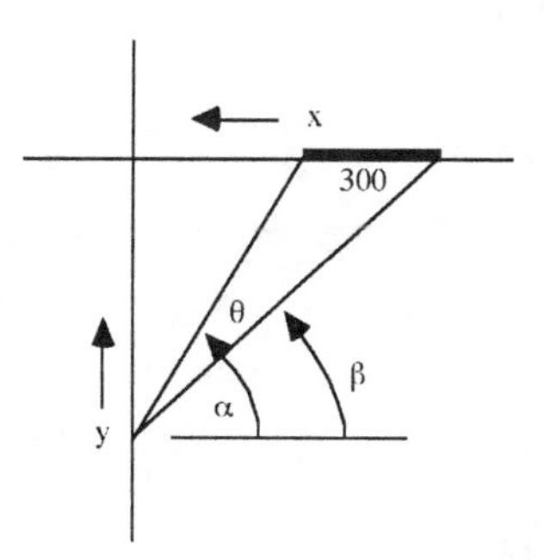

In[13]:= `{y=19*t-37.6,x=57.2-27*t,L=300/5280};`

The length of the ship must be expressed in miles to be compatible with the other units.

Greek letters are typically used to denote angles, and they **can be used in a Mathematica notebook** as well. It is not clear how to do this by looking at the input cells below (only the Greek letter itself appears), so pay attention to the next line. **To enter α in an input cell, type \[Alpha] in an input or text cell.** The symbol (\) is a backslash; it is not the same as a division bar (/) that would be used in, for example 1/2. **As soon as a closing bracket is placed on \[Alpha], this symbol disappears and is replaced by α.** Greek letters can also be capitalized. For example, to **enter the Greek letter γ (gamma) or its capital Γ** into an input or text cell, **type \[Gamma], or \[CapitalGamma]** respectively.

In[14]:= `{α =ArcTan[-(y/x)],β =ArcTan[-(y/(x+L))]}`

Out[14]= $\left\{-\arctan\left[\frac{-37.6+19\ t}{57.2-27\ t}\right],\ -\arctan\left[\frac{-37.6+19\ t}{57.2568-27\ t}\right]\right\}$

In[15]:= $\theta = \alpha - \beta$

Out[15]= $-\arctan\left[\frac{-37.6+19\ t}{57.2-27\ t}\right] + \arctan\left[\frac{-37.6+19\ t}{57.2568-27\ t}\right]$

In[16]:= `Plot[θ ,{t,0,10}];`

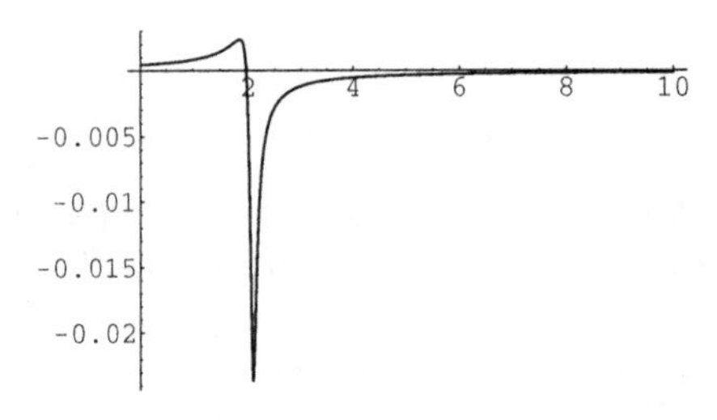

In[17]:= `slope=Simplify[D[θ ,t]]`

Out[17]= $(0.000990409\ (-2.09395+t)\ (-1.86395+t))/((4.30468-4.1474\ t+t^2)\ (4.29872-4.14459\ t+t^2))$

In[18]:= `FindRoot[slope==0,{t,1.9,2.2}]`

Out[18]= $\{t \to 2.09395\}$

In[19]:= `t0=t/.%`

Out[19]= 2.09395

In[20]:= `MaxAngle=N[Abs[θ /.t->t0]]`

Out[20]= 0.0236336

In[21]:= `DegreeMax=N[(MaxAngle*180)/Pi]`

Out[21]= 1.3541

In[22]:= `f=x^2+y^2`

Out[22]= $(57.2-27\ t)^2+(-37.6+19\ t)^2$

This is the square of the distance between the ship and the observer. It will be a minimum at the same time that the distance itself will be a minimum, but it is an easier expression to differentiate.

In[23]:= `eq=D[f,t]==0`

Out[23]= $-54\ (57.2-27\ t)+38\ (-37.6+19\ t) == 0$

```
In[24]:= Solve[eq,t]
Out[24]= {{t → 2.07229}}
```

```
In[25]:= t1=t/.%
Out[25]= {2.07229}
```

```
In[26]:= MinDistance=Sqrt[f/.t->t1]
Out[26]= {2.1687}
```

```
In[27]:= MaxAngleDistant=Sqrt[f/.t->t0]
Out[27]= 2.28349
```

Notice that $t0 > t1$, so the distance is a minimum, before the viewing angle is a maximum.

5.4 Exponential Growth and Decay

The differential equation

$$\frac{dy(t)}{dt} = ky(t)$$

is so common in mathematics and its applications, that its solution

$$y(t) = Ce^{kt},$$

where C is a constant of integration, quickly becomes a familiar if not memorized formula to most students of mathematics. If the most common initial condition $y(0) = y_0$ is given, then C takes the form $C = y_0$. As much as any other problem, this classic differential equation justifies the introduction of the exponential function into the central core of mathematics.

Example 5.6 *A radioactive sample with a half life of* 117.537 *years must remain in protective storage until less than* 0.005 *grams of the substance are left. If 50 grams of the substance were put into a storage locker, how long will this take?*

```
In[1]:= A[t_]:=50*Exp[k*t]
```

```
In[2]:= FindRoot[A[117.537]==25,{k,0}]
Out[2]= {k → -0.00589727}
```

```
In[3]:= k=k/.%
Out[3]= -0.00589727
```

```
In[4]:= A[t]
Out[4]= 50 e^(-0.00589727 t)
```

```
In[5]:= FindRoot[A[t]==0.005,{t,200}]
Out[5]= {t → 1561.79}
```

```
In[6]:= t0=t/.%
```

```
Out[6]= 1561.79
```

We check the amount after $t0$ years.

```
In[7]:= A[t0]
Out[7]= 0.0050001
```

Newton's Law of Cooling is a principle that leads to a differential equation very similar to the classic equation $\frac{dy}{dt} = ky$ just discussed, and it is frequently presented in calculus courses at this time. This principal states that the **rate at which a body cools is proportional to the difference between the temperature of the body and the temperature of the surrounding medium**. Translated into mathematics, the equation takes the form

$$\frac{dT(t)}{dt} = K(T(t) - A),$$

where $T(t)$ is the temperature of the body at time t, A is the constant temperature of the surrounding medium, and K is the constant of proportionality—a constant which depends on the physical properties of the medium.

This differential equation quickly turns into the well known $\frac{dy}{dt} = Ky$, if we just let $y(t) = T(t) - A$. As a consequence, we can easily specify the solution to *Newton's Law of Cooling* as

$$T(t) = A + Ce^{Kt},$$

where C is a constant of integration, whose value is determined by the initial condition.

Example 5.7 *A steel beam is taken out of a furnace and placed in an environment of* $72°$ *Fahrenheit. After 10 minutes the beam has a temperature of* $1287°$, *and after 15 minutes, its temperature is* $986°$. *If we must wait until its temperature is* $100°$ *before we can continue working with it, then how long must we wait? What was the temperature of the beam when it came out of the furnace?*

```
In[8]:= Clear["Global`*"]

In[9]:= T[t_]:=72+a*Exp[k*t]

In[10]:= {eq1=T[10]==1287,eq2=T[15]==986}
Out[10]= {72 + a e^(10 k) == 1287, 72 + a e^(15 k) == 986}

In[11]:= Solve[{eq1,eq2},{a,k}]
```

Output omitted. Actually, the output gives a solution. It just looks complicated and inappropriate. We could use numerical techniques, but start up values for a and k are hard to estimate. Even with reasonable values, the method fails to converge.

```
In[12]:= FindRoot[{eq1,eq2},{a,2000},{k,-1}]
FindRoot :: cvnwt :
  Newton's method failed to converge to the
     prescribed accuracy after 15 iterations.
```

```
Out[12]= {a → 1368.55, k → -0.967226}
```

Mathematica could use our help. The above equation to not hard to solve exactly, even with pencil and paper. Here is an interesting example of a **system of equations which is easier to solve exactly than it is to solve by a decimal approximation**.

```
In[13]:= Solve[eq1,a]
```

Out[13]= $\{\{a \to 1215\ e^{-10\ k}\}\}$

```
In[14]:= a=a/.%[[1]]
```

Out[14]= $1215\ e^{-10\ k}$

```
In[15]:= eq2
```

Out[15]= $72 + 1215\ e^{5\ k} == 986$

```
In[16]:= Solve[eq2,k]
Solve::ifun: Inverse functions are being used
    by Solve, so some solutions may not be found.
```

Out[16]= $\{\{k \to -\frac{1}{5}\ \log[\frac{1215}{914}]\}\}$

```
In[17]:= k=k/.%[[1]]
```

Out[17]= $-\frac{1}{5}\ \log[\frac{1215}{914}]$

```
In[18]:= a
```

Out[18]= $\frac{1793613375}{835396}$

```
In[19]:= InitialTemp=N[T[0]]
Out[19]= 2219.02
```

This is the initial temperature in degrees Fahrenheit

```
In[20]:= FindRoot[T[t]==100,{t,25}]
Out[20]= {t → 76.2225}
```

```
In[21]:= wait=t/.%[[1]]
Out[21]= 76.2225
```

This is how long it takes, in minutes, to cool to 100°.

```
In[22]:= T[10]
Out[22]= 1287
```

This just checks to see if the answer is what it should be.

The study of differential equations is a vast and important branch of applied mathematics. We usually do not memorize solutions to differential equations as we did in this section, but we use instead, the theory and techniques presented in a course in differential equations taken soon after completing the calculus sequence.

Mathematica, not surprisingly, has some powerful commands for solving differential equations. A brief introduction to differential equations is often presented towards the end of a calculus sequence, and so, it is included as well in this manual—although in a different form. In order to keep size and cost to a minimum, it is not included in the paper form of this manual, but an introduction to differential equations—Chapter 14 of this manual—will **soon be available at a web site on the internet**.

5.5 The Hyperbolic Functions and Their Inverses

The hyperbolic functions have the same names in Mathematica as they do in traditional mathematics. Their inverses, just like the inverses of the trigonometric functions, are denoted by attaching the prefix "Arc" to the function name. There are many interesting manipulations involving these functions that can be done very effectively with Mathematica.

Example 5.8 *Derive the differentiation formula for* $\sinh^{-1}(x)$ *based on formulas for the exponential and logarithmic functions.*

Here is the differentiation formula.

```
In[23]:= D[ArcSinh[x],x]
```

Out[23]= $\frac{1}{\sqrt{1+x^2}}$

We begin our derivation of this formula, by turning the *arcsinh* function into a more familiar *sinh* function.

```
In[24]:= eq1=y==ArcSinh[x];
```

```
In[25]:= Solve[eq1,x]
```

Out[25]= $\{\{x \to \sinh[y]\}\}$

```
In[26]:= eq2=x==(x/.%[[1]])
```

Out[26]= $x == \sinh[y]$

All input statements serve as examples of how to write our own input statements. As another example, notice the structure of the input statement used to create the equation *eq*2.

The function $\sinh(y)$ is defined in terms of exponentials. We let Mathematica make the transformation to exponentials for us with the **new command TrigToExp[ ]**. Not surprisingly, **there is also a ExpToTrig[ ] command**.

```
In[27]:= eq3=TrigToExp[eq2]
```

Out[27]= $x == \frac{1}{2}(-e^{-y} + e^{y})$

Now that we have the equation in terms of exponential functions, we solve this for y in terms of x.

```
In[28]:= Solve[eq3,y]
```

```
Solve :: ifun : Inverse functions are being used
    by Solve so some solutions may not be found.
```

Out[28]= $\{\{y \to \log[x - \sqrt{1+x^2}]\}, \{y \to \log[x + \sqrt{1+x^2}]\}\}$

There are two solutions, but the term %[[1]] involves the logarithm of an expression which is clearly always negative.

```
In[29]:= y=(y/.%[[2]])
```

Out[29]= $\log[x + \sqrt{1+x^2}]$

```
In[30]:= yp=D[y,x]
```

Out[30]= $\dfrac{1 + \frac{x}{\sqrt{1+x^2}}}{x + \sqrt{1+x^2}}$

```
In[31]:= Simplify[yp]
```

Out[31]= $\dfrac{1}{\sqrt{1+x^2}}$

5.6 Indeterminate Forms and L'Hopital's Rule

If f and g are differentiable functions on an open interval containing $x = a$, except possibly at a itself, and if $h = f/g$ is a $(0/0)$ or (∞/∞) indeterminate form at $x = a$, then, according to *L'Hopital's Rule*

$$\lim_{x \to a} h(x) = \lim_{x \to a} \frac{f(x)}{g(x)} = \lim_{x \to a} \frac{f'(x)}{g'(x)},$$

provided the limit on the right exists in a finite or infinite sense. Variations of this rule can be stated for left and right hand limits and for limits with $a = \pm\infty$.

Mathematica is capable of computing most limits without any help from us. It's response to a limit request, however, can be somewhat unsettling—the only output is a number. In most other problems, even if we do not verify the result with an alternate computation, we are, at least, involved enough in the computation to judge whether or not an answer seems reasonable. In problems involving indeterminate forms, it is usually hard to tell, on the face of it, whether the value of a limit seems reasonable. To be reassured, an alternate calculation is called for, and *L'Hopital's Rule* is a useful tool to use in the verification.

At first glance, it might seem that more elementary methods could be used to judge whether or not the value of a limit seems reasonable. If we want supportive evidence that Mathematica's answer to $lim_{x \to a} h(x)$ is correct, why not simple evaluate $h(x)$ for a few values of x close to $x = a$? This frequently will work, but if h is of indeterminate form at $x = a$, then $h(x)$ is computationally unstable for values of x close to $x = a$. In the indeterminate form $(0/0)$, for example, when the numerator and denominator become small enough, round off errors frequently become sizable. Recall what happen in Example 2.4 when we tried this verification approach.

Example 5.9 *Compute the following limit and verify the answer with an alternate calculation.*

$$\lim_{x \to 0} \frac{6\sin(x) - 6x + x^3}{2x^5}.$$

This limit is easily evaluated directly using Mathematica's limit command. To verify the answer, notice that the expression is clearly a (0/0) indeterminate form, and so *L'Hopital's Rule* applies. After we apply the rule, we have another (0/0) indeterminate form, and so *L'Hopital's Rule* can be applied again and again, until we arrive at a fraction, which is no longer a (0/0). At that point, the value of the limit becomes obvious.

```
In[32]:= f=(6*Sin[x]-6*x+x^3)/(2*x^5); Limit[f,x->0]
```

Out[32]= $\frac{1}{40}$

```
In[33]:= f1=D[Numerator[f],x]/D[Denominator[f],x]
```

Out[33]= $\frac{-6 + 3\ x^2 + 6\ \cos[x]}{10\ x^4}$

```
In[34]:= f2=D[Numerator[f1],x]/D[Denominator[f1],x]
```

Out[34]= $\frac{6\ x - 6\ \sin[x]}{40\ x^3}$

```
In[35]:= f3=D[Numerator[f2],x]/D[Denominator[f2],x]
```

Out[35]= $\frac{6 - 6\ \cos[x]}{120\ x^2}$

```
In[36]:= f4=D[Numerator[f3],x]/D[Denominator[f3],x]
```

Out[36]= $\frac{\sin[x]}{40\ x}$

```
In[37]:= f5=D[Numerator[f4],x]/D[Denominator[f4],x]
```

Out[37]= $\frac{\cos[x]}{40}$

At this point, the limit is no longer a 0/0 indeterminate form and its value is obviously 1/40.

In the next example, we make a low level use of the command Part[ ], which has been used before, but without explanation. It would be worthwhile to add this command to our vocabulary. **The input expression Part[*expr*, *j*] returns the *j th* term in the expression *expr*.** For a product, $p = ab$, of terms, Part[p,1] return a, and Part[p,2] returns b, as you would expect. We will discuss this command in more detail when it is used in a more involved way.

Example 5.10 *Compute the following limit and verify the answer with an alternate calculation.*

$$\lim_{x \to 0+} (1 + \sin(x))^{\cot(x)}$$

This limit is a 1^∞ indeterminate form. To use *L'Hopital's Rule* on a function f which has a 1^∞ indeterminate form, we first write f in the form $f = exp(ln(f))$ and then consider the limit of $p = ln(f)$. Using properties, of logarithms, p can be expressed as a (0/0) or (∞/∞) indeterminate form.

```
In[38]:= f=(1+Sin[x])^Cot[x];Limit[f,x->0,Direction->-1]
```

```
Out[38]= E
```

```
In[39]:= p=Log[f]
Out[39]= log[(1 + sin[x])^cot[x]]
```

Most of the properties of logarithms do not hold for complex valued expressions. It turns out that each term in this expression has a complex valued interpretation, and unless it is told otherwise, Mathematica always makes allowances for complex values. Recall that the **PowerExpand[ ]** command makes the **assumptions necessary to carry out the simplification**. This command must be used cautiously, because mistakes can be made if the assumptions are not valid.

```
In[40]:= p1=PowerExpand[p]
Out[40]= cot[x] log[1 + sin[x]]
```

Before we can use *L'Hopital's Rule*, the expression $p1$ must be turned into a $(0/0)$ or (∞/∞) fraction. We try to create the expression

$$\cot(x)\ln(1+\sin(x)) = \frac{\ln(1+\sin(x))}{tan(x)},$$

but Mathematica immediately simplifies this fraction and returns the original expression. (Thank you Mathematica.) To prevent this from happening, we create this fraction, and differentiate its numerator and denominator at the same time.

```
In[41]:= p2=D[Part[p1,2],x]/D[Tan[x],x]
              cos[x]^3
Out[41]= ------------
          1 + sin[x]
```

Notice that this is no longer an indeterminate form. Its limit value is obviously 1, and this verifies Mathematica's answer of e for the original limit.

```
In[42]:= p3=Limit[p2,x->0]
Out[42]= 1
```

```
In[43]:= answer=E^p3
Out[43]= E
```

5.7 Exercise Set

Some New and Some Old Mathematica features and Commands

Apart[ ]	Clear[ ]	ComplexExpand[ ]
delayed assignment: $(a := b)$	Expand[ ]	ExpToTrig[ ]
FindRoot[ ]	FullSimplify[ ]	Get[ ] or <<
ImplicitPlot[ ]	LeftRiemannDisplay[ ]	MidRiemannDisplay[ ]
Module[ ]	Part[ ]	PowerExpand[ ]
Range[ ]	Remove[ ]	RightRiemannDisplay[ ]
Show[ ]	Sum[ ]	substitution: $(expr/.x->a)$
TrigToExp[ ]	Table[ ]	Together[ ]
<<Graphics`ImplicitPlot`		

1. Show that $f(x) = 1 + \log_2(x+2)$ and $g(x) = 2^{x-1} - 2$ are inverse functions of each other. Demonstrate with an appropriate graph the symmetry between the curves. Choose a plot range of an appropriate size, and use the same scale on both axes.

2. Show that $f(x) = \frac{1}{300}x^2 - \frac{3}{200}x^2 + \frac{13}{400}x + 7$ is one-to-one for $\infty \leq x \leq \infty$. Find a formula for its inverse $g(x) = f^{-1}(x)$, and demonstrate the symmetry between the curves $y = f(x)$ and $y = g(x)$ with an appropriate graph. Choose a plot range of an appropriate size, and use the same scale on both axes.

3. Show that $f(x) = x^3 - 6x^2 + 15x + 7$ is one-to-one on $(-\infty, \infty)$ and find a formula for its inverse $g(x) = f^{-1}(x)$. By computing derivatives directly, and then evaluating them as decimals, demonstrate, for $p = g(3)$ and $q = f(8)$, that

$$g'(3) = \frac{1}{f'(p)}, \; g'(q) = \frac{1}{f'(8)}.$$

4. Show that $f(x) = x/\sqrt{x^2+4}$ is one-to-one on $(-\infty, \infty)$ and find a formula for $g(x) = f^{-1}(x)$. Show that $f(g(x)) = x$ and $g((f(x)) = x$ for all real x. Demonstrate with an appropriate graph the symmetry between the curves $y = f(x)$ and $y = g(x)$. Show that for every a,

$$g'(b) = \frac{1}{f'(a)} \text{for } b = f(a).$$

5. Find the largest interval on which $f(x) = x^2 - e^x$ is one-to-one. You may want to look at a graph for guidance, but don't rely on your eye sight. See how effectively you can use the differential calculus to establish this one-to-one'ness. If $g(x)$ denotes its inverse, approximate $g'(7)$ as a decimal.

6. Find a decimal expansion for e which is correct to 30 decimal places.

7. Evaluate the following as decimals. Notice their similarities or differences.

a) $e^{3/2}$	b) $\sqrt{e^3}$	c) $\frac{1}{e^{\sqrt{2}}}$	d) $e^{-\sqrt{2}}$
e) $\ln(7^2)$	f) $2\ln(7)$	g) $\ln(8+9)$	h) $\ln(8) + \ln(9)$
i) $\ln(8 \times 9)$	j) $\ln(\frac{5}{3})$	k) $\ln(5) - \ln(3)$	l) $\frac{\ln(5)}{\ln(3)}$

8. Differentiate the following. Express your answer first as an expression using D[ ] and then as a function using Derivative[ ] or the *primed notation* we use to denote a derivative in a pencil and paper environment. Notice the answers. Based on your knowledge of basic differentiation formulas, are the answers correct?
a) $f(x) = e^{x^2+5x+1}$ b) $g(x) = \ln(1 + \cos(2x))$

9. Differentiate the function $f(x) = (\ln(1 + \ln(1 + x^2)))^2$. Does the answer look correct? Answer this question one way or the other by using the definition of a derivative as a limit of a difference quotient.

10. Set up a command (module) for estimating the value of $ln(x) = \int_1^x \frac{1}{t} dt)$ with a Riemann sum obtained by dividing the interval $[1, x]$ into 40 subintervals of equal length and always using left end points of intervals to establish heights of rectangles. Use the command to estimate the value of $ln(0.1)$, $ln(10)$, $ln(100)$.

11. Use implicit differentiation and the D[ ] command to find the slope of the line tangent to the graph of $x^2 + y^2 = e^{x+y}$ at the point on the graph corresponding to $x = -1/2$. Plot the equation along with the tangent line. Is there more than one point on the curve

corresponding to $x = -1/2$? (Hint: Recall that replacing y by $y[x]$ tells Mathematica that y depends on x.)

12. Find the absolute maximum and absolute minimum and where they occur, for the function f defined by

$$f(x) = \frac{5x^7}{e^x + e^{-x}}.$$

Express your answers as decimal numbers with 10 digits of accuracy. Include some evidence (not just visual evidence) that larger (or more negative) values will not be found outside of your plot window.

13. Evaluate the following exactly. Simplify your answers.
a) $\log_7(16807)$ b) $\log_{16807}(7)$ c) $\tanh(\ln(e))$
d) $\mathrm{sech}(\ln(17))$ e) $\arctan(-\sqrt{3})$ f) $\sin(\arctan(17))$
g) $\arcsin(\sin(\frac{78}{5}\pi))$ h) $\tanh(\mathrm{arcsinh}(13))$

14. Convert arctanh(3/4) to a form involving evaluations of natural logarithms. A direct approach gives an answer, but little else. Is it correct? You could evaluate both numbers as decimals, but this approach would only show that the two values are approximately the same. Verify with an alternate, more mathematical approach. Begin with the equation $tanh(y) = 3/4$, where y represents the answer.

15. Compute a decimal approximation (10 digits) for each of the following.
a) $\log_8(2.45)$ b) $\sqrt{\arctan(26)}$ c) $e^{\cos(5.3)}$ d) $\sqrt{2\sqrt{2}}$

16. Differentiate the following functions and simplify your answers if appropriate.
a) $f(x) = \arctan(-\sqrt{x})$ b) $g(t) = \frac{t^5 ln(t^2+1)}{(ln(t))^2}$ c) $h(w) = w^{\cos(w)}$

17. Find the points of intersection of $y = x^3$ and $y = e^{x/600}$. How do you know that there is more than one? A convincing analytical answer to this question is essential.

18. Evaluate the following integrals, simplify the answers, and verify that the answers are correct with an alternate calculation.
a) $\int \frac{\cos(3x)\sin(3x)}{4+\cos^2(3x)}dx$ b) $\int_{-1}^{3} \frac{e^{2x}}{9+e^{4x}}dx$
c) $\int_{-4}^{-2} \frac{x}{1-x^2}dx$ d) $\int \frac{1}{\sqrt{4-x^2}}dx$
e) $\int \frac{x}{\sqrt{4-x^2}}dx$ f) $\int \frac{x^2}{\sqrt{4-x^2}}dx$

19. Verify the identity $\arctan\frac{1+x}{1-x} = \arctan(x) + \frac{\pi}{4}$. (Hint: Use the idea that if if $F(x)$ and $G(x)$ are both antiderivatives of the same function, then they differ by a constant.)

20. Derive the differentiation formula for arctanh(x) based on formulas for the exponential and logarithmic functions.

21. Find the area (as a decimal number) of the region bounded by the curves

$$y = x^3, \text{ and } y = 5 - e^{-x}.$$

Verify the answer with an alternate calculation.

22. Find the area of the region bounded by the y-axis and the curves $y = \cos(x)$ and $y = -0.9 + \frac{3}{x-1}$. The two curves here actually bound infinitely many regions. Consider only the one region which is also bounded by the $y-$axis.

23. A movie theater has a 20 foot tall screen mounted on its front wall. The bottom of the screen is 20 feet above the floor, and the entire floor is flat and horizontal. How far from the front wall should a person sit in order to maximize the size of the viewing angle of the screen ? What is the maximum viewing angle in degree measure?

24. When an environmental group first began to monitor the population of a threatened species of bird, there were an estimated 8,500,000 of the birds in the country. That was 50 years ago. The results of a current study suggest that the population has dwindled to 1,900,000. Emergency laws designed to save the bird from extinction are automatically placed in effect if and when the population drops to 500,000. Assume that over the lifetime of the study, the population has been subject to the same law of exponential decay. How much time remains before an emergency is declared?

25. The owner of a small pond is introducing bass into the pond for the first time. The pond is big enough and diverse enough to sustain a maximum population of 10,000 adult bass. It is expected that the population of adult bass will grow at a rate which is proportional to the difference between the maximum sustainable population and the current population. The pond is initially seeded with 50 adult bass. Three years later the population is estimated to be 1400. What will be the population be after fifteen years ? How long will it take to reach a population which is 95% of it maximum population ?

26. A certain bacteria in the human body is capable doubling its population every 45 minutes. If there were 20 bacteria in the initial contaminant, what will the population be 24 hours later ? Assume that the population changes at a rate proportional to the size of the population.

27. Compute the following limits if they exist, and use *L'Hopital's Rule* to verify the answers. Let a, b, and c denote arbitrary constants. Recall our failed attempt to verify the first limit in Example 2.4.

a) $\lim_{x\to 1} \frac{2\cos(x-1)+x^2-2x-1}{(x-1)^4}$ b) $\lim_{x\to\infty}(1+\frac{2}{x})^x$

c) $\lim_{x\to 0^+} \sin(x)\ln(x)$ d) $\lim_{x\to 0^+}(\cos(ax)+\sin(bx))^{(c/x)}$

Project: Best seat in the theater

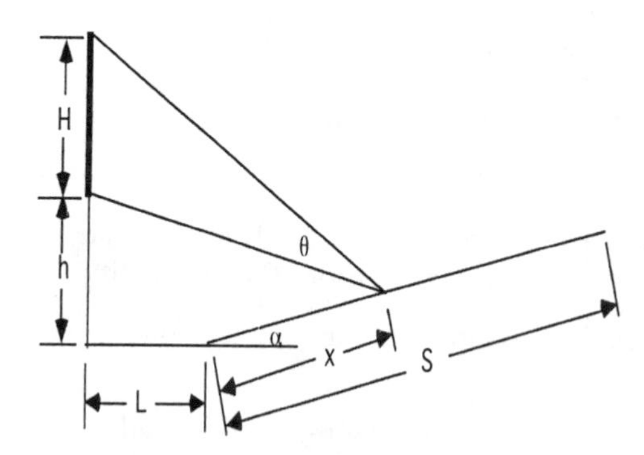

The above diagram describes the basic shape of a movie theater. A screen of height H feet is placed on a wall h feet above a level floor. The floor remains horizontal for the first L feet from that wall and then begins to rise at an incline of angle a as shown. This incline, of course, is where the seats are placedg its length from the first row of seats to the last row in back is S. How far up this incline should we sit in order to have the best view of the movie? The best view, naturally, is one where θ is as large as possible.

Now suppose we plan to see a movie in a theater where $H = 30$, $h = 10$, $L = 20$, $S = 60$, $\alpha = 20°$, where the lengths are all in feet. Determine the value x which corresponds to a maximum viewing angle and find this angle θ. If the rows of seats are placed 3 feet apart with the first row right at the base of the incline, then what row should we sit in?

Suppose management plans to build a new theater where all of the dimensions would be kept the same except for α, which would be changed to $\alpha = 16.5°$. Determine how much this would change the viewing angle by computing the new values of x and θ.

Chapter 6

Techniques of Integration

Neither Mathematica nor mathematics will ever supply us with enough tools to integrate, in closed form, all of the functions we would like to integrate. This is not for want of trying. We defined the function $\ln(x)$ in order to solve the problem $\int \frac{1}{x}\,dx$, and this same strategy could be used for other integrals just as well. If an antiderivative for the integral $\int f(x)\,dx$ cannot be found, simply pick a number a, define the function $F(x) = \int_a^x f(t)\,dt$, study the properties of F so that it becomes a usable special function, and then add it to a list of known special functions. This would allow us to solve not only $\int f(x)\,dx$, but a host of related integrals as well. Unfortunately, this is an endless task. It turns out that no matter how many special functions we add to this list, we can always break outside of this list by integrating functions within the list. This means that there will always be "standard" functions with "unknown" antiderivatives.

The transcendental functions introduced in the last chapter complete the list of functions commonly referred to as the *elementary special functions* of mathematics. These functions include: the power function $p(x) = x^r$ for a rational number r, the trigonometric functions and their inverses, the various logarithmic and exponential functions, and the hyperbolic functions and their inverses. The functions, which we are able to access or express (in worldly terms) consist of all the combinations of these special functions using finitely many compositions and finitely many algebraic operations. Certainly this is a large list, but as we stated in the last paragraph, it is not large enough.

Mathematica has added a few more exotic functions to this list, and so it can evaluate a few more integrals. Furthermore, its integration command is essentially complete. If an antiderivative is expressible in terms of these special functions, Mathematica will almost always find it. If it cannot, then it is likely that we will fail as well. There may be exceptions to this general statement, but the exceptions are probably too rare to be worth pursuing.

Since Mathematica is willing to just give us answers, should we just accept them, without any concern over how answers evolve? Mathematica makes this possible, and it is appropriate behavior, at least some of the time. There is, however, a need for balance. Some knowledge of the techniques of integration is essential for a genuine understanding of calculus. It helps to pay attention to basic ideas, if we want to use mathematics effectively and creatively. At the very least, we must be able to judge whether our answers are correct and appropriate, and this too, requires some grounding in fundamentals. To build this strong background, the time honored practice of integrating with pencil and paper is still an effective learning experience.

In this chapter, the pencil and paper integration techniques of a traditional calculus course are the main focus of attention. To enhance our pencil and paper work, these same techniques are mimicked on a Mathematica notebook. This allows us to focus attention of the techniques, rather than the computations involved, and it allows us to do more involved and interesting problems. Along the way, we will introduce several new Mathematica commands and hopefully learn more about classical integration techniques as well.

Since Mathematica will be able to evaluate, without any help from us, the integrals we consider in this chapter, we have to turn our work here into somewhat of a "game." To begin with, **we will have to use the inert form InertInt[ ] of the integration command, in order to suppress premature evaluation**, otherwise Mathematica will just compute the answer directly. Our objective here is to manipulate a given integral, just like we do with pencil and paper techniques, until it appears as an entry on a simple table of integrals (or as a linear combination of entries on the table). This table of integrals consists of the 15 or so basic antiderivatives, which are simple consequences of basic derivatives. We may as well give a name to this strategy for evaluating integrals, so that we don't have to repeat all of these instructions every time an integral is to be evaluated in this fashion. With this in mind, let us simply call any approach which evaluates integrals in this way an **elementary integral approach**.

The **commands we will be using do not appear in Mathematica**. We already entered three user-defined integration commands, InertInt[ ], IntBySubs[ ], and BasicInt[ ], on a separate file called "MyLibrary." These commands, and a few more, which will be added to this file, will be used throughout this chapter.

The Part[ ] command will also turn out to be very useful in our work, so we take this opportunity to explain more about this very interesting command. It has another very convenient form. If an expression is assigned the name p, then **either Part[p, j] or p[[j]]** can be used to return the jth term in p. We have been using this notation for some time to extract elements from a list. Here are some examples to shed light on this most useful command.

```
In[44]:= p=a+b+c+d+e+f;q=a*b*c*d*e*f;
         r=a*b*c+d+e+f;s=(ab+cd)*(e+f);
```

What do you see when you look at r? It is a sum of 4 terms, the first of which has subterms. What do you see when you look at s? It is a product of 2 terms, both of which have subterms. We can access individual terms as follows:

```
In[45]:= {Part[p,3],Part[q,2],Part[r,3],Part[s,1]}
Out[45]= {c, b, e, ab + cd}
```

```
In[46]:= {p[[3]],q[[2]],r[[3]],s[[1]]}
Out[46]= {c, b, e, ab + cd}
```

The first term, s[[1]] in s is a sum of 2 terms. We can use either notation to extract individual terms from this sum, but notice that a use of the Part[ ] command, probably makes it look more readable.

```
In[47]:= {s[[1]][[2]],Part[s[[1]],2]}
Out[47]= {cd, cd}
```

If A is a sum [product, list, etc.] of terms, then A[{$j_1, j_2, \ldots, j_n$}] represents the sum [product, list, etc.] of the terms $j_1, j_2, \ldots, j_n$ from A. This feature of the Part[ ] command can be very useful. Whatever the structure of A is, a combination of the **same kind is returned**.

```
In[48]:= {p[[{1,3}]],q[[{2,5,6}]],r[[{1,4}]]}
Out[48]= {a + c, b e f, a b c + f}
```

```
In[49]:= a=f[x,y]^2
Out[49]= f[x,y]^2
```

```
In[50]:= {a[[1]],a[[2]]}
Out[50]= {f[x,y],2}
```

```
In[51]:= Part[a[[1]],2]
Out[51]= y
```

The next example is included because it is easy to make a mistake when subtraction is involved. As you can see, Mathematica doesn't recognize subtraction as an operation on a par with addition. It prefers to think of this subtraction as $a + (-1)b$. Mathematica sees the second term here as $-b$. To extract the term b itself, just identify it as the second term of $(-1)b$.

```
In[52]:= p=c-d;
```

```
In[53]:= {p[[2]],Part[p[[2]],2]}
Out[53]= {-d, d}
```

6.1 Integration by Substitution Revisited

This topic was studied in Chapter 3, but now our list of special functions is much larger, and so we are able to integrate many more functions.

Suppose we wish to evaluate an integral $\int p(x)\,dx$. Perhaps Mathematica cannot evaluate the integral, or perhaps we just want to understand how Mathematica arrived at its answer. If we can identify the integral as being of the form $\int f(g(x))g'(x)\,dx$, then we can let $u = g(x)$ and

$$\int p(x)\,dx = \int f(g(x))g'(x)\,dx = \int f(u)\frac{du}{dx}\,dx = \int f(u)\,du.$$

If the integral $\int f(u)\,du$ happens to be an elementary or simple integral, then it is an easy matter to finish the problem.

It is time to **load the commands InertInt[ ], IntBySubs[ ], BasicInt[ ]** from our special file that we named "MyLibrary" (The commands are defined on page 49, if the "MyLibrary" file is not available.) **We don't have to enter them onto the current notebook**. Just open the file, put the mouse pointer into each of the three input cells and hit the Enter Key. The commands will not appear on our current notebook, but they will be entered, nevertheless.

Example 6.1 *Evaluate the integral* $\int \frac{e^{3x}}{9+e^{6x}}dx$ *using Mathematica's integration command directly. Verify the answer, by using an elementary integral approach.*

The number on the first input cell is 4 because **we first entered the three special commands** in the way we as described above.

```
In[4]:= f:=Exp[3x]/(9+Exp[6x]);
```

```
In[5]:= a=Integrate[f,x]
```

```
Out[5]= -1/9 arctan[3 e^(-3 x)]
```

```
In[6]:= ix=InertInt[f,x]
```

Out[6]= $\text{int}\left[\frac{e^{3\,x}}{9+e^{6\,x}}, d[x]\right]$

We let $u = e^{3x}$ because the integral looks like the elementary integral $\int \frac{1}{a^2+x^2}\,du$.

```
In[7]:= iu=IntBySubs[ix,Exp[3x],u]
```

Out[7]= $\text{int}\left[\frac{1}{3\,(9+e^{6\,x})}, d[u]\right]$

```
In[8]:= iu1=iu/.Exp[6x]->u^2
```

Out[8]= $\text{int}\left[\frac{1}{3\,(9+u^2)}, d[u]\right]$

```
In[9]:= iu2=BasicInt[iu1]
```

Out[9]= $\frac{1}{9}\ \arctan\left[\frac{u}{3}\right]$

Now we must return to x.

```
In[10]:= ans=iu2/.u->Exp[3x]
```

Out[10]= $\frac{1}{9}\ \arctan\left[\frac{e^{3\,x}}{3}\right]$

Is Mathematica's answer correct?

```
In[11]:= D[a,x]
```

Out[11]= $\frac{e^{-3\,x}}{1+9\,e^{-6\,x}}$

```
In[12]:= Simplify[%]
```

Out[12]= $\frac{e^{3\,x}}{9+e^{6\,x}}$

In the next example we define the command **CompleteSquare[ ]**. This is another command, which **does not live in Mathematica, so it should be included in our file "MyLibrary"**. Like the other commands in this file, it can be accepted and used on face value. The main part of the formula, however, might look like the familiar way to complete a square on $f(x) = ax^2 + bx + c$.

Example 6.2 *Evaluate the integral* $\int \frac{5}{\sqrt{7-3x-2x^2}}$ *using Mathematica's integration command directly. Verify the answer using an elementary integral approach.*

```
In[13]:= f=5/Sqrt[7-3*x-2*x^2];
```

```
In[14]:= a=Integrate[f,x]
```

Out[14]= $-\frac{5\ \arcsin\left[\frac{-3-4\ x}{\sqrt{65}}\right]}{\sqrt{2}}$

```
In[15]:= ix=InertInt[f,x]
```

Out[15]= $\text{int}\left[\frac{5}{\sqrt{7-3\ x-2\ x^2}}, d[x]\right]$

```
In[16]:= CompleteSquare[f_,x_]:=Module[{a,b,c},
         {c,b,a}=CoefficientList[f,x];
         a*(x+b/(2*a))^2+(4*a*c-b^2)/(4*a)]
```

```
In[17]:= poly=7-3*x-2*x^2
```

Out[17]= $7-3\ x-2\ x^2$

```
In[18]:= poly1=CompleteSquare[poly,x]
```

Out[18]= $\frac{65}{8}-2\ \left(\frac{3}{4}+x\right)^2$

```
In[19]:= p2=f/.poly->poly1
```

Out[19]= $\frac{5}{\sqrt{\frac{65}{8}-2\ \left(\frac{3}{4}+x\right)^2}}$

```
In[20]:= ix=InertInt[p2,x]
```

Out[20]= $\text{int}\left[\frac{5}{\sqrt{\frac{65}{8}-2\ \left(\frac{3}{4}+x\right)^2}}, d[x]\right]$

```
In[21]:= iu=IntBySubs[ix,Sqrt[2]*(x+3/4),u]
```

Out[21]= $\text{int}\left[\frac{5}{\sqrt{2}\ \sqrt{\frac{65}{8}-2\ \left(\frac{3}{4}+x\right)^2}}, d[u]\right]$

```
In[22]:= iu1=iu/.x+3/4->u/Sqrt[2]
```

Out[22]= $\text{int}\left[\frac{5}{\sqrt{2}\ \sqrt{\frac{65}{8}-u^2}}, d[u]\right]$

```
In[23]:= iu2=BasicInt[iu1]
```

Out[23]= $\frac{5\ \arcsin\left[2\ \sqrt{\frac{2}{65}}\ u\right]}{\sqrt{2}}$

In[24]:= `ans=iu2/.u->Sqrt[2]*(x+3/4)`

Out[24]= $\dfrac{5\ \arcsin\left[\frac{4\ \left(\frac{3}{4}+x\right)}{\sqrt{65}}\right]}{\sqrt{2}}$

In[25]:= `ans1=Simplify[ans]`

Out[25]= $\dfrac{5\ \arcsin\left[\frac{3+4\ x}{\sqrt{65}}\right]}{\sqrt{2}}$

Example 6.3 *Use Mathematica's integration command to evaluate the integral*

$$\int \sin^4(x)\cos^7(x)dx$$

directly. Verify the answer using an elementary integral approach.

A substitution of the form $u = \sin(x)$ or $u = \cos(x)$ would seem appropriate. Notice, however, that if we try $u = \sin(x)$ then one power of the cosine function can be used for the differential, and the remaining $\cos^6(x)$ can easily be turned into $(1 - \sin^2(x))^3$ and hence into $(1-u^2)^3$. A substitution of the form $u = \cos(x)$ would not work. One power of the sine function would be needed for the differential, and the remaining 3 powers of $\sin(x)$ could not be turned (without radicals) into an expression involving $u = \cos(x)$.

In[26]:= `f=Sin[x]^4*Cos[x]^7`

Out[26]= $\cos[x]^7\ \sin[x]^4$

In[27]:= `a=Integrate[f,x]`

Out[27]= $\frac{1}{1182720}(16170\ \sin[x] - 2310\ \sin[3\ x] - 2541\ \sin[5\ x] - 165\ \sin[7\ x] + 385\ \sin[9\ x] + 105\ \sin[11\ x])$

In[28]:= `ix=InertInt[f,x]`

Out[28]= $\text{int}\left[\cos[x]^7\ \sin[x]^4,\ d[x]\right]$

In[29]:= `iu=IntBySubs[ix,Sin[x],u]`

Out[29]= $\text{int}\left[u^4\ \cos[x]^6,\ d[u]\right]$

In[30]:= `iu1=iu/.Cos[x]->Sqrt[1-u^2]`

Out[30]= $\text{int}\left[u^4\ \left(1-u^2\right)^3,\ d[u]\right]$

In[31]:= `iu2=int[Expand[iu1[[1]]],d[u]]`

Out[31]= $\text{int}\left[u^4 - 3\ u^6 + 3\ u^8 - u^{10},\ d[u]\right]$

In[32]:= `iu3=BasicInt[iu2]`

Out[32]= $\frac{u^5}{5} - \frac{3\ u^7}{7} + \frac{u^9}{3} - \frac{u^{11}}{11}$

In[33]:= `ans=iu3/.u->Sin[x]`

Out[33]= $\frac{\sin[x]^5}{5} - \frac{3\ \sin[x]^7}{7} + \frac{\sin[x]^9}{3} - \frac{\sin[x]^{11}}{11}$

This answer is different from the one we obtained when we integrated directly. We could try to turn one answer into the other or to turn both answers into some common third answer, but this can become tedious and tricky. We opt for the easiest (though less insightful) approach below. We have used this method before to show that two answers are equivalent, and it is worth remembering.

```
In[34]:= Simplify[a-ans]
Out[34]= 0
```

6.2 Integration by Parts

When we integrate by parts (by hand), we start with an integral, and choose u (the part which gets differentiated) and dv (the part which gets integrated). Actually, all we really have to choose is either u or dv, since the value of one is implied by the choice for the other. Consequently, our command **IntByParts[ ] has just two arguments. The first specifies an (inert) integral, and the second, a choice for u.** An integration by parts command does not exist in Mathematica, so we define our own command below. **This command should be added to our "MyLibrary" file.** The first entry in this command must be an inert integral entered by using the InertInt[ ] command.

To start our notebook, we first **enter the commands InertInt[ ], IntBySubs[ ], BasicInt[ ]** from the file "MyLibrary".

```
In[4]:= IntByParts[p_,u_]:=Module[{x,f,dv,v},
        x=Part[Part[p,2],1];f=Part[p,1]; dv=f/u;
         v=Integrate[dv,x];u*v+InertInt[-v*D[u,x],x]]
```

The integration by parts formula

$$\int u\,dv = uv - \int v\,du = uv + \int (-v)du$$

is clearly seen in the definition of this command. The second form of the formula is used in the definition, because it turns out to be slightly more efficient to have a positive sign in front of the final (inert) integral.

Example 6.4 *Evaluate the integral $\int x^3 e^{5x} dx$ directly, using Mathematica's integration command. Verify the answer using an elementary integral approach.*

```
In[5]:= f=x^3*Exp[5x]
Out[5]= e^(5 x) x^3
```

In[6]:= a=Integrate[f,x]

Out[6]= $e^{5\ x} \left(-\frac{6}{625} + \frac{6\ x}{125} - \frac{3\ x^2}{25} + \frac{x^3}{5}\right)$

```
In[7]:= ix=InertInt[f,x]
Out[7]= int[e^(5 x) x^3, d[x]]
```

```
In[8]:= ix1=IntByParts[ix,x^3]
```

Out[8]= $\frac{1}{5} e^{5x} x^3 + \text{int}\left[-\frac{3}{5} e^{5x} x^2, d[x]\right]$

```
In[9]:= ix2=ix1[[1]]+IntByParts[ix1[[2]],x^2]
```

Out[9]= $-\frac{3}{25} e^{5x} x^2 + \frac{1}{5} e^{5x} x^3 + \text{int}\left[\frac{6}{25} e^{5x} x, d[x]\right]$

```
In[10]:= ix3=ix2[[{1,2}]]+IntByParts[ix2[[3]],x]
```

Out[10]= $\frac{6}{125} e^{5x} x - \frac{3}{25} e^{5x} x^2 + \frac{1}{5} e^{5x} x^3 + \text{int}\left[-\frac{6 e^{5x}}{125}, d[x]\right]$

```
In[11]:= ix4=ix3[[{1,2,3}]]+BasicInt[ix3[[4]]]
```

Out[11]= $-\frac{6 e^{5x}}{625} + \frac{6}{125} e^{5x} x - \frac{3}{25} e^{5x} x^2 + \frac{1}{5} e^{5x} x^3$

Example 6.5 *Evaluate the integral $\int e^{5x}\cos(2x)dx$ using Mathematica's integration command directly. Verify the answer using an elementary integral approach.*

```
In[12]:= f=Exp[5x]*Cos[2x]
```

Out[12]= $e^{5x} \cos[2x]$

```
In[13]:= a=Integrate[f,x]
```

Out[13]= $\frac{1}{29} e^{5x} (5 \cos[2x] + 2 \sin[2x])$

```
In[14]:= ix=InertInt[f,x]
```

Out[14]= $\text{int}\left[e^{5x} \cos[2x], d[x]\right]$

```
In[15]:= ix1=IntByParts[ix,Cos[2x]]
```

Out[15]= $\frac{1}{5} e^{5x} \cos[2x] + \text{int}\left[\frac{2}{5} e^{5x} \sin[2x], d[x]\right]$

```
In[16]:= ix2=ix1[[1]]+IntByParts[ix1[[2]],Sin[2x]]
```

Out[16]= $\frac{1}{5} e^{5x} \cos[2x] + \text{int}\left[-\frac{4}{25} e^{5x} \cos[2x], d[x]\right] + \frac{2}{25} e^{5x} \sin[2x]$

Notice how the original integral has returned. Far from being a wasted effort; a solution is at hand. The solution, however, is no longer a calculus problem, but rather a purely algebraic one. We **replace the original integral, in $ix2$ by an unassigned name u ($u =$ answer)**, set up the equation $u = ix2$, and solve the resulting equation for u.

```
In[17]:= eq=u==ix2[[{1,3}]]-4*u/25
```

Out[17]= $u == -\frac{4u}{25} + \frac{1}{5} e^{5x} \cos[2x] + \frac{2}{25} e^{5x} \sin[2x]$

```
In[18]:= s=Solve[eq,u]
```

Out[18]= $\{\{u \to \frac{1}{29} e^{5x} (5 \cos[2x] + 2 \sin[2x])\}\}$

```
In[19]:= ans=u/.s[[1]]
```

Out[19]= $\frac{1}{29} e^{5x} (5 \cos[2x] + 2 \sin[2x])$

6.3 Reduction Formulas

Most Tables of Integrals include a large number of formulas known as reduction formulas. These are formulas, such as the one in the next example, which express an integral in terms of an "easier form"—usually a lower power— of a similar integral. Mathematica is a useful tool that can be used to establish many of these formulas.

Example 6.6 *Verify the reduction formula*

$$\int \csc^n(x)dx = -\frac{1}{n-1}\frac{\cos(x)}{\sin^{n-1}(x)} + \frac{n-2}{n-1}\int \csc^{n-2}(x)dx$$

```
In[1]:= Integrate[Csc[x]^n,x]
```

The answer is inappropriate and so it is omitted. We could replace $\csc(x)$ by $1/\sin(x)$ and try again, but Mathematica's answer would still be inappropriate. It turns out that **we can do a better job than Mathematica** by manipulating the inert integral with our special commands. We use the second form for this integral, since it is easier to manipulate sines and cosines than it is to work with $\csc(x)$.

```
In[2]:= f=Sin[x]^(-n)
```

When we integrate by parts, there are several strategies to consider in our choice of u and dv. One of these strategies involves looking for a part of the integral which is easily integrated. Since this is true of $\csc^2(x)$, that would prompt us to choose $dv = \csc^2(x)$. With this choice of dv, $u = \csc^{n-2}(x)$ becomes he rest of the integral, by default. Remember that with our integration by parts command, all we have to specify is u.

We begin by entering the commands InertInt[ ], IntBySubs[ ],IntByParts[ ],BasicInt[ ], from the file "MyLibrary."

```
In[7]:= ix=InertInt[f,x]
```

Out[7]= $\text{int}[\sin[x]^{-n}, d[x]]$

```
In[8]:= ix1=IntByParts[ix,Sin[x]^-(n-2)]
```

Out[8]= $\text{int}\left[(2-n) \cos[x]^2 \sin[x]^{-n}, d[x]\right] - \cos[x] \sin[x]^{1-n}$

```
In[9]:= ix2=ix1/.Cos[x]^2->1-Sin[x]^2)
```

Out[9]= $\text{int}\left[(2-n) \sin[x]^{-n} \left(1-\sin[x]^2\right), d[x]\right] - \cos[x] \sin[x]^{1-n}$

Notice that the original integral would reappear (as part of the inert integral) if we expanded the product $\sin^{-n}(x)(1-\sin^2(x))$. We could ask Mathematica to do this, but it is just as easy to skip this step and go on to the next. Once the original integral reappears, the problem can be **solved without calculus**, by simply solving an equation.

This situation is similar to one that occurred in Example 6.5. We **replace the original integral, in $ix2$ by an unassigned name u (u = answer)**, set up the equation $u = ix2$, and solve the resulting equation for u. This equation can be set up directly, by just looking at the above output.

```
In[10]:= eq=u==(2-n)*u-(2-n)*InertInt[Sin[x]^(-n+2),x]+ix2[[2]]
Out[10]= u == (2 - n) u - (2 - n) int[sin[x]^(2-n), d[x]] - cos[x] sin[x]^(1-n)
```

```
In[11]:= s=Solve[eq,u]
Out[11]= {{u -> -((2 - n) int[sin[x]^(2-n), d[x]] + cos[x] sin[x]^(1-n))/(-1 + n)}}
```

6.4 Inverse Trigonometric Substitutions

In this section, we consider integrals involving expressions of the form $\sqrt{x^2+a^2}$, $\sqrt{x^2-a^2}$, $\sqrt{a^2-x^2}$. Frequently, integrals involving other powers of these basic quadratic expressions can be evaluated using the same techniques. The integrals are, of course, easily evaluated using Mathematica's Integrate[ ] command directly, but it is not uncommon for answers to bear little resemblance to the original integrand. Why does an answer take a particular form? To answer this question, our work continues in the same spirit as it has in the other sections of this chapter. We use Mathematica to mimic what we would have to do to evaluate the integral with "pencil and paper" techniques.

The germ of the idea is to use the substitutions and identities,

$$x = a\sin(\theta), \text{ and } 1-\sin^2(\theta) = \cos^2(\theta), \text{ on the expression } a^2 - x^2,$$

$$x = a\tan(\theta), \text{ and } 1+\tan^2(\theta) = \sec^2(\theta), \text{ on the expression } a^2 + x^2,$$

$$x = a\sec(\theta), \text{ and } \sec^2(\theta)-1 = \tan^2(\theta), \text{ on the expression } x^2 - a^2,$$

to turn these basic quadratic expressions into perfect squares. This allows us to cancel the radical—a major simplification. The resulting trigonometric integral can then be evaluated by standard techniques.

The approach is really just integration by substitution, and so our special IntBySubs[ ] command will easily handle the conversion over to a trigonometric integral. Our past substitutions inside the IntBySubs[ ] command have been of the form $u = g(x)$ (a new variable as a function of the old variable). With the current technique, we consider substitutions of the form $x = g(\theta)$ (the old variable as a function of a new variable). To enter this into the IntBySubs[ ] command, we must use it in the form $\theta = g^{-1}(x)$.

This command automatically converts the differential dx over to the new differential $d\theta$. We could ignore these details, but to make this a good learning experience, we should at least make a mental note of the relationships

$x = a\sin(\theta)$	$x = a\tan(\theta)$	$x = a\sec(\theta)$
$dx = a\cos(\theta)\,d\theta$	$dx = a\sec^2(\theta)\,d\theta$	$dx = a\sec(\theta)\tan(\theta)\,d\theta$

between the variables and the differentials.

Example 6.7 *Evaluate the following integral using Mathematica's integration command directly. Verify the answer using the above approach.*

$$\int \frac{a^2-x^2}{x^4}\,dx$$

Recall that we can enter the symbol θ in an input cell, by typing \[Theta]. Alternately, pull down the File Menu, then the Palettes Menu and click on BasicTypesetting. To avoid all of these complications, we use u instead of θ.

In[1]:= **f=Sqrt[a^2-x^2]/x^4**

Out[1]= $\frac{\sqrt{a^2 - x^2}}{x^4}$

In[2]:= **ans1=Integrate[f,x]**

Out[2]= $\left(-\frac{1}{3\ x^3} + \frac{1}{3\ a^2\ x}\right) \sqrt{a^2 - x^2}$

Enter the commands InertInt[ ], IntBySubs[ ], BasicInt[ ], from the file "MyLibrary."

In[6]:= **ix=InertInt[f,x]**

Out[6]= $\text{int}\left[\frac{\sqrt{a^2 - x^2}}{x^4}, \text{d}[x]\right]$

In[7]:= **iu=IntBySubs[ix,ArcSin[x/a],u]**

Out[7]= $\text{int}\left[\frac{a\ \sqrt{a^2 - x^2}\ \sqrt{1 - \frac{x^2}{a^2}}}{x^4}, \text{d}[u]\right]$

In[8]:= **iu2=iu/.x->a*Sin[u]**

Out[8]= $\text{int}\left[\frac{\csc[u]^4\ \sqrt{1 - \sin[u]^2}\ \sqrt{a^2 - a^2\ \sin[u]^2}}{a^3}, \text{d}[u]\right]$

In[9]:= **g=Simplify[iu2[[1]]]**

Out[9]= $\frac{\sqrt{\cos[u]^2}\ \sqrt{a^2\ \cos[u]^2}\ \csc[u]^4}{a^3}$

In[10]:= **g1=PowerExpand[g]**

Out[10]= $\frac{\cot[u]^2\ \csc[u]^2}{a^2}$

In[11]:= **iu3=InertInt[g1,u]**

Out[11]= $\text{int}\left[\frac{\cot[u]^2\ \csc[u]^2}{a^2}, \text{d}[u]\right]$

In[12]:= **iu4=BasicInt[iu3]**

Out[12]= $-\frac{\cot[u]^3}{3\ a^2}$

In[13]:= **ans2=iu4/.u->ArcSin[x/a]**

Out[13]= $-\frac{a\ \left(1 - \frac{x^2}{a^2}\right)^{3/2}}{3\ x^3}$

In[14]:= **Together[ans1]**

Out[14]= $\dfrac{\sqrt{a^2 - x^2}\ (-a^2 + x^2)}{3\ a^2\ x^3}$

6.5 Rational Functions

Mathematica can compute integrals of rational functions directly. It uses the same approach of partial fraction decomposition that would be used to compute the integral of a rational function by hand. As you recall, one of the steps in this process involves factoring the denominator of the rational function into a product of linear and irreducible quadratic terms.

According to the *Fundamental Theorem of Algebra*, every polynomial of degree one or more can be expressed as a product of linear and quadratic terms, but finding the factors is not always an easy matter. In Section 5.1, we mentioned that there is no formula for finding the roots of a general polynomial of degree n, for any $n \geq 5$. For the very same reason, there is no formula for factoring polynomials of degree $n \geq 5$ (factoring and finding roots of polynomials are equivalent).

In spite of this mathematical barrier, Mathematica is quite good at factoring polynomials which have nice factors, and so one can expect Mathematica to be reasonably successful at integrating quite a few rational functions. Nevertheless, the next example was chosen with some care. The denominator of a randomly chosen rational function of this size would probably not have such nice factors.

In[1]:= **f=(6x^6-25x^5+71x^4-235x^3+317x^2-582x+468)/ (2x^5-7x^4+16x^3-56x^2+32x-112)**

Out[1]= $\dfrac{468 - 582\ x + 317\ x^2 - 235\ x^3 + 71\ x^4 - 25\ x^5 + 6\ x^6}{-112 + 32\ x - 56\ x^2 + 16\ x^3 - 7\ x^4 + 2\ x^5}$

In[2]:= **Integrate[f,x]**

Out[2]= $-2\ x + \frac{3\ x^2}{2} - \frac{3}{4+x^2} - \frac{7}{2}\ \arctan\left[\frac{x}{2}\right] + \frac{3}{2}\ \log[-7 + 2\ x] + \frac{3}{2}\ \log\left[4 + x^2\right]$

If it is desired, the **Apart[ ] command** can be used to find the partial fraction decomposition of a rational function $r(x)$. As you can see, however, Mathematica can compute the integral directly without additional input. The degree of the numerator of $r(x)$ does not have to be less than the degree of the denominator of $r(x)$, as it does when doing the method by hand. Mathematica first does the necessary long division.

In[3]:= **Apart[f]**

Out[3]= $-2 + 3\ x + \dfrac{3}{-7 + 2\ x} + \dfrac{6\ x}{(4 + x^2)^2} + \dfrac{-7 + 3\ x}{4 + x^2}$

Look at this decomposition. It justifies the value for the above integral.

Mathematica has several related commands, that can be used to gain further insight into this decomposition. For example, f is not proper—the degree of its numerator is not less than the degree of the denominator. We could determine the quotient and remainder from a **long division process** by using the commands **PolyQuotient[ ] and PolyRemainder[ ]**.

In[4]:= **up=Numerator[f];down=Denominator[f];**

In[5]:= **q=PolynomialQuotient[up,down,x]**

```
Out[5]= -2 + 3 x
```

```
In[6]:= r=PolynomialRemainder[up,down,x]
Out[6]= 244 - 182 x + 109 x^2 - 35 x^3 + 9 x^4
```

As a consequence of the long division process, we have the following equivalent form for f. This is followed by a factorization of the denominator of the proper fraction (the numerator doesn't factor). All this sheds further light on the partial fraction decomposition of this particular rational function.

```
In[7]:= f1=q+r/down
```

$$\text{Out[7]= } -2+3\,x+\frac{244-182\,x+109\,x^2-35\,x^3+9\,x^4}{-112+32\,x-56\,x^2+16\,x^3-7\,x^4+2\,x^5}$$

```
In[8]:= f2=f1[[{1,2}]]+Factor[f2[[3]]]
```

$$\text{Out[8]= } -2+3\,x+\frac{244-182\,x+109\,x^2-35\,x^3+9\,x^4}{(-7+2\,x)\,(4+x^2)^2}$$

6.6 Improper Integrals

The command Integrate[ ] can be used to compute directly improper integrals of the form

$$\int_a^{\infty} f(x)dx,\ \int_{-\infty}^{b} f(x)dx,\ \int_{-\infty}^{\infty} f(x)dx.$$

It can also compute improper integrals of the form

$$\int_a^b f(x)dx,$$

where f is discontinuous at one or more points in the interval [a,b]. Mathematica **automatically treats the integral as being improper, if f is discontinuous at either one of the two end points $x = a$ or $x = b$** of the integration interval. However, Mathematica **should be told, if there are discontinuities strictly between a and b.**

Here is a sample of Mathematica responses to a few different improper integrals.

```
In[1]:= Integrate[10/(x^2+4)^2,{x,-Infinity,Infinity}]
```

$$\text{Out[1]= } \frac{5\,\pi}{8}$$

```
In[2]:= Integrate[x/(x^2+4),{x,-Infinity,0}]
```

We omit Mathematica's lengthy response. This integral does not converge. To see why, we compute the integral again, using the definition of an improper integral.

```
In[3]:= p=Integrate[x/(x^2+4),{x,t,0}]
```

$$\text{Out[3]= } \log[2]-\frac{1}{2}\log\left[4+t^2\right]$$

```
In[4]:= Limit[p,t->-Infinity]
```

```
Out[4]= ∞
```

Mathematica automatically checks the end points for discontinuities and treats the integral as improper if they are found. In the next example, the integrand has a discontinuity midway between the limits of integration. In this case, to be on the safe side, Mathematica should be told. Notice where the discontinuity at $x = 0$ is entered into the integration command.

```
In[5]:= Integrate[1/x^(2/3),{x,-1,0,8}]
Out[5]= 6 - 3 (-1)^(1/3)
```

```
In[6]:= Expand[%]
Out[6]= 6 - 3 (-1)^(1/3)
```

Is the answer 6+3=9?

```
In[7]:= N[%]
Out[7]= 4.5 - 2.59808 i
```

This improper integral exists, and it certainly is not complex valued (not according to our interpretation). **Did Mathematica make a mistake?** Not really! It depends on what interpretation one gives to a negative number raised to a fractional power. Recall that Mathematica **always evaluates** a^r **as a complex number, when** $a < 0$ **and the rational number** $r = p/q$ **is not an integer**. It is easy to be frustrated by Mathematica's behavior on matters of this nature, but do not blame Mathematica—it is a necessary inconvenience.

In the present case, it is easy to overcome the problem, by writing $x^{2/3}$ in an equivalent form (x^2 is always positive).

```
In[8]:= g=1/(x^2)^(1/3)
                1
Out[8]= ----------
        (x^2)^(1/3)
```

```
In[9]:= Integrate[g,{x,-1,0,8}]
Out[9]= 9
```

This real valued calculation is more reliable than the first one, but it confirms the value we expected from the first, where we used -1 as a value for $(-1)^{1/3}$. For further confirmation:

```
In[10]:= {Integrate[g,{x,-1,0}],Integrate[g,{x,0,8}]}
Out[10]= {3, 6}
```

In order to integrate an expression involving $\sqrt[3]{x}$, **we use the If[ ] command to first define a real valued cubed root.**

```
In[11]:= g=If[x<0,-1/(-x)^(1/3),1/x^(1/3)]
```

```
In[12]:= Integrate[g,{x,-1,0,8}]
         9
Out[12]= -
         2
```

The integrand in the next example has a discontinuity at its left end point $x = 0$. No special instructions are necessary; Mathematica automatically treats it as an improper integral.

```
In[13]:= Integrate[1/Sin[x^2],{x,0,1}]
```

Out[13]= $\int_0^1 \csc[x^2]\,dx$

```
In[14]:= NIntegrate[1/Sin[x^2],{x,0,1}]
```

Mathematica's elaborate warning, about this calculation not working out, is omitted. A following large number is then given as output.

Out[14]= $1.2406551784\ 10^{3498}$

What does this mean? Does this integral converge or diverge ?.

Frequently, we can tell if an improper integral converges or diverges, by comparing it to an improper integral of known convergence or divergence. We compare $1/\sin(x^2)$ to $1/x$ by looking at the graphs of $sin(x^2)$ and x.

```
In[15]:= Plot[{Sin[x^2],x},{x,0,1}];
```

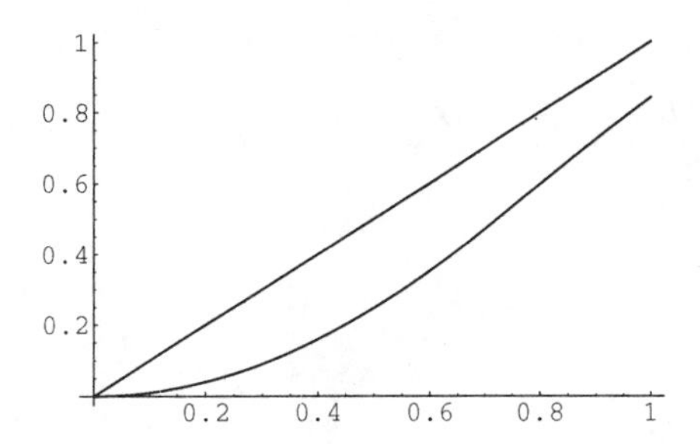

This plot reveals that $0 < sin(x^2) < x$ for $0 \leq x \leq 1$, which means that

$$\frac{1}{sin(x^2)} > \frac{1}{x}\ (0 \leq x \leq 1),$$

and so

$$\int_0^1 \frac{1}{x}dx = \infty \quad \Rightarrow \quad \int_0^1 \frac{1}{\sin(x^2)} = \infty.$$

Example 6.8 *A large lake is initially seeded with 2,000 small mouth bass. It is anticipated that the population will grow at the rate $R(t) = (200 + 10t^3)e^{-0.14t}$ fish per year t years after the initial seeding. What population does the lake eventually level off to?*

This problem could be handled as a simple antidifferentiation problem, but it could just as well be thought of as the answer to an improper integral. At time t, if dt is a "small" interval of time, $R(t)$ is essentially constant, so $R(t)dt$ determines the increase in population during this time interval. We simply "add up" all of the little increases over time.

```
In[16]:= ;R[t_]:=(200+10t^3)*E^(-0.14t)
```

```
In[17]:= pop=2000+Integrate[R[t],{t,0,Infinity}]
Out[17]= 159613.
```

Example 6.9 *Find the arc length of the curve $f(x) = x * \sin(1/x)$ $(0 < x \leq 1)$ and the curve $g(x) = x^2 * \sin(1/x)$ $(0 < x \leq 1)$, if they exist.*

The formula used to evaluate arc length of the curve $y = h(x)$ $(a \leq x \leq b)$ is

$$L(h) = \int_a^b \sqrt{1 + (h'(x))^2}\, dx.$$

In both of the curves in this example, as $x \to 0^+$, $1/x \to \infty$, and so $\sin(1/x)$ begins to oscillate with greater and greater frequency between $+1$ and -1. Look at the graphs of the two curves in the following work session. Would you ever expect that one of them has infinite length? Our eyes can be deceiving.

```
In[18]:= f=.;g=.;f[x_]=x*Sin[1/x];g[x_]=x^2*Sin[1/x];
```

```
In[19]:= Plot[f[x],{x,0,1}];
```

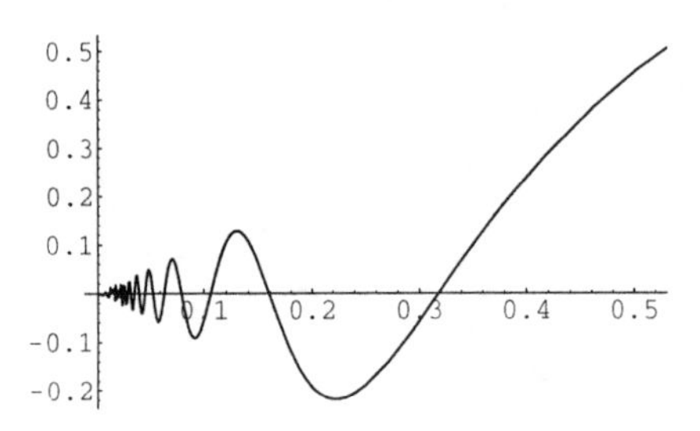


```
In[20]:= Plot[g[x],{x,0,1}];
```

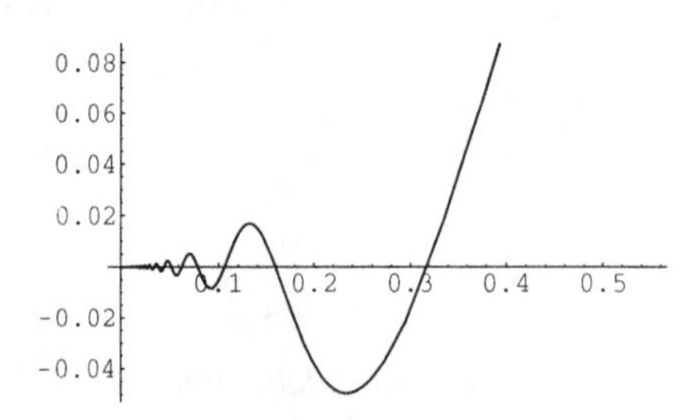


```
In[21]:= lf=Sqrt[1+(f'[x])^2]
```

Out[21]= $\sqrt{1 + \left(-\frac{\cos\left[\frac{1}{x}\right]}{x} + \sin\left[\frac{1}{x}\right]\right)^2}$

```
In[22]:= lg=Sqrt[1+(g'[x])^2]
```

Out[22]= $\sqrt{1 + \left(-\cos\left[\frac{1}{x}\right] + 2\ x\ \sin\left[\frac{1}{x}\right]\right)^2}$

Both lf and lg are discontinuous at $x = 0$, and so their integrals are definitely improper integrals. It is interesting to look at the graphs of these expressions before we attempt to compute their integrals.

```
In[23]:= Plot[lf,{x,0,1}];
```

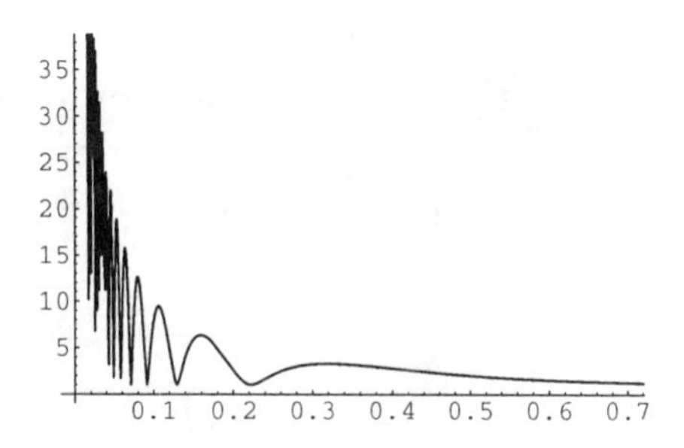


```
In[24]:= Plot[lg,{x,0,1}];
```

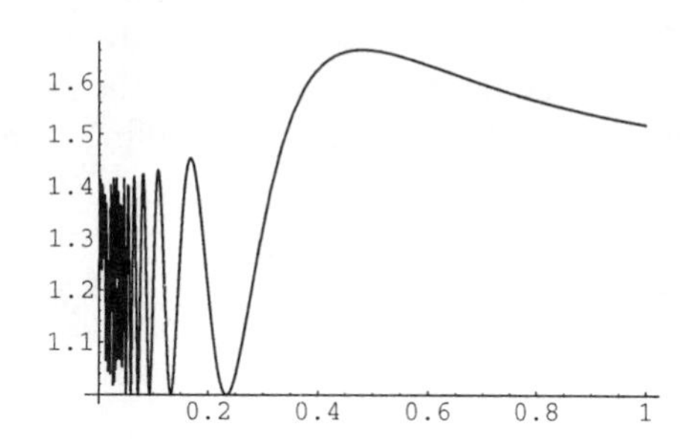

Finally, we attempt to compute the arc lengths of the curves $y = f(x)$ and $y = g(x)$ on the interval [0,1].

```
In[25]:= Lf=NIntegrate[lf,{x,0,1}]
```

Mathematica's elaborate warning is omitted. It then gives a value, which is not to be trusted. This calculation is not good evidence, but it turns out, that this integral diverges; the length of the curve $y = f(x)$ $(0 < x \leq 1)$ is infinite.

```
Out[25]= 15.2409
```

```
In[26]:= Lg=NIntegrate[lg,{x,0,1}]
```

Mathematica's elaborate warning for this calculation is also omitted. It gives a value, which is not to be trusted, but as it turns out, this integral converges; the length of the curve $y = g(x)$ $(0 < x \leq 1)$ is reasonably close to the value Mathematica gives after its warning.

```
Out[26]= 1.46938
```

It is hard to resolve the numerical problems and get more definitive answers. Integrating instead from $x = a$ to $x = 1$, where a is a small, nonzero number, say $a = 0.0001$, would seem like a reasonable strategy, but it does not help. Look up the NIntegrate[ ] command in the Help File. **There are many options available to use when numerical problems**

are encountered, but none of them seem to help in this particular situation. There are packages worth looking at (From the Help File, click the Add-Ons button, then Standard Packages, then NumericalMath), but we shall not pursue the matter any further.

6.7 Exercise Set

Some New and Some Old Mathematica features and Commands

Apart[ ]	Assumptions-¿	BasicInt[ ]
Clear[ ]	CompleteSquare[ ]	Expand[ ]
ExpToTrig[ ]	FullSimplify[ ]	If[ ]
Infinity	IntByParts[ ]	IntBySubs[ ]
InertInt[ ]	Integrate[ ]	Module[ ]
NIntegrate[ ]	Part[ ]	PolyQuotient[ ]
PolyRemainder[ ]	PowerExpand[ ]	Range[ ]
Sum[ ]	TrigToExp[ ]	Together[ ]

Recall the alternate form for the Part[ ] command: $p[[j]]$ is the jth term of p, and $p[\{j_1, j_2, \ldots, j_n\}]$ is the combination of terms $j_1, j_2, \ldots, j_n$ from p

In problems 1 through 14, evaluate the integral directly, and then verify your answer by mimicking a pencil and paper approach—what we have been calling an elementary integral approach.

1. $\int \sin^3(x)\cos^6(x)\,dx$

2. $\int \sin^2(x)\cos^7(x)\,dx$

3. $\int \frac{x^3}{\sqrt{4-x^2}}\,dx$

4. $\int \frac{p^3}{\sqrt{4-p^4}}\,dp$

5. $\int \sin(5x)\cos(2x)\,dx$ (Integrate by parts twice.)

6. $\int \tan^5(4y)\sec^7(4y)\,dy$

7. $\int \tan^4(4y)\sec^8(4y)\,dy$

8. $\int (5+3x^2)\sqrt{x^2+8}\,dx$

9. $\int \frac{\sqrt{a^2+x^2}}{x^4}\,dx$

10. $\int \frac{1}{(x^2+8)^3}\,dx$

11. $\int (2s^2+7s-8)e^{-5s}\,ds$

12. $\int \frac{\sin(2x)}{9+\cos^4(x)}\,dx$

13. $\int \sin(t)\ln(\cos(t))\,dt$

14. $\int \frac{x^7-12x^3+21x^2+36}{x^5+3x^4+4x^3+12x^2}\,dx$

In the remaining problems, use Mathematica freely. without restrictions.

15. Evaluate the following integrals. Verify that your answer is correct using the easiest method.

a) $\int \ln(x + \sqrt{a^2 + x^2})dx$
b) $\int \sqrt[7]{1 + \sqrt[3]{x}}\, dx$.
c) $\int (a^2 + \sqrt[3]{x})^n\, dx$, $(n \neq -1)$

16. Evaluate the following improper integral directly. Confirm your answer by evaluating it a second time, using the definition of an improper integral as a certain limit of an antiderivative.

$$\int_1^\infty \frac{x^2 + 25}{x(x^2 + 4)^2}\, dx$$

17. Evaluate the improper integral $\int_1^\infty (ax^2 + bx + c)e^{-px}\, dx$ where a, b, c, p are constants, with $p > 0$. Notice that the integral clearly diverges for $p < 0$. (Mathematica's optional integration "rule" Assumptions− >*an inequality* will be needed in this problem.)

18. Use Mathematica to establish the reduction formula

$$\int x^m (\ln(x))^n dx = \frac{x^{m+1}(\ln(x)^n}{m+1} - \frac{n}{m+1} \int x^m (\ln(x))^{n-1} dx (m \neq -1).$$

19. Use Mathematica to establish the following reduction formula. Try to get Mathematica to do most of the work.

$$\int \cos^m(x) \sin^n(x) dx =$$

$$\frac{\cos^{(m-1)}(x) \sin^{(n+1)}(x)}{m+n} + \frac{m-1}{m+n} \int \cos^{(m-2)}(x) \sin^n(x) dx.$$

This is not an easy problem. Integrate by partsg try to get the original integral to reappear. Aggressive substitution may be necessary to get Mathematica to cooperate. Look up the option Expand[expr,*pattern*] in the Help File to avoid unwanted expansion. It will help to use the command Part[*expr*, *j*]=*expr*[[*j*]].

20. Evaluate the improper integral $\int_{-a}^{a} \sqrt{\frac{a+x}{a-x}}\, dx$, where a is a positive constant. Mathematica gives an answer, but it warns us that the integral may not converge. Is the answer correct? Use an alternate strategy.

21. The *Horn of Gabriel* is the surface generated by revolving the curve $y = 1/x$ $(1 \leq x < \infty)$ about the x-axis. Show that the surface area of the *Horn of Gabriel* is infinite, but that the volume enclosed by the horn $(1 \leq x < \infty)$ is finite. As the saying goes, *"You can fill the horn with paint, but you can't paint the surface of the horn."*

Chapter 7

Sequences and Series

Imagine the large supply of functions we can create and use with our current knowledge of mathematics. This collection would include the power function, $p(x) = x^r$, where r is a real constant, all of the polynomial and rational functions, the trigonometric, logarithmic, exponential, hyperbolic functions and all of their inverses. It would include all of the combinations of these functions using finitely many compositions and finitely many algebraic operations. These are the functions, which we have the power to know. We may need Mathematica's help, but we can evaluate them, graph them, manipulate them, and study their behavior.

The collection is large, and we can certainly use these functions to create a variety of complex mathematical models. Yet, one of the truly fascinating aspects of mathematics is that this supply of accessible functions occupies only a small corner in a much larger warehouse of "nice," "well-behaved" functions that play a significant role in mathematics and its applications. Most of the functions in this warehouse live in a world that is currently beyond our reach.

Examples come from a variety of mathematical situations. We have already mentioned that antiderivatives of continuous functions are frequently (actually, usually) beyond our reach, even though the *Fundamental Theorem of Calculus* says that they always exist. The same can be said about solutions (of the form $y = f(x)$) to some fairly simple equations in x and y. The same can be said about inverses to some functions, even simple functions like $f(x) = x + \sin(x)$, for example. Solutions to differential equations, which are so important to mathematical applications, are frequently hard to describe.

In all of these situations, and others as well, answers to very practical problems can turn out to exist, yet be beyond our ability to describe in terms of familiar functions. How are we to reach these functions, and study their properties? Certainly this is an important issue, if it can be addressed.

In this chapter we study ideas, which will ultimately provide us with a means to get a handle on all of the functions we usually need in mathematics, including those that may be quite inaccessible otherwise. It turns out that functions in this list can be described, surprisingly enough, by expressions that look and behave very much like the simplest functions, namely polynomials

$$p(x) = \sum_{j=0}^{n} a_j x^j, \text{ or } p(x) = \sum_{j=0}^{n} a_j (x-a)^j.$$

The only difference is that we must allow the degree of the polynomial to go to infinity. In the process, "long polynomials"

$$p(x) = \sum_{j=0}^{\infty} a_j x^j, \text{ or } p(x) = \sum_{j=0}^{\infty} a_j (x-a)^j$$

called **power series** and **Taylor series** are created.

Realizing a function as a power series gives us an important means to study the behavior of the function. It turns out that power series behave very much like ordinary polynomials, although this requires a careful analysis. Frequently, the coefficients, a_n ($n = 1, 2, \ldots$), for a particular power series can all be evaluated, and this gives us a very specific way to formulate the function represented by the power series.

This is not an easy topic, and there are many issues in this chapter that must be discussed along with power series. Power series and Taylor series are introduced in the last section in this chapter, after all the other details are considered. Since it is the most important issue, some text books prefer to discuss power series—especially Taylor series—first, followed by all of the other details. In that case, read a portion of the last section of this chapter first. It may help to postpone the material on power series until later, since this is a bit more abstract. Mathematica has a command, Series[ ], for producing a Taylor series. You can read the material concerning this command, along with the example, without going through the rest of the chapter first.

7.1 Mathematica's Limit Command

We have used this command before, and it will be used in almost every problem in this chapter. It seemed like a well behaved, straightforward command when we used it before, but this reputation is not well deserved. It deserves its own section in this manual, so that we can deal with its surprising problems before we study the mathematical topics of this chapter. Mathematica can sometimes compute the most complicated limits, and it can fail to compute the simplest limits.

One of the problems is not Mathematica's fault. When we call for a calculation of the form Limit[$a[n]$, $n-$ >Infinity], **Mathematica assumes that n is a real number** tending to infinity, but in this chapter, **n will always be an integer** tending to infinity. This can cause unexpected problems. For example, $(-1)^n$ simply alternates between $+1$ and -1 as n passes through the positive integers, but $(-1)^n$ is complex valued for every other real number n. We will have to remain alert to complications of this sort throughout this chapter.

Here are a few examples to show how the limit command can fail.

```
In[1]:= Limit[x^x,x->Infinity]
Out[1]= ComplexInfinity
```

Why is the output "ComplexInfinity," and not just ∞, as it is in the next computation?

```
In[2]:= Limit[2^x,x->Infinity]
Out[2]= ∞
```

Think about the definition of $\tanh(x)$ in terms of exponentials, and it follows readily that $\tanh(x) \to 1$ as $x \to \infty$. Mathematica, however, has a hard time with this. This is followed by another limit, which should easily evaluate to 0, but Mathematica is unable to perform the calculation.

```
In[3]:= Limit[Tanh[x],x->Infinity]
Out[3]= Limit[tanh[x], x → ∞]
```

```
In[4]:= f=10^x/x^x
Out[4]= 10^x x^-x
```

```
In[5]:= Limit[f,x->Infinity]
Out[5]= Limit[10^x x^-x, x → ∞]
```

Fortunately, these problems, and others we may encounter, can all be overcome. **Mathematica has an enhanced Limit[ ] command that resides in the Calculus`Limit` package**. This package must be loaded using the prefix operator (<<) before the enhanced command is available.

In this section, however, we show that **we can do just as well without an enhanced limit command**. A few suggestions may help. **Simplify expressions as much as possible, before you take a limit**. Consider using the commands Simplify[ ], FullSimplify[ ], Expand[ ], and PowerExpand[ ]. Remember that the **PowerExpand[ ] command makes assumptions** on the variables. Be careful when you use it!.

Here is a more important suggestion. **Problems with the computation $\lim_{x\to x_0} f(x)$ seem more pronounced when $f(x)$ contains variable expressions as exponents.** This problem can be overcome by using the properties of logarithms and the idea that

$$\text{If } \lim_{x\to x_0} \ln(f(x)) = L, \text{ then } \lim_{x\to x_0} f(x) = e^L.$$

By simplifying $\ln(f(x))$, the variable powers can be brought out of the power position. The PowerExpand[ ] command performs simplifications like

$$\ln\left(b(x)^{p(x)}\right) = p(x)\ln(b(x)).$$

Once the variable powers are gone, the limit command will frequently work. **This strategy will be used frequently to compute limits.**

L'Hopital's Rule might also help, when Mathematica fails to evaluate a limit. Sometimes, just setting up an expression as a product, or a fraction, is all Mathematica needs to compute a limit.

Here are solutions to the two unresolved limits entered above,

```
In[6]:= g=TrigToExp[Tanh[x]]

         -e^-x + e^x
Out[6]= ------------
          e^-x + e^x
```

```
In[7]:= Limit[g,x->Infinity]
Out[7]= 1
```

```
In[8]:= f1=PowerExpand[Log[f]]
Out[8]= x log[10] - x log[x]
```

```
In[9]:= f2=Factor[f1]
Out[9]= x (log[10] - log[x])
```

```
In[10]:= Limit[f2,x->Infinity]
Out[10]= -∞
```

It follows that the limit of f as $x \to \infty$ is $e^{-\infty} = 0$, as expected.

7.2 Sequences

While the concept of a sequence is needed for the development of the rest of the chapter, It is also a concept, which is interesting in its own right. Mathematically, a sequence is just a function whose domain is the set of of positive integers or a subset of that set. It follows that the standard methods for defining functions in Mathematica can be used to define sequences just as well. Of course, this means that notation such as $\{a_n\}$ for denoting a sequence will have to be dropped in favor of functional notation when Mathematica is being used. If a denotes the sequence (function),

$$\left\{\frac{(-1)^n}{n^2}\right\},$$

then the nth term of the sequence and the convergence or divergence of the sequence can be determined in the expected manner.

```
In[1]:= a[n_]:=(-1)^n/n^2
```

```
In[2]:= {a[7],a[20],a[201]}
```

Out[2]= $\left\{-\frac{1}{49}, \frac{1}{400}, -\frac{1}{40401}\right\}$

```
In[3]:= Limit[a[n],n->Infinity]
```

Out[3]= $\text{Limit}\left[\frac{(-1)^n}{n^2}, n \to \infty\right]$

We talked about problems with the alternating sign $(-1)^n = \pm 1$ in the previous section (it is complex valued when n is not an integer). Try to anticipate this problem and drop the alternating sign before a limit is computed.

```
In[4]:= Limit[1/n^2,n->Infinity]
Out[4]= 0
```

It follows that the sequence converges to 0. If we had obtained any number other than 0 in the above limit computation, then the sequence would have been divergent, because of the alternating ± 1.

The terms of a sequence can be displayed in a list by using the Table[ ] command. We have used this command before. The index inside of this command is a local variable. To make an issue of this, we begin the next statement by assigning a value to n.

```
In[5]:= n=7
```

This assignment has no effect on the next computation, because n is a local variable.

```
In[6]:= s=Table[(2n+1)/(n+3),{n,1,20}]
```

Out[6]= $\left\{\frac{3}{4}, 1, \frac{7}{6}, \frac{9}{7}, \frac{11}{8}, \frac{13}{9}, \frac{3}{2}, \frac{17}{11}, \frac{19}{12}, \frac{21}{13}, \frac{23}{14}, \frac{5}{3}, \frac{27}{16}, \frac{29}{17}, \frac{31}{18}, \frac{33}{19}, \frac{7}{4}, \frac{37}{21}, \frac{39}{22}, \frac{41}{23}\right\}$

```
In[7]:= s[[n]]
```

Out[7]= $\frac{3}{2}$

It is customary in mathematics to denote a sequence with notation such as $\{a_n\}$. In Mathematica, curly brackets are used to define lists. The elements of a list are ordered, so this is natural notation to use for that of a sequence.

Since we will be more interested in **infinite sequences**, they will usually be defined as **Mathematica functions**, and the Table[ ] command will used mainly to evaluate and list some of the terms of a sequence.

Example 7.1 *Show that the sequence* $\{a_n\}$ *defined by* $a_n = \frac{n^{100}}{100^n}$ *is eventually a decreasing sequence which converges to zero. At what point in the sequence does it become decreasing? Determine an* N *such that* $0 < a_n < 10^{-8}$ *for all* $n > N$.

We begin by evaluating, for the sake of interest, the first few terms of the sequence.

```
In[8]:= a=.; a[n_]:=n^100/100^n
```

```
In[9]:= Table[N[a[n],4],{n,1,10}]
```

Out[9]= $\{0.01, 1.26765\,10^{26}, 5.15378\,10^{41}, 1.60694\,10^{52}, 7.88861\,10^{59}, 6.53319\,10^{65}, 3.23448\,10^{70}, 2.03704\,10^{74}, 2.65614\,10^{77}, 1.\,10^{80}\}$

Does this sequence diverge to infinity? We attempt to compute the limit of the sequence, and we immediately run into problems with the limit command that we mentioned earlier. A direct attempt to compute a limit, and other attempts along the way are worth trying, but we do not record our failures below.

```
In[10]:= a1=Log[a[n]]
```

Out[10]= $\log\left[100^{-n}\ n^{100}\right]$

```
In[11]:= a2=PowerExpand[a1]
```

Out[11]= $-n\ \log[100] + 100\ \log[n]$

Mathematica has an easier time with products and quotients. Notice that this next step is essentially what we would do, if we were going to set up the problem for an application of *L'Hopital's Rule*.

```
In[12]:= a3=n*Expand[a2/n]
```

Out[12]= $n\left(-\log[100] + \frac{100\ \log[n]}{n}\right)$

```
In[13]:= Limit[a3,n->Infinity]
```

Out[13]= $-\infty$

Remember, this value is the limit of the logarithm of the sequence, so the original sequence **converges to** $e^{-\infty} = 0$. As we can see, **merely listing the first 10 terms, and witnessing the phenomenal growth, did not suggest the long term pattern of values**.

Since the sequence is defined as a Mathematica function, we can study the sequence by plotting the function. We do this next. This seems to show that the sequence does eventually decrease, starting with approximately the 22nd term. This is a useful approach, but not the most rigorous. Our eyes cannot really establish, with any certainty, that the graph continues to decrease forever. Thus, to show the decreasing nature of this sequence more rigorously, we follow this plot with a determination of the sign of the first derivative.

```
In[14]:= Plot[a[n],{n,1,40},PlotRange->{0,2*10^90}];
```

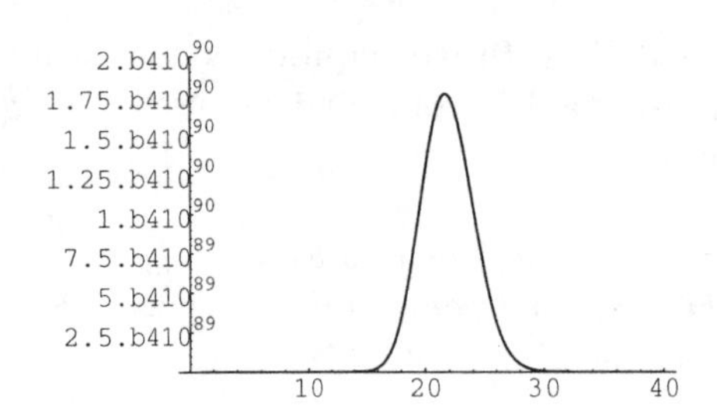

In[15]:= `ap=D[a[n],n]`

Out[15]= $100^{1-n}\ n^{99} - 100^{-n}\ n^{100}\ \log[100]$

In[16]:= `ap=Together[ap]`

Out[15]= $-100^{-n}\ n^{99}\ (-100 + n\ \log[100])$

The term *ap* will be 0, when the $4th$ term (-1 is the first term) is 0. The output would look complicated, if we set the entire expression $ap = 0$

In[17]:= `s=Solve[ap[[4]]==0,n]`

Out[17]= $\{\{n \to \frac{100}{\log[100]}\}\}$

In[18]:= `n0=(n/.s[[1]]);N[p]`

Out[18]= 21.7147

It follows that the derivative *ap* is negative for $n > 22$, and so the sequence decreases for $n > 22$. For the sake of interest, we mention that we could have computed the next larger integer directly with the following new command. Another command in the same family is introduced as well.

In[19]:= `{Ceiling[n0],Floor[n0]}`

Out[19]= {22, 21}

The **commands, Ceiling[ ] and Floor[ ]**, are two of several related integer valued commands, with **Ceiling[x] returning the smallest integer greater than or equal to x**. These commands are not really essential, but are mentioned so that you may use them if you wish.

The easiest way to solve the inequality $a_n < 10^{-8}$ is to plot this pair of expressions. The horizontal and vertical ranges can be adjusted, by trial and error, until they are appropriate.

In[20]:=
```
Plot[{a[n],10^(-8)},{n,80,150},
PlotRange->{0,2*10^(-8)}];
```

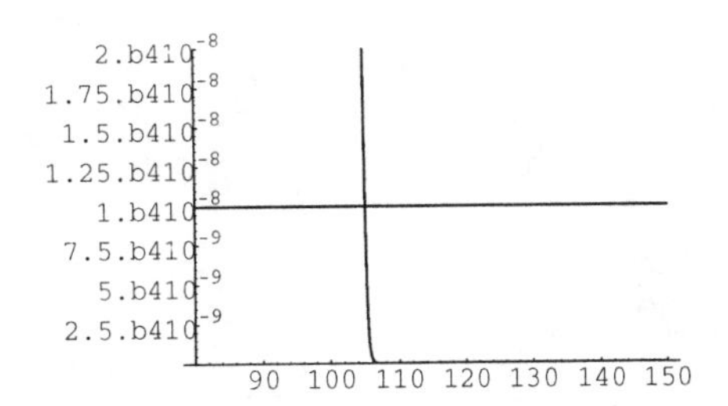

From the graph, it looks like $a_n < 10^{-8}$ for $n > 105$. Alternately or in conjunction with the above plot, we can find this point with the FindRoot[ ] command. A plot still helps. It can be used to find appropriate start-up values and to make a solution seem more believable.

Mathematica has difficulty solving the equation $a_n = 10^{-8}$. This is interesting! The terms involved are too small. **One way to overcome this problem is to solve the equivalent equation** $\ln(a_n) = \ln(10^{-8}) = -8$. Taking the logarithm of both sides, turns small numbers into numbers close to -8.

```
In[21]:= FindRoot[a2==-8,{n,50}]

Out[21]= {n → 102.212}
```

A sequence is just a function whose domain is a set of integers. Frequently, however, the formula used to define a sequence $\{a_n\}$ is well defined, not just for integer values of n, but also for all (or most) real numbers n as well. We took advantage of this when we plotted the sequence in the last example, treating it essentially as a function of a real number. Some sequences, on the other hand, cannot be extended in such a way. Mathematical terms like $(-1)^n$ and $n!$ are quite simple when n is an integer, but when n is not an integer, they represent very advanced mathematical structures that are frequently complex valued and which have no place in our setting. Any sequence involving terms such as these cannot be differentiated or plotted as we did the sequence in the last example. We can, however, still use Mathematica to plot the sequence over its integer domain. We show the technique by plotting the sequence

$$\{\frac{e^n n!}{\sqrt{n} n^n}\}$$

in the next Mathematica work session.

```
In[22]:= a[n_]:=E^n*n!/Sqrt[n]/n^n
```

```
In[23]:= s=Table[a[n],{n,1,20}];
```

```
In[24]:= ListPlot[s];
```

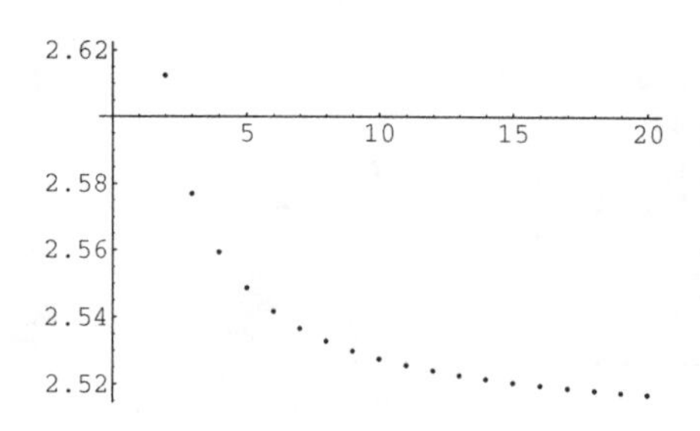

When $s = \{a_1, a_2, \ldots, a_n\}$ is a list of real numbers, the ListPlot[ ] command plots the points $\{(1, a_1), (2, a_2), \ldots, (n, a_n)\}$. When s is a list of ordered pairs, it plots the corresponding points.

Example 7.2 *Show that the sequence $\{a_n\}$ defined by $a_n = (\frac{n+3}{7+n})^{2n}$ converges and find its limit. Verify its limit by an alternate means.*

```
In[25]:= a[n_]:=((n+3)/(n+7))^(2n)
```

```
In[26]:= Table[N[a[n],4],{n,1,30}]
Out[26]= {0.25, 0.0952599,
    0.046656, 0.0268932, 0.0173415, 0.0121224, 0.00899927, 0.00699557,
    0.00563771, 0.00467615, 0.00397028, 0.0034364, 0.00302231, 0.0026942,
    0.00242938, 0.0022122, 0.0020316, 0.00187955, 0.00175015, 0.00163894,
    0.00154253, 0.00145831, 0.00138419, 0.00131856, 0.00126009, 0.00120773,
    0.00116061, 0.00111801, 0.00107933, 0.00104408}
```

Does the sequence converge slowly to zero?

```
In[27]:= Limit[a[n],n->Infinity]
```
Out[27]= $\frac{1}{e^8}$

```
In[28]:= N[%]
Out[28]= 0.000335463
```

As we can now see, the limit is small, but definitely nonzero.

To gain a deeper understanding of this limit, we could use *L'Hopital's Rule* to verify its value. We did similar work in Example 5.9 and Example 5.10. Using this technique on a sequence $\{a_n\}$ is no different than using it on a function $f(x)$. One simply treats the variable n as a continuous variable similar to the variable x.

```
In[29]:= b=PowerExpand[Log[a[n]]]
Out[29]= 2 n (log[3 + n] - log[7 + n])
```

```
In[30]:= {up=2*b[[3]],down=1/n}
```
Out[30]= $\{2\ (\log[3+n] - \log[7+n]),\ \frac{1}{n}\}$

```
In[31]:= b1=D[up,n]/D[down,n]
```

Out[31]= $-2\ n^2 \left(\frac{1}{3+n} - \frac{1}{7+n}\right)$

```
In[32]:= b2=Together[b1]
```

Out[32]= $-\frac{8\ n^2}{(3+n)\ (7+n)}$

```
In[33]:= b3=D[Numerator[b2],n]/D[Denominator[b2],n]
```

Out[33]= $-\frac{16\ n}{10+2\ n}$

```
In[34]:= b4=D[Numerator[b3],n]/D[Denominator[b3],n]
Out[34]= -8
```

This verifies that the original limit is e^{-8}.

The sequences that we have worked with so far have all been explicitly defined. In other words, they have all been of the form $\{a_n\}$ where a_n is given as a formula in n. A sequence can also be defined by expressing the nth entry in the sequence in terms of some (or all) of the previous $n-1$ entries of the same sequence. Such a sequence is said to be **recursively defined**. It is a more natural way of defining some sequences. Frequently, when a sequence is defined recursively, it is in a form ideally suited for listing values numerically in the most efficient way.

For example, suppose $\{b_n\}$ is the sequence of monthly balances that would result if \$20,000 is borrowed at 9% annual interest compounded monthly, with a monthly payment of \$400. The next balance would be the previous balance plus interest minus the payment. In the most natural way, without any need to look up a financial formula, it is easy to conclude that

$$b_{n+1} = (1 + \frac{.09}{12})b_n - 400, b_0 = 20000.$$

This defines the sequence $\{b_n\}$ recursively. Knowing b_0, we can use this formula to compute b_1 and then, in turn, $b_2, b_3, \ldots$. Since it is computationally so easy to go from any one entry to the next, this would be the most efficient way to list, numerically, the whole sequence of balances.

Computations such as these, are carried out using **Mathematica's programming tools**. A recursively defined sequence, however, can be set up is a very special way, which is exceedingly simple.

```
In[35]:= b=.; b[0]=20000;b[n_]:=(1+.09/12)*b[n-1]-400
```

This sets up the sequence as a program. The program runs when we call for a value, by entering, for example $b[60]$ in an input cell. **Do not do this!** Actually, the above statement is perfectly valid, but it would take long time to run, because **Mathematica does not remember any of its values**. In order to compute $b[60]$, it needs the value of, for example, $b[21]$, many times, and each time it needs this value, it must run a part of the program all over again to find it. There is a **much faster way to run this program**.

```
In[36]:= b=.; b[0]=20000;b[n_]:=b[n]=(1+.09/12)*b[n-1]-400
```

Notice the **use of the assignment operator** ($=$). To compute $b[60]$, we still need to compute $b[j]$ for $j < 60$, but as soon as a value for a particular j is computed, it is stored under the name $b[j]$, so it only has to be computed once. The program will now run very quickly. Here are some sample values.

```
In[37]:= {b[10],b[30],b[60]}
Out[37]= {17413.9, 11624.3, 1143.97}
```

There are some **unexpected assignment consequences**. Normally, we can unassign a name, such as b, by a simple input statement of the form (**b = .**), which is another form of the Unset[ ] command. In the present case, however, **we have also assigned values to $b[0], b[1], \ldots, b[60]$**, and all of these names must be unassigned as well. The subtleties, are not worth laboring over, so suffice it to say, that the **Clear[ ] command is a stronger form of unassignment**. It also unassigns the values of $b[j]$ for every j.

```
In[38]:= Clear[b]
```

Writing programs in Mathematica is of central importance, and the current problem provides an opportunity to at least plant a seed for future growth. Any problem involving **repetitive calculation** can probably be done **more efficiently** by doing the work inside of a **program**. While programs can sometimes become quite involved, many situations call for really simple programs. We will not develop programming skills at this time, but we take this opportunity to **introduce the Do[ ] command**, with a simple example.

```
In[39]:= Do[Print["Sin"[j*x]==Expand[Sin[j*x],Trig->True]],{j,2,6}]
Sin[2 x] == 2 cos[x] sin[x]
Sin[3 x] == 3 cos[x]^2 sin[x] - sin[x]^3
Sin[4 x] == 4 cos[x]^3 sin[x] - 4 cos[x] sin[x]^3
Sin[5 x] == 5 cos[x]^4 sin[x] - 10 cos[x]^2 sin[x]^3 + sin[x]^5
Sin[6 x] == 6 cos[x]^5 sin[x] - 20 cos[x]^3 sin[x]^3 + 6 cos[x] sin[x]^5
```

The steps of a program are separated by (;) semicolons, even though they don't seem to appear in the above statement, so the output is not normally displayed. This is why we entered the **Print[ ]** command. The double quotes are used to turn "Sin" into a string. Without the double quotes, the output would be "True, True,...,True."

Look up the Do[ ] command in the Help File. It will lead to several other basic programming commands for the interested reader.

Example 7.3 *A very well-known sequence, called the Fibonacci sequence, is defined recursively by*

$$x_{n+2} = x_{n+1} + x_n, \quad x_1 = x_2 = 1.$$

List the first 30 terms of the Fibonacci sequence. Determine the 400th term, x_{400}.

This sequence was first formulated 800 years ago to describe the populations of successive generations of breeding rabbits. Since that time, it has found its way into many mathematical problems. In addition, the Fibonacci sequence is found as a number pattern in a surprisingly large number of diverse settings in nature. It is such an interesting sequence that a mathematical journal, The Fibonacci Quarterly, is devoted to its study.

```
In[40]:= x=.; x[0]=0;x[1]=1;x[n_]:=x[n]=x[n-1]+x[n-2]
```

```
In[41]:= Table[x[n],{n,1,30}]
Out[41]= {1, 1, 2, 3, 5, 8, 13, 21, 34, 55, 89, 144, 233, 377, 610,
          987, 1597, 2584, 4181, 6765, 10946, 17711, 28657,
          46368, 75025, 121393, 196418, 317811, 514229, 832040}
```

Mathematica cannot compute $x[400]$ directly. It requires **too many programming steps**, a number which is set by the **value of $RecursionLimit**. We could change this value, but it is just as easy to simply compute an intermediate value. It doesn't take too many steps to go from $x[200]$ to $x[400]$.

```
In[42]:= x[200]
Out[42]= 280571172992510140037611932413038677189525
```

```
In[43]:= x[400]
Out[43]= 17602368064501396646822694539241125077038438330449219l
           8867259928965753450442160196755
```

7.3 Series

The Sum[ ] command was introduced on page 42, and it will be used rather centrally through the remainder of this chapter. This is a good time to remind ourselves about **Mathematica's penchant for being exact**! Sums can be exceedingly complicated, and we should try to anticipate this, before we hit the enter key. To find the exact value of Sum[$1/j, \{j, 1, 2000\}$], for example, we would first have to find a common denominator. The sum would be a fraction a/b, where a and b each have roughly one thousand digits.

Since the arithmetic and symbolic complications quickly get out of hand, it is often **advisable to use the NSum[ ] command** rather than the Sum[ ] command. The combination N[Sum[ ]] just calls for the command NSum[ ].

Now, let us begin the topic of this section. To gain a good understanding of the concept of a series, it helps to start with its definition. A series is a sequence—**a sequence**—a sequence $\{s_n\}$ of partial sums. Each term in this sequence is much more manageable than the whole series, because each term s_n is a finite sum. We already have a good understanding of finite sums, and the previous section gave us a background in sequences. So now our attention will be focused on this sequence $\{s_n\}$. Does it converge or diverge? If it converges, can we compute its limit? Here are some examples, where we use Mathematica to shift our attention from the series itself to its sequence of partial sums. We start with the example

$$\sum_{j=1}^{\infty} \frac{1}{j^3}.$$

```
In[44]:= ps[n_]:=N[Sum[1/j^3,{j,1,n}]]
```

Rather than list the first 150 terms in the sequence of partial sums, we list just every 5 *th* term. Notice how this is done as an option inside the Table[ ] command. This option is available with many Mathematica commands.

```
In[45]:= Table[ps[n],{n,1,150,5}]
Out[45]= {1., 1.19029, 1.19828, 1.20022, 1.20098, 1.20135, 1.20155, 1.20168, 1.20177,
          1.20183, 1.20187, 1.2019, 1.20192, 1.20194, 1.20196, 1.20197, 1.20198,
          1.20199, 1.202, 1.202, 1.20201, 1.20201, 1.20202, 1.20202, 1.20202,
          1.20203, 1.20203, 1.20203, 1.20203, 1.20203}
```

The above list of terms from the sequence of partial sums would suggest that, roughly speaking

$$\sum_{j=1}^{\infty} \frac{1}{j^3} \approx 1.20203.$$

The above sequence was not computed very efficiently. To compute, for example, the 29th term, psum(29), in the sequence, Mathematica added up all 29 items that appear in psum(29). Then, to compute psum(30), it duplicated most of its work by adding, essentially all over again, the 30 items that appear in psum(30). It would have been easier for Mathematica to use the term psum(29) that was just computed in the previous step, and then simply add $1/30^2$ to psum(29) to get psum(30).

Here is a simple program that compute the first 1000 terms of this sequence more efficiently. The program is very fast, because of the efficient computation we discussed above. No output is displayed. (Program steps are separated by the equivalent of semicolons.) Try this program yourself, and notice how quickly it is executed. A table command then displays every $20th$ term in the output, but this also is very fast.

```
In[46]:= s[1]=1;Do[s[j]=N[s[j-1]+1/j^3],{j,2,1000}]
```

```
In[47]:= Table[s[n],{n,1,1000,20}]
```

The output was omitted to save space.

Before we finish our discussion of this series, we show that Mathematica is able to calculate its value directly. We should not expect such success with just any series. It is easy to ask too much out of Mathematica in evaluating series, and frequently their values cannot even be expressed in closed form or, in other words, in terms of familiar mathematical expressions.

```
In[48]:= Clear[s]; Sum[1/j^3,{j,1,Infinity}]
Out[48]= Zeta[3]
```

This answer is correct, but in a calculus course, it is **not an acceptable way of expressing answers**. When we get such exotic answers, we should always **look for more acceptable ways to express answers**. This is not the appropriate place to discuss the so-called **Riemann Zeta function**,, but, since we have an exact answer, let's take the opportunity to compare its decimal value with the values we generated above with our sequence of partial sums.

```
In[49]:= N[%]
Out[49]= 1.20206
```

We have been discussing the idea that a series is, by definition, just a sequence, the sequence of partial sums. One special example, namely, the series $\sum_{k=1}^{\infty}(-1)^k$, is particularly good for shedding light on this matter. This series may appear to be very puzzling, until it is written as a sequence of partial sums. Take note!

```
In[50]:= psum[n_]:=Sum[(-1)^j,{j,1,n}]
```

```
In[51]:= Table[psum[n],{n,1,10}]
Out[51]= {-1, 0, -1, 0, -1, 0, -1, 0, -1, 0}
```

Obviously, this oscillating pattern between -1 and 0 continues forever, so that the sequence, and hence, by definition, the series itself, diverges. The point to be made here, is that the practice of using the definition of a series to express it as a sequence of partial sums, can, at least in some instances, increase our understanding of a series.

7.4 Positive Term Series

The *Comparison Test* (*Inequality Comparison Test*) and the *Limit Comparison Test* are used to determine the convergence or divergence of a given positive term series, by comparing it to another well chosen positive term series of known convergence or divergence. The *Inequality Comparison Test* is not very suitable for Mathematica involvement. Using this test is more a matter of logical argument, than it is a computation. On the other hand, given Mathematica's ability to compute limits, the *Limit Comparison Test* can be used quite effectively.

To use a comparison test, we need a collection of positive term series of known convergence or divergence. While any positive term series, once it is known to converge or diverge, can be put into this collection, it is usually restricted, in a classroom setting, to a certain **"standard set"** consisting of all of the geometric series, together with all of the so-called p-series. The well known convergent series $\sum_{n=0}^{\infty} 1/n!$ will also be placed in our standard set.

Deciding what to compare a given series to involves "educated guess work." Since Mathematica's limit command is so easy to use, we can make guesses with less hesitance, knowing that it will not take much work to perform the test.

Example 7.4 *Determine whether the following series converges or diverges.*

$$\sum_{n=1}^{\infty} \frac{\sqrt{n^3+2}}{2^n}$$

The key idea here is that 2^n gets large so fast that it quickly overwhelms the numerator and makes it somewhat insignificant. This would suggest that it behaves somewhat like the series

$$\sum_{n=1}^{\infty} (\frac{1}{2})^n$$

which is a convergent geometric series. If so, then our series probably converges. Unfortunately, this geometric series is smaller than the given series and so it cannot be used in the *Comparison Test.* It is easily seen that the *Limit Comparison Test* fails for pretty much the same reason. What we need is to compare the given series to a somewhat larger convergent series. This is accomplished by using any larger geometric series which still converges. The following limit shows that the series converges.

```
In[1]:= a[n_]:=Sqrt[n^3+2]/2^n; b[n_]:=(2/3)^n;
```

It is natural to make a direct attempt to evaluate the limit, but we know that Mathematica frequently fails to deliver. To save space, we do not record our failed attempts.

```
In[2]:= r=PowerExpand[Log[a[n]/b[n]]]
```

Out[2]= $-n\,(-\log[3]+\log[4])+\frac{1}{2}\log[2+n^3]$

```
In[3]:= r1=n*Expand[r/n]
```

Out[3]= $n\left(\log[3]-\log[4]+\frac{\log[2+n^3]}{2\,n}\right)$

```
In[4]:= Limit[r1,n->Infinity]
```

Out[4]= $-\infty$

It follows that

$$L = \lim_{n\to\infty} \frac{a_n}{b_n} = e^{-\infty} = 0.$$

Consequently, the series $\sum a_n$ converges, because $\sum b_n$ converges,

Example 7.5 *Determine whether the following series converges or diverges.*

$$\sum_{n=1}^{\infty} \frac{(2n)!}{(n!)^2}$$

The ratio test is always the most promising test to use in series involving factorials. (Recall that the Greek letter ρ is entered by typing \[Rho].) **The FullSimplify[ ] command was needed to simplify the factorials**. Take note!

```
In[5]:=  a=.; a[n_]:=(2n)!/(n!)^2;
```

```
In[6]:= r=FullSimplify[a[n+1]/a[n]]
                 2
Out[6]= 4 - -----
              1 + n
```

```
In[7]:= ρ=Limit[r,n->Infinity]
Out[7]= 4
```

Since $\rho > 1$ it follows that the series diverges.

Once it is known that a series converges, a partial sum can be used to approximate the value of the series. This approximation, however, is of little value unless the error between the exact sum and approximate partial sum can be estimated. For example, the series

$$\sum_{j=2}^{\infty} \left(\frac{1}{\ln(\sqrt{j})}\right)^{\ln(\sqrt{j})}$$

can be shown to converge (see Problem 12 in the exercises), but it converges so slowly that the 10,000*th* term is still approximately 0.009 in order of magnitude. We would obviously have to add tens of thousands of terms to come close to the exact value of the series.

By way of contrast, there are positive term divergent series whose partial sums get large at an almost imperceptible rate. There are many anecdotes written about how slowly the partial sums of the harmonic series get large.

> Suppose that 10 million years ago, one of our ancestors began to add up the terms of the harmonic series at the rate of 100 terms per second, and that the task was continued, without fail, each second thereafter until the present time, with the duty to continue the additions passed on smoothly from generation to generation. One can show that, at the present time, the sum of all these terms would be somewhere between 37 and 39, continuing, nevertheless, to increase, in a very slow race, to infinity. (see Problem 13 in the exercises.)

If $\sum_{k=1}^{\infty} b_k$ is a convergent series with sum s and if s_n denotes its nth partial sum, then the nth error term is defined to be

$$e(n) = \mid s - s_n \mid = \mid \sum_{k=1}^{\infty} b_k - \sum_{k=1}^{n} b_k \mid = \mid \sum_{k=n+1}^{\infty} b_k \mid .$$

Deciding how far out one needs to go in the sequence of partial sums of a convergent series in order to get a finite sum which is a good approximation for the whole series is not always an easy matter. Whenever this error term can be measured, it deserves our attention.

One of the important characteristics of the *Integral Test* is that it can be used not only to show convergence, but also to measure the size of the error between the exact sum and a partial sum. Given a series $\sum_{k=1}^{\infty} b_k$, suppose that f is a nonincreasing real valued continuous function on the interval $[1, \infty]$ such that $f(k) = b_k$ for each k. Basically the same idea used to prove the *Integral Test* can be used to prove that if the series converges then the error term satisfies

$$e(n) \leq \int_n^{\infty} f(x)dx.$$

The nonincreasing nature of the function f, in the *Integral Test* is a critical condition, but it can be relaxed. Like most tests for convergence, words like "for all" can usually be replaced by "eventually," and so it is it is only necessary to show that f is eventually nonincreasing. This complicates the error term only slightly. Using the same argument used to prove the *Integral Test* one can show that as long as f is nonincreasing for all $x \geq N$, the error term still satisfies

$$e(n) \leq \int_n^{\infty} f(x)dx \text{ for } n \geq N.$$

Example 7.6 *Show that the series* $\sum_{k=1}^{\infty} \frac{k^3}{e^k}$ *converges. Approximate the sum with an error not exceeding* 10^{-8}.

Based on growth patterns for the exponential function, it may be clear that the underlining function $f(x) = x^3/e^x$ is eventually decreasing. This may not, however, be clear to everyone, and we cannot use the *Integral Test* until the decreasing nature of this function is established. It is easy enough to see this, if it is not already clear, by plotting $f(x)$. More rigorously, ask Mathematica to factor $f'(x)$, to see that $f'(x) < 0$ for $x > 3$. This step is not included, and we proceed immediately with the *Integral Test*. Since the improper integrals $\int_1^{\infty} f(x)dx$ and $\int_n^{\infty} f(x)dx$ $(n \geq 1)$ either all converge or all diverge, there is no need to establish convergence of the series as a separate issue. If the integral involved in any error term converges, then the series converges, and so we turn our attention immediately to this error term.

```
In[8]:= f[x_]:=x^3*Exp[-x]
```

```
In[9]:= error=Integrate[f[x],{x,n,Infinity}, Assumptions->{n>0}]
Out[9]= e^-n (6 + 6 n + 3 n^2 + n^3)
```

```
In[10]:= Plot[{error,10^(-8)},{n,1,50},PlotRange->{0,2*10^(-8)}];
```

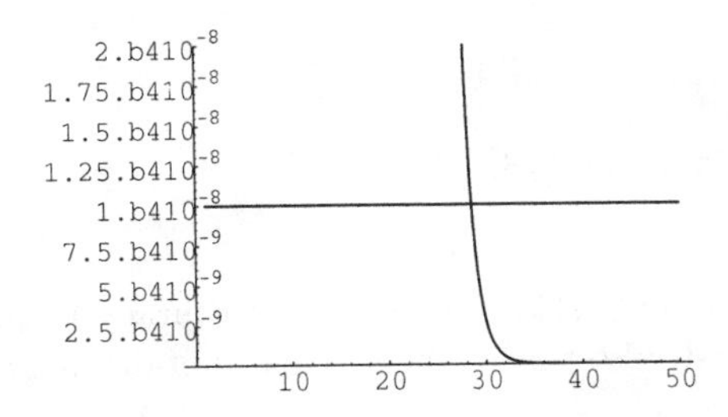

The error term will clearly be small enough for $n = 30$.

```
In[11]:= s=N[Sum[f[k],{k,1,30}],9]
Out[11]= 6.00651279
```

Interestingly enough, our approximate sum of $s = 6.0065127$ is correct to this many decimal places, even though we had to sum very few terms in the series to reach this degree of accuracy.

7.5 Alternating Series and Absolute Convergence

The error term can also be estimated quite easily for an alternating series. According to the *Alternating Series Test*, if $\{a_k\}$ is a nonincreasing sequence of positive numbers which converges to 0, then the alternating series $\sum_{k=1}^{\infty}(-1)^{k+1}a_k$ converges, and for each n, the error term satisfies

$$e(n) \leq a_{n+1}.$$

If the sequence $\{a_k\}$ is nonincreasing only for $k \geq N$, then it is easy to see that the error term still satisfies

$$e(n) \leq a_{n+1} \text{ for } n \geq N.$$

Example 7.7 *Show that the sequence* $\sum_{k=1}^{\infty}(-1)^k \frac{1}{k^2+2^k}$ *converges and approximate its sum with an error not exceeding* 10^{-8}

This series clearly converges by the *Alternating Series Test*, and so we consider the error term immediately.

```
In[12]:= a=.; a=1/(k^2+2^k);
```

```
In[13]:= Plot[{a,10^(-8)},{k,1,50},PlotRange->{0,2*10^(-8)}];
```

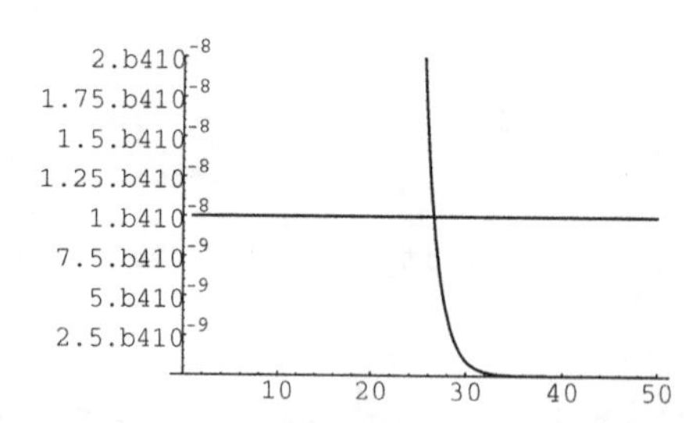

The error term will clearly be small enough for $n = 27$.

```
In[14]:= s=N[Sum[(-1)^k*a,{k,1,28}],9]
Out[14]= -0.247076768
```

The approximate sum is $s = -0.2470767$. This much accuracy is quite striking considering that we only added the first 27 terms of the series to obtain this approximation.

Example 7.8 *Determine whether the series* $\sum_{n=1}^{\infty}(-1)^n \left(n^{-1+\frac{1}{n}}\right)$ *converge absolutely, conditionally, or diverges.*

When problems like this are solved by hand, the series under consideration is usually studied with some care before a decision is made regarding which convergence test to use. The labor involved in using these tests is often great enough that there is a reluctance to use a test without a reasonable expectation of a successful outcome. With Mathematica, on the other hand, the computations are no cause for concern, and so there is no real need to be so cautious before a test is used.

We first consider the absolute value series and compare it to the $p = 2$ series. This is seen to be inconclusive. The series is then compared to the Harmonic Series, and this test is successful. It follows that the series does not converge absolutely.

```
In[15]:= a=n^(1/n-1);
```

```
In[16]:= Limit[a/(1/n^2),n->Infinity]
Out[16]= ∞
```

```
In[17]:= Limit[a/(1/n),n->Infinity]
Out[17]= 1
```

According to the *Alternating Series Test*, the series will converge if $\{n^{\frac{1}{n}-1}\}$ is a decreasing sequence which converges to 0. The last limit we computed above clearly implies that the sequence converges to 0 (and then some). Showing that the sequence decreases can be done with a plot. We choose to do it somewhat more rigorously by showing that its derivative is always negative.

```
In[18]:= ap=D[a,n]
```

Out[18]= $\left(-1+\frac{1}{n}\right) n^{-2+\frac{1}{n}} - n^{-3+\frac{1}{n}} \log[n]$

```
In[19]:= ap1=Together[ap]
```

Out[19]= $-n^{-3+\frac{1}{n}} (-1+n+\log[n])$

The left hand term (along with the negative sign) (more precisely, $ap1[[\{1,2\}]]$) is clearly negative, and the last term (the parenthetical expression $ap1[[3]]$) is clearly always positive. The derivative, therefore, is always negative, and so the sequence decreases. It follows that the series converges conditionally.

Example 7.9 *Determine whether the series* $\sum_{n=1}^{\infty}(-1)^n \frac{n^{\ln(n)}}{2^{\sqrt{n}}}$ *converges absolutely, conditionally, or diverges.*

Using the *Root Test* on the absolute value series is seen to be inconclusive. This is followed by a comparison with the $p = 2$ series, which turns out to be successful.

```
In[20]:= a=.; a[n_]:=n^Log[n]/2^Sqrt[n];
```

```
In[21]:= r=a[n]^(1/n); r1=PowerExpand[Log[r]]
```

Out[21]= $\frac{-\sqrt{n}\log[2]+\log[n]^2}{n}$

```
In[22]:= Limit[r1,n->Infinity]
```

Out[22]= 0

It follows that $\rho = e^0 = 1$, so the test is inconclusive.

```
In[23]:= r=a[n]/(1/n^2);r1=PowerExpand[Log[r]]
```
Out[23]= $-\sqrt{n}\ \log[2] + \log[n]\ (2 + \log[n])$

```
In[24]:= r2=Sqrt[n]*Expand[r1/Sqrt[n]]
```

Out[24]= $\sqrt{n}\left(-\log[2] + \frac{2\ \log[n]}{\sqrt{n}} + \frac{\log[n]^2}{\sqrt{n}}\right)$

```
In[25]:= Limit[r2,n->Infinity]
```
Out[25]= $-\infty$

This means that

$$\lim_{n\to\infty} \frac{a_n}{1/n^2} = e^{-\infty} = 0,$$

so the given series converges, because the $p = 2$ series converges. Consequently the series converges absolutely.

7.6 Power Series

Example 7.10 *Find the exact interval of convergence of the power series* $\sum_{n=1}^{\infty} \frac{n^{\sqrt{n}}}{2^n} x^n$

The radius of convergence is found by using the *Ratio* or *Root Test*. These tests can only be used on positive term series or, in other words, to test for absolute convergence. In this example the *Root Test* is much easier to use.

```
In[1]:= a[n_]:=n^Sqrt[n]/2^n;X=Abs[x];
```

```
In[2]:= r=PowerExpand[(a[n]*X^n)^(1/n)]
```

Out[2]= $\frac{1}{2}\ n^{\frac{1}{\sqrt{n}}}\ \text{Abs}[x]$

```
In[3]:= ρ=Limit[r1,n->Infinity]
```

Out[3]= $\frac{\text{Abs}[x]}{2}$

The series converges for $|\, x \,| < 2$ (where $\rho < 1$) and diverges for $|\, x \,| > 2$ (where $\rho > 1$). It clearly diverges at $x = \pm 2$, and so the exact interval of convergence is the open interval $(-2, 2)$.

Example 7.11 *Find the exact interval of convergence of the power series*

$$\sum_{n=2}^{\infty} \frac{\ln(n)^2}{n} (x-3)^n.$$

We use the *Ratio Test* to establish the interval of convergence in this example.

```
In[4]:= a[n_]:=Log[n]^2/n;X=Abs[x-3];
```

```
In[5]:= r=a[n+1]X^(n+1)/(a[n]*X^n)
```

$$\text{Out[5]= } \frac{\text{n Abs}[-3+\text{x}]\ \log[1+\text{n}]^2}{(1+\text{n})\ \log[\text{n}]^2}$$

```
In[6]:= ρ=Limit[r,n->Infinity]
Out[6]= Abs[-3 + x]
```

It follows that the series converges for $\rho = |x-3| < 1$, and diverges for $\rho = |x-3| > 1$. At $x = 4$, the series is the positive term series $\sum_{n=2}^{\infty} a[n]$ is clearly larger than the harmonic series, and so it diverges. At $x = 2$, the series is the alternating series $\sum_{n=2}^{\infty}(-1)^n a(n)$. To prove that this series converges by the *Alternating Series Test*, we show next that $\{a(n)\}$ is a decreasing sequence which converges to zero.

```
In[7]:= Limit[a[n],n->Infinity]
Out[7]= 0
```

```
In[8]:= ap=D[a[n],n]
```

$$\text{Out[8]= } \frac{2\ \log[\text{n}]}{\text{n}^2} - \frac{\log[\text{n}]^2}{\text{n}^2}$$

```
In[9]:= ap1=Factor[ap]
```

$$\text{Out[9]= } -\frac{(-2+\log[\text{n}])\ \log[\text{n}]}{\text{n}^2}$$

The derivative *ap* is clearly negative for all $n > e^2$, and so the sequence $\{a(n)\}$ eventually decreases. It follows that the series converges, and so the exact interval of convergence is the half open interval $[2, 4)$.

Mathematica has a command for computing Taylor series. **If f is a function which has derivatives of all order in some open interval containing $x = a$, and if n is a positive integer, the command**

t=Series[f[x],{x,a,n}];

computes the Taylor series of f centered at $x = a$, and displays the result as

$$t := f(a) + f'(a)(x-a) + \frac{f''(a)}{2!}(x-a)^2 + \cdots + \frac{f^{(n)}(a)}{n!}(x-a)^n + O[x^{n+1}].$$

The term $O[x^{n+1}]$ (read big Oh of x^{n+1}) represents all the terms in the series of order $(n+1)$ or larger. To a certain extent, **this display is merely a visual display and cannot be used in subsequent Mathematica computations until it is altered.** Mathematica, understandably, has a hard time dealing with the term $O[x^{n+1}]$. Unless, this display is the end result, **most of the time it will be necessary to turn it into a legitimate polynomial before it can be used** in any further work.

The command **Normal[ ] converts a Taylor series to a polynomial, by dropping the $O[x^{n+1}]$ term. The result is the Taylor polynomial of order n for f centered at $x = a$.**

If f has derivatives of order $k = 1, 2, \ldots, n$ in some open interval containing $x = a$, then the Taylor polynomial of order m for f at $x = a$ exists for each $m \leq n$. However, if f is not infinitely differentiable in some open interval containing $x = a$, then the Taylor series for f at $x = a$ clearly does not exist. In this case, an attempt to evaluate Series$[f[x], \{x, a, j\}]$ (for any j) will result in strange output, which is clearly wrong. In addition, **Series[ ] is actually a call for a series that is somewhat more general than a Taylor series.** It is **possible to get output** from Series$[f[x], \{x, a, n\}]$, even when **f has a discontinuity at $x = a$.** This is not a Mathematica mistake, but it is **not acceptable output in a calculus course.**

Example 7.12 *Find the Taylor polynomials of* $f(x) = \frac{12x}{x^2+36}$ *of order* $n = 5, 8, 11$ *centered at* $x = 0$. *Plot all three of them together with the function* f *on the interval* $[-10..10]$. *Use Taylor's Theorem to determine the error involved in approximating* $f(x)$ *by its Taylor polynomial of order 11, on the interval* $[-5, 5]$ *and on the interval* $[-2, 2]$.

```
In[10]:= f=12x/(x^2+36)
```

Out[10]= $\frac{12\,x}{36 + x^2}$

```
In[11]:= t5=Series[f,{x,0,5}]
```

Out[11]= $\frac{x}{3} - \frac{x^3}{108} + \frac{x^5}{3888} + O[x]^6$

We must now turn this series into a legitimate polynomial. For the remaining two Taylor polynomials, we combine these commands to compute the final polynomials in fewer steps.

```
In[12]:= t5=Normal[t5]
```

Out[12]= $\frac{x}{3} - \frac{x^3}{108} + \frac{x^5}{3888}$

```
In[13]:= t8=Normal[Series[f,{x,0,8}]]
```

Out[13]= $\frac{x}{3} - \frac{x^3}{108} + \frac{x^5}{3888} - \frac{x^7}{139968}$

```
In[14]:= t11=Normal[Series[f,{x,0,11}]]
```

Out[14]= $\frac{x}{3} - \frac{x^3}{108} + \frac{x^5}{3888} - \frac{x^7}{139968} + \frac{x^9}{5038848} - \frac{x^{11}}{181398528}$

We now plot, f, along with its three Taylor polynomials. We could identify which one is the graph of f, by plotting it separately. Polynomials eventually always get large, and it turns out that $|f(x)| < 1$ for all x. This should help identify which one is the graph of f.

```
In[15]:= Plot[{f,t5,t8,t11},{x,-10,10},PlotRange->{-2,2}];
```

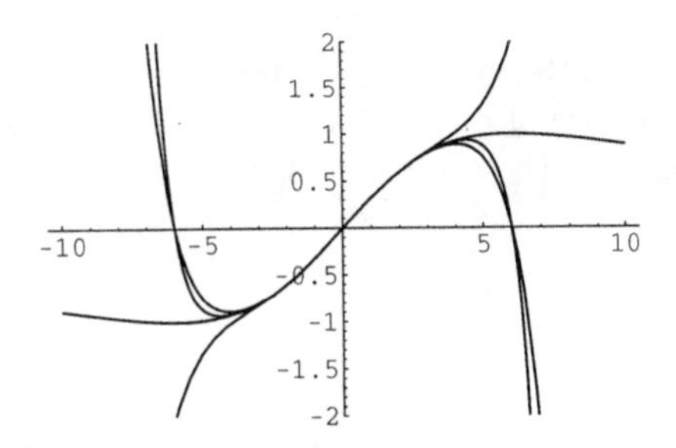

In the last plot command, we specified a vertical range, rather than allow Mathematica to select this range for us. Since polynomials get large rather quickly, Mathematica would have selected a rather large vertical scale which would have produced too flat of a picture. Instead, we picked a scale by considering only the values of the original function f.

By looking at this graph, we can see, visually at least, how close the Taylor polynomials are to the graph of the function f that they represent. To get a more reliable estimate of the error term involved, we consider the **remainder term** in *Taylor's Theorem*. If $t_n(x)$ is the Taylor polynomial of order n for f based at $x = a$, then the error term is defined by

$$e_n(x) = \mid f(x) - t_n(x) \mid .$$

If f has $n+1$ derivatives on an interval I centered at $x = a$, then *Taylor's Theorem* implies that

$$e_n(x) < \frac{M}{(n+1)!} \mid x - a \mid^{n+1} \text{ for all } x \text{ in } I,$$

where M is the maximum value of $\mid f^{(n+1)}(x) \mid$ for x in I. As a practical issue, there is usually no advantage gained in computing M accurately, since clearly any number larger than M can be used in place of M. Using a number larger than M would generate a slightly weaker error term, but the effect is usually so slight that it is of no consequence. Consequently, the usual practice is to use some easily obtained number known to be larger than every value of $\mid f^{(n+1))}(x) \mid$ for x in I. As we finish the above example, notice how a value for M is easily selected, visually, off of a graph.

```
In[16]:= fp12=D[f,{x,12}];
```

```
In[17]:= Plot[fp12,{x,-10,10},PlotRange->{-.5,.5}];
```

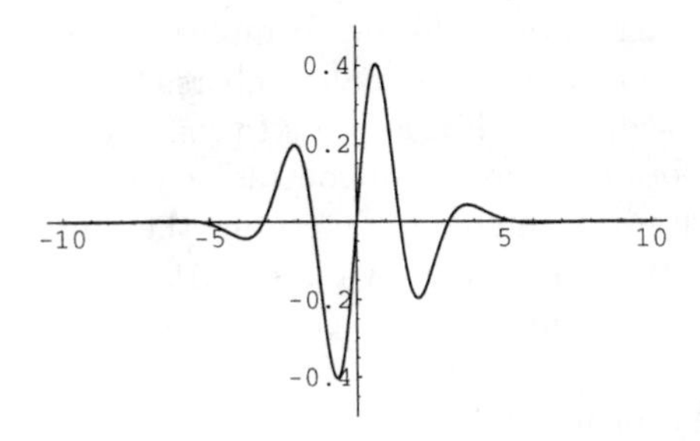


```
In[18]:= M=.4
Out[18]= 0.4
```

We let $e5$ and $e2$ denote the error terms on the interval [-5,5] and [-2,2], respectively, corresponding to the Taylor polynomial of order 12 for $f(x)$ at $x = 0$. Actually, The values we compute next are upper bounds for the error terms.

```
In[19]:= e5=M*5^12/12!
Out[19]= 0.203875
```

```
In[20]:= e2=M*2^12/12!
Out[20]= 3.42045 10^-6
```

Notice how much better the estimate is on $[-2, 2]$ than it is on $[-5, 5]$.

7.7 Exercise Set

Some New and Some Old Mathematica features and Commands

Apart[ ]	Ceiling[ ]	Clear[ ]
Do[ ]	Expand[ ]	ExpToTrig[ ]
Factor[ ]	Floor[ ]	FindRoot[ ]
FullSimplify[ ]	Get[ ] or <<	ImplicitPlot[ ]
Limit[ ]	ListPlot[ ]	Log[ ]
Normal[ ]	NSum[ ]	Part[ ]
PowerExpand[ ]	Print[ ]	Range[ ]
Series[ ]	Show[ ]	Simplify[ ]
Sum[ ]	TrigToExp[ ]	Table[ ]
Together[ ]	<<Graphics`ImplicitPlot`	

Recall the alternate form for the Part[ ] command: $p[[j]]$ is the $j\,th$ term of p and $p[\{j_1, j_2, \ldots, j_n\}]$ is the combination of terms $j_1, j_2, \ldots, j_n$ from p

The command Sum[ ] calls for an exact evaluation. It can quickly lead to results, which are more complicated than desired.

1. A sequence $\{a_n\}$ is defined by each of the following formulas. In each case list the first 50 terms of the sequence as decimal numbers. Use Mathematica's limit command to determine the limit of the sequence, if it converges.
a) $a_n = \frac{n}{2^n}$ for $n = 1, 2, \ldots$ b) $a_n = \frac{n^2}{10n+100}$ for $n = 1, 2, \ldots$

2. A sequence $\{a_n\}$ is defined by each of the following formulas. In each case list every tenth term up to the $500\,th$ term of the sequence as decimal numbers. By looking at this list, does the sequence appear to converge or diverge. If it appears to converge, try to predict its limit ? Compute a_{1000} as a decimal. Does this confirm or contradict your previous conclusions. If it converges, use Mathematica's limit command to determine the limit of the sequence. If it converges or if it diverges to infinity, then, purely by guess work, find a term in the sequence which is "close" to its limit value. (You may have to go quite far out in the sequence to find such a term.)
a) $a_n = (n + (-1)^n \sqrt{n}) \sin(\frac{1}{n})$ for $n = 1, 2, \ldots$
b) $a_n = \frac{(\ln(n))^{\ln(n)}}{n^5}$ for $n = 2, 3, \ldots$

3. Show that the sequence in Problem 2b is eventually an increasing sequence. Find the **exact** term (pump up the accuracy) in the sequence where it changes from a decreasing to an increasing sequence?

4. A sequence $\{a_n\}$ is defined by each of the following formulas. In each case determine the limit of the sequence, if it converges, or state that the sequence diverges. Some of the sequences may have obvious limits which can be evaluated without Mathematica, by a simple observation. Identify such sequences, and determine their limits, before using Mathematica. If the sequence does not have an obvious limit, identify its indeterminate form before you attempt to compute its limit.

a) $a_n = \sec(\frac{1}{n})^{5n^2}$ for $n = 1, 2, \ldots$

b) $a_n = (1 + \tanh(n))^{1/n}$ for $n = 1, 2, \ldots$

c) $a_n = (1 - \tanh(n))^{1/n}$ for $n = 1, 2, \ldots$

d) $a_n = \frac{n^{2n+1}}{(n+1)^n}$ for $n = 1, 2, \ldots$

5. The sequence $\{a_n\}$ in problem 4c is quite interesting. Evaluate terms in the sequence as decimals, and try to find a term far enough out in the sequence to be "close" to the sequence's limit value. Prepare, however, to be frustrated in this endeavor. Is Mathematica's limit computation correct? To remove any lingering doubt, use *L'Hopital's Rule,* to verify the value of the limit found in Problem 4c.

Why is Mathematica's decimal evaluator performing in this way? Surprisingly enough, Mathematica's behavior here is clear and predictable, and in order to understand the limitations of technology, it is important to pursue this question. Devise a way to express a_n as a decimal, in such a way that the decimal value of a_n gets close to the limit of the sequence, as n gets large. Use Mathematica's Table command to list the terms a_n in this manner for n from 10 to 50. Recall that the command N[A,k] evaluates A with k digits of acuracy.

6. Let L denote the limit of the sequence $\{a_n\}$ in Problem 4c. Find an N such that $| a_n - L |< 10^{-3}$ for all $n > N$. This turns out to be a nontrivial problem. First try to find N with a graphical approach. If this fails to be useful, try to find N by experimenting numerically with different values of n. Why do these methods fail so dramatically? How can this failure be avoided? One approach that works involves converting the hyperbolic tangent to its exponential form. This can be done by using its definition or by **using the TrigToExp[ ] command** to convert an expression to its **exponential form**. Once you find an integer n for which a_n is sufficiently close, how do you know that all subsequent terms in the sequence will also be sufficiently close? It may help to establish that the sequence $\{a_n\}$ (in its converted form) is either increasing or decreasing.

7. Let $\{x_n\}$ be the sequence defined recursively by $x_n = (1 - \frac{20}{n} + \frac{2}{n^2})x_{n-1}$ for $n = 2, 3, \ldots$ and $x_1 = 3$. List the first 50 values of x_n as decimals. Observe the initial chaos and how the terms eventually settle down.

8. Let $\{b_n\}$ be the sequence defined recursively by

$$b_n = b_{n-1} + (-1)^n \frac{2n-1}{n^2 - n}, \quad b_1 = 1.$$

List the first 50 terms exactly. Plot the points (j, b_j) $(j = 1, \ldots, 50)$, using the ListPlot[ ] command. Does the sequence appear to converge?

9. Suppose that one month before the first day of your college education, your parents put a lump sum of cash in an account earning 8% annual interest compounded monthly. You are allowed to draw out $1300 per month from this account for living and educational expenses, with your first draw occurring one month after this lump sum deposit. The account is set up so that the balance is 0 after exactly 48 monthly draws. What is the value of the initial deposit into this account ?

Let $\{b_n\}$ be the sequence of monthly balances. Define this sequence recursively, guess at an initial deposit and compute $b[48]$. Adjust your initial guess until $b[48]$ is (almost) 0.

Alternately, here is a much more efficient approach. **Use the Simplify command when you set up (define) your recursive sequence.** Let b[0]=p (the unknown initial deposit). Solve the equation $b[48] = 0$ for p.

Regardless of which approach you use, list all 48 of the monthly account balances once you have settled the matter of the initial deposit. Take note once this problem is finished, that by defining the sequence of balances recursively, complicated finance formulas were avoided. No special formulas were needed other than a basic monthly interest and withdrawal statement.

10. It is easy to ask Mathematica to do too much when using the Sum command. Remember that Mathematica will always be exact unless it is allowed to make a decimal approximation. In this problem, demonstrate the output, probably undesirable, resulting from an exact calculation of $\sum_{n=1}^{500} \frac{1}{n^2}$ Explain the form of this output, and what Mathematica did to generate it.

11. Some of the mystery surrounding the notion of a series can be removed by appreciating that a series is just a sequence—the sequence of partial sums. Viewing a series in this way may seem unnecessary in calculations, but to fully understand what a series is, as a concept, it helps to pay some attention to the definition of a series. With this in mind, use Mathematica to set up the sequence of partial sums of the series $\sum_{j=0}^{\infty} \frac{1}{n!}$. Use the evalf command when you define this sequence. List the first 20 terms of this sequence. (The output should be a sequence of decimal numbers.) Increase the number of digits of accuracy to 15, and repeat these calculations. (Recall that this can be done by entering **15 as an optional second argument in N[ ].**) You will soon know, if you do not already know, that this series converges to e. Compare the 20 terms in this sequence to a decimal expansion for e.

12. Determine whether the following series converge absolutely, conditionally, or diverge (p denotes a fixed real number).

a) $\sum_{j=1}^{\infty} \frac{j^4}{1.2^j}$ b) $\sum_{j=2}^{\infty} \frac{100^{2j}}{j!}$
c) $\sum_{j=2}^{\infty} (1 - \sqrt[j]{j})^j$ d) $\sum_{j=2}^{\infty} (\sqrt{4 + j^2} - j)$
e) $\sum_{j=2}^{\infty} (\frac{1}{\ln(j)})^{\ln(j)}$ f) $\sum_{j=2}^{\infty} (\frac{1}{\ln(\sqrt{j})})^{\ln(\sqrt{j})}$
g) $\sum_{j=3}^{\infty} (\frac{1}{\ln(\ln(j))})^{\ln(\ln(j))}$ h) $\sum_{j=2}^{\infty} (\frac{1}{j \ln(j)}$
i) $\sum_{j=2}^{\infty} (\frac{1}{j(\ln(j))^2}$ j) $\sum_{j=2}^{\infty} (\frac{1}{\ln(j))})^p$
k) $\sum_{j=1}^{\infty} \frac{(j+1)^j}{j^j j!}$ l) $\sum_{j=1}^{\infty} \frac{(j!)^3}{j\sqrt{j}}$

13. The inequality used to prove the *Integral Test* can be used to prove the inequality

$$\ln(n+1) < \sum_{j=1}^{n} \frac{1}{j} < 1 + \ln(n).$$

Use this inequality to establish the claim made in the anecdote about the harmonic series on page 132

14. Show that the series $\sum_{k=2}^{\infty} \frac{1}{k(\ln(k))^6}$ converges. Determine an n such that the error term $e(n) < 10^{-4}$. Use this to approximate the sum of the series with this much accuracy.

15. Show that the series $\sum_{k=0}^{\infty} \frac{(-4)^k}{k!}$ converges. Determine an n such that the error term $e(n) < 10^{-8}$. Use this to approximate the sum of the series with this much accuracy.

16. Find the exact interval of convergence of each of the power series.

a) $\sum_{n=2}^{\infty} \frac{\ln(n)}{n^2 4^n} x^n$ b) $\sum_{n=2}^{\infty} \ln(n)^{\ln(n)} (x+3)^n$

17. For each of the following, compute the Taylor polynomial of order 8 based at the given point a. Compare each Taylor polynomial with the function it represents, by plotting both of them on the same graph over an appropriate interval.
a) $\cos(x)$, $a = 0$ b) e^x, $a = 0$ c) $\ln(x)$, $a = 1$ d) $\sqrt{x}$, $a = 9$

18. Let $f(x) = (12x^2 - 17x - 5)\sin(3x)$. Determine the Taylor polynomial of order $n = 9$ for $f(x)$ based at $x = 0$. Plot $f(x)$ along with its Taylor polynomial on the interval $-1 \leq x \leq 1$ Use Taylor's remainder formula to estimate the error involved in approximating $f(x)$ by its Taylor polynomial. Compare this estimate to the actual difference between $f(x)$ and its Taylor polynomial by plotting this difference function on the interval $-1 \leq x \leq 1$.

19. Repeat problem 16 using the Taylor polynomial of order 15 instead.

20. Let $f(x) = x\cos(\sqrt{x})$. Determine the Taylor polynomial of order $n = 3$ for $f(x)$ based at $x = 50$. Plot $f(x)$ along with its Taylor polynomial on the interval $30 \leq x \leq 70$. Use Taylor's remainder formula to estimate the error involved in estimating $f(x)$ by its Taylor polynomial. Compare this estimate to the actual difference between $f(x)$ and its Taylor polynomial by plotting this difference function on the interval $30 \leq x \leq 70$.

21. Compute the Taylor polynomial of order 8 of $f(x) = (x + a)^8$ based at $x = 0$, but before you do, anticipate what the outcome should be in terms of more elementary structures. Use Mathematica to verify your claim.

22. Let $f(x) = \frac{1}{1-x}$. Without computing the Maclaurin series for $f(x)$, anticipate what you would expect it to be. Verify your hunch by computing the series and displaying all of the terms of order $n \leq 15$. Write the result as an equation named eq. Replace x by $-x^2$ in *eq* and integrate both sides. Verify that the result represents the Maclaurin series for $\arctan(x)$. Replace x by $-x$ in *eq* and integrate both sides. Verify that the result represents the Maclaurin series for $\ln(1 + x)$. A Taylor or Maclaurin series for a function $f(x)$ does not have to converge to $f(x)$. In these two cases, convergence to corresponding function is immediate. Why? What are intervals of convergence?

23. Express the polynomial $f(x) = 8x^5 - 6x^4 + 27x^2 + 32x + 12$ as a polynomial in powers of $(x + 20)$.

24. The function $f(x)$ defined below appears to be discontinuous at $x = 0$, but the discontinuity turns out to be removable. Starting with a Maclaurin series for $\cos(x)$, find a power series which represents $f(x)$ for all $x \neq 0$. Then define $f(0)$ so that f is continuous at $x = 0$.

$$f(x) = \frac{2\cos(x^2) + x^4 - 2}{x^8}$$

Use Mathematica's Series[ ] command to compute a power series expansion for $f(x)$ directly and compare it to the above approach. As we pointed out earlier, the output of the command Series[ ] is more a "display" than it is a valid mathematical combination of terms. The "Big-Oh" term appearing in the output frequently prevents the kind of manipulation desired. When this happens, one way to resolve the problem is to convert the Taylor expression into a legitimate polynomial of fairly high degree. If you wish, you can even add a symbol to this polynomial which represents the "Big-Oh" term which was dropped when you converted to a polynomial. You will have an easier time manipulating the individual terms in this expression.

Problems 25, 26, and 27, are interesting applications of Taylor series and Taylor polynomials. Some important functions of mathematics cannot be expressed in elementary terms, even though they are known to exist. This is an event, which happens frequently in mathematics. How do we get a "handle" on such a function? The answer is, of course, through a

Taylor series. Hopefully, these problems will demonstrate, in a convincing way, the practical importance of this major topic.

25. The equation $x^3 - y^3 = 64xy$, defines y implicitly as one or more functions of x. Plot the equation. The top half of the inner loop of this curve is the graph of one of the functions, which are defined implicitly by the equation. Let $f(x)$ denote this function of x, and let J denote the interval over which $f(x)$ is defined. Can we find a formula for $f(x)$? We know that Mathematica will solve a cubic equation exactly, but the solution is easily seen to be too complicated to be of much use.

Let $p_3(x)$ denote the Taylor polynomial of $f(x)$ of order 3 based at a point centrally located on the interval J. Use the D[ ] command to differentiate the original equation implicitly several times to determine the coefficients of the Taylor polynomial $p_3(x)$ as decimal numbers. Plot the original equation and $y = p_3(x)$ together to see how good of an approximation $p_3(x)$ is to the function $f(x)$ on the interval J.

26. The function $f(x) = 2x + \sin(x)$ has a derivative, which is everywhere positive, and so $f(x)$ is globally one-to-one. Its inverse,$f^{-1}(x)$, exists, but it is impossible to find a formula for it in terms of the familiar functions of mathematics. Find instead the Taylor polynomial $p_5(x)$ of order 5 for $f^{-1}(x)$ based at $x = 0$. Use the D[ ] command to differentiate the equation $x = f(y)$ several times implicitly to determine the coefficients of the Taylor polynomial $p_5(x)$. Plot $f^{-1}(x)$ and $p_5(x)$ together to see how good of an approximation $p_5(x)$ is to $f^{-1}(x)$ on an interval centered about the origin. Choose a large enough interval to show the effective range of the approximation. The inverse function will have to be plotted in the form of the equation $x = f(y)$ using the ImplicitPlot[ ] command.

27. Consider the differential equation $\frac{dy}{dx} = (\arctan(x))^2$, with initial condition $y(4) = 13$. There is a unique solution $y = y(x)$ to this equation according to the*Fundamental Theorem Of Calculus* but Mathematica cannot find its formula. It makes sense, then, to look for a polynomial $y = p(x)$ which could be used as an approximate solution. What would we expect of such a polynomial? Certainly we would want $p(4) = 13$, and for all x in some interval containing $x = 4$, we would want not only $p(x)$ to be close to $y(x)$, but also $p'(x)$ to be close to $y'(x) = (arctan(x))^2$. With this in mind, find a polynomial $p(x)$ such that $p(4) = 13$ and

$$|y(x) - p(x)| < 0.0001, \ |y'(x) - p'(x)| < 0.0001, \text{ for } 3 \leq x \leq 5.$$

Hint: If $p'(x)$ and $y'(x)$ are close on the interval, then $p(x)$ and $y(x)$ will also be close on the interval. (See a proof of this following the hint.) Using plotting tools and Mathematica's Series[ ] command, find a polynomial which is close enough to $(\arctan(x))^2$. The rest is easy! What would be a good point to use as a base for the Taylor polynomial?

Proof of the inequality: This is a consequence of the *Mean Value Theorem*. Let $G(x) = y(x) - p(x)$. Then $G(4) = 0$ (since $y(4) = 13$ and $p(4) = 13$). By the *Mean Value Theorem*, for each x in $[3, 5]$ there is a c between x and 4 such that

$$\begin{aligned} |y(x) - p(x)| &= |G(x) - G(4)| = |G'(c)|\,|x - 4| \\ &= |y'(c) - p'(c)|\,|x - 4| < 0.001|x - 4| < 0.001. \end{aligned}$$

Chapter 8

Plane Curves and Polar Coordinates

In this chapter we discuss several different types of curves in the plane and several different methods that can be used to describe planar curves.

8.1 Conic Sections

A conic section is a curve which can be described by intersecting a double cone with a plane. A circle, parabola, ellipse, and hyperbola can all be realized in this way, along with certain so-called "degenerate" cases such as a line, a pair of lines, or a point. Every conic section is the graph of an equation of the form

$$ax^2 + bxy + cy^2 + dx + ey + f = 0, \tag{8.1}$$

and conversely, any such second degree equation is a conic section (or an empty set).

A second degree equation or any other equation in two variables for that matter, can easily be graphed using Mathematica's **ImplicitPlot[ ] command, which resides in the Graphics`ImplicitPlot` package**. A graph will not, however, reveal the exact location of, for example, the foci or vertices. The geometric information associated with a conic section is found by first putting equation 8.1 into a standard form. If b, the coefficient of the xy term is 0, the equation can be put into its standard form by first completing a square on the x and/or y terms. If b, the coefficient of the xy term is not 0, then a rotation is performed to transform the original equation into a new second degree equation in which the new "b-like" term is 0. Once this coefficient is 0, the new equation can be put into its standard form by completing squares.

Working with these equations is easy enough conceptually, but the computations can be very tedious if they are done by hand. A Mathematica notebook is an excellent environment to carry out the computational details.

With Mathematica's ImplicitPlot command it is easy to see the rotational effect on the graph of an equation of the form 8.1 where the xy-term $b \neq 0$. In the next Mathematica work session we graph the equation $x^2 + 4xy + 5y^2 + 4x - 6 = 0$.

```
In[1]:= <<Graphics'ImplicitPlot'
```

```
In[2]:= eq=x^2+4x*y+5y^2+4x-6==0
Out[2]= -6 + 4 x + x^2 + 4 x y + 5 y^2 == 0
```

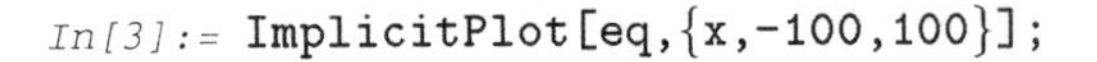

```
In[3]:= ImplicitPlot[eq,{x,-100,100}];
```

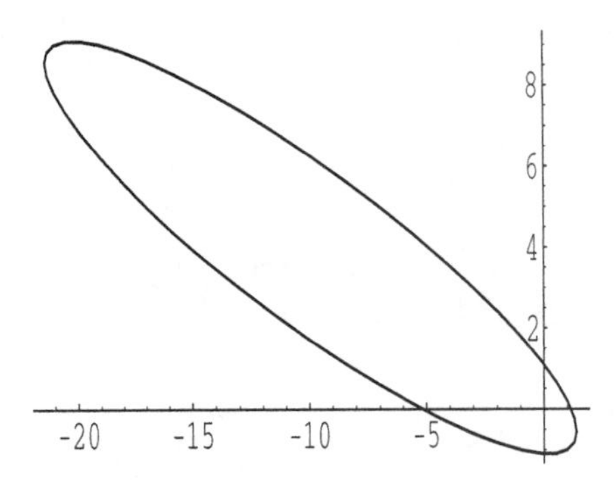

There are two forms for the ImplicitPlot[ ] command, and the form used above, **where a range for only one variable, in this case x, is specified**, is ideally suited for plotting conic sections. The choice of this variable determines the horizontal axis. **The second form of this command, namely ImplicitPlot[eq, $\{x, a, b\}$, $\{y, c, d\}$], is an entirely different mechanism for plotting equations**. It is very useful, but not for plotting conic sections. It will be introduced, when it is needed.

Before we begin our work, **the new Collect[ ] and Coefficient[ ] commands are introduced**. The following short, meandering work session (which serves no other purpose) is used to demonstrate how these useful commands work. The Coefficient[ ] command, which speaks for itself, occurs only briefly, at the very end.

```
In[4]:= ex=Expand[(a*x+b*y+c)^2+(d*x+e*y+f)^2]
Out[4]= c^2 + f^2 + 2 a c x + 2 d f x + a^2 x^2 + d^2 x^2 + 2 b c y + 2 e f y +
          2 a b x y + 2 d e x y + b^2 y^2 + e^2 y^2
```

```
In[5]:= ex1=Collect[ex,{y,x}]
Out[5]= c^2 + f^2 + (2 a c + 2 d f) x + (a^2 + d^2) x^2 +
          (2 b c + 2 e f + (2 a b + 2 d e) x) y + (b^2 + e^2) y^2
```

Notice that the above input statement **does not quite produce what we want**. The problem is resolved by telling Mathematica to collect, not only with respect to x and y, **but also with respect to $x * y$**.

```
In[6]:= ex2=Collect[ex,{x,y,x*y}]
Out[6]= c^2 + f^2 + (2 a c + 2 d f) x + (a^2 + d^2) x^2 + (2 b c + 2 e f) y +
          (2 a b + 2 d e) x y + (b^2 + e^2) y^2
```

This command has the ability to **collect terms in a variety of other ways**. To demonstrate this, we first set up an expression.

```
In[7]:= ex3=Expand[(a+b*x+c*x^2)*(e+f*x)]
Out[7]= a e + b e x + a f x + c e x^2 + b f x^2 + c f x^3
```

```
In[8]:= ex4=ex3/.{x->(x+1)}
Out[8]= a e + b e (1 + x) + a f (1 + x) + c e (1 + x)^2 + b f (1 + x)^2 + c f (1 + x)^3
```

```
In[9]:= Collect[ex4,x+1]
```

```
Out[9]= a e + (b e + a f) (1 + x) + (c e + b f) (1 + x)^2 + c f (1 + x)^3
```

```
In[10]:= Coefficient[ex2,x*y]
Out[10]= 2 a b + 2 d e
```

Example 8.1 *Graph the conic section* $x^2 + 4x + 6y^2 - 24y + 2 = 0$ *and find the coordinates of the center, foci and vertices.*

```
In[11]:= eq=x^2+4x+6y^2-24y+2==0
Out[11]= 2 + 4 x + x^2 - 24 y + 6 y^2 == 0
```

The package Graphics`ImplicitPlot` was loaded earlier, so there is no need to load it again. However, we do need the command, CompleteSquare( ) that we created in Chapter 6 and placed it in our personal library. To use this command, we must open up that notebook and either enter the command directly from that notebook, or paste it into this work session.

```
In[12]:= ImplicitPlot[eq,{x,-8,4}];
```

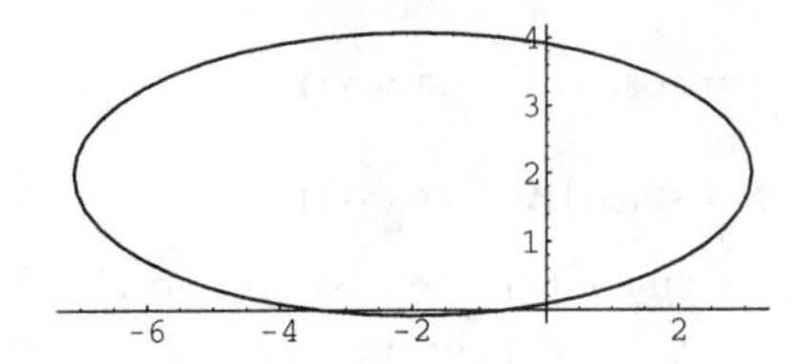

```
In[13]:= CompleteSquare[f_,x_]:=Module[{a,b,c},{c,b,a}=
         CoefficientList[f,x];a*(x+b/(2*a))^2+(4*a*c-b^2)/(4*a)]
```

```
In[14]:= eq1=CompleteSquare[eq[[1]],x]==0
```

$$\text{Out[14]=}\ (2 + x)^2 + \frac{1}{4}\left(-16 + 4\left(2 - 24\,y + 6\,y^2\right)\right) == 0$$

```
In[15]:= eq2=CompleteSquare[eq1[[1]],y]==0
```

$$\text{Out[15]=}\ \frac{1}{24}\left(-576 + 24\left(-2 + (2 + x)^2\right)\right) + 6\,(-2 + y)^2 == 0$$

```
In[16]:= eq3=Collect[eq2[[1]],{2+x,-2+y}]==0
Out[16]= -26 + (2 + x)^2 + 6 (-2 + y)^2 == 0
```

```
In[17]:= eq4=eq3[[1]]+26==26
Out[17]= (2 + x)^2 + 6 (-2 + y)^2 == 26
```

```
In[18]:= eq5=eq4[[1]]/26==1
```

$$\text{Out[18]=}\ \frac{1}{26}\left((2 + x)^2 + 6\,(-2 + y)^2\right) == 1$$

```
In[19]:= eq6=Collect[eq5[[1]],{2+x,-2+y}]==1

Out[19]= 1/26 (2 + x)^2 + 3/13 (-2 + y)^2 == 1
```

```
In[20]:= {a=N[Sqrt[26]],b=N[Sqrt[13/3]],c=N[Sqrt[a^2-b^2]]}
Out[20]= {5.09902, 2.08167, 4.65475}
```

```
In[21]:= ct={-2,2}
Out[21]= {-2, 2}
```

```
In[22]:= {f1={-2+c,2},f2={-2-c,2}}
Out[22]= {{2.65475, 2}, {-6.65475, 2}}
```

```
In[23]:= {v1={-2+a,2},v2={-2-a,2}}
Out[23]= {{3.09902, 2}, {-7.09902, 2}}
```

The transformation equations

$$x = \cos(\alpha)X - \sin(\alpha)Y$$

$$y = \sin(\alpha)X + \cos(\alpha)Y$$

give us the relationship between the xy-coordinate system and the XY-coordinate system obtained by rotating the xy-coordinate system through an angle α as shown in the figure.

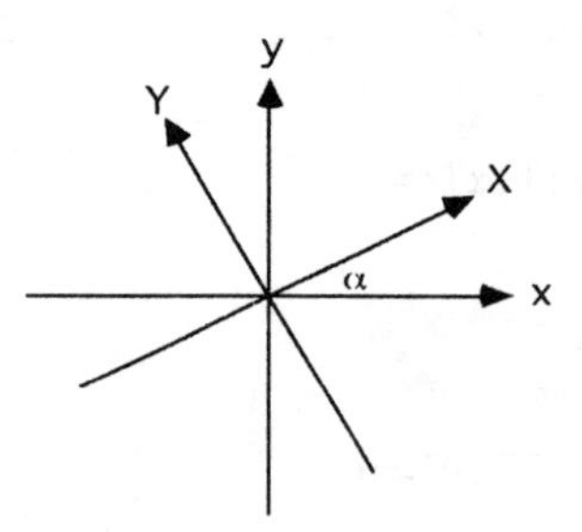

A point P in the plane which has coordinates (x, y) in the xy-coordinate system would have coordinates (X, Y) in the rotated XY-coordinate system. These equations transform equation 8.1 into an equation of the form

$$AX^2 + BXY + CY^2 + DX + EY + F = 0. \tag{8.2}$$

A rotation is used to transform an equation of the form 8.1 into an equation of the form 8.2 where $B = 0$. Then the equation can be identified and its foci, vertices, etc., can be evaluated. Of course the graph of 8.1 is readily available without any of this work by using the implicitplot( ) command, but the foci, vertices, etc. can not be computed without an actual rotation. Recall that $B = 0$ if α is chosen to satisfy

$$\alpha = \frac{1}{2}\cot^{-1}(\frac{a-c}{b}) = \frac{1}{2}\tan^{-1}(\frac{b}{a-c}). \tag{8.3}$$

While it is tedious to show this result by hand, it is a pleasant and interesting Mathematica exercise. (see Exercise 3 at the end of the chapter)

The so-called discriminant of a second degree equation is unchanged by a rotation, regardless of the angle α. That is to say, under a rotation,

$$b^2 - 4ac = B^2 - 4AC.$$

Using this result, one can tell at a glance, whether an equation of the form 8.1 is an parabola, ellipse, or hyperbola, although one has to allow for certain degenerate possibilities. This result can be very useful, even when Mathematica's plotting tools are used to graph a second degree equation. Without a hint on where to look, it can sometime be very hard to find the complete graph. **A graph being viewed on a computer screen might be a parabola or half of a hyperbola. How do you know**? Fortunately, it is easy to know the basic type of curve before a plot is attempted.

To show that the discriminant is unchanged by a rotation is a tedious computation to do by hand, but it is another interesting Mathematica exercise. (See Exercise 2 at the end of the chapter.)

Example 8.2 *Graph the conic section defined by the equation*

$$2x^2 - y^2 + 4xy - 2x + 3y = 6,$$

and find the coordinates of its center, foci, and vertices.

This problem can be done in a somewhat more straightforward manner, at least initially, if it is done in a decimal environment. One simply evaluates α as a decimal. The transformation equations, which can then be set up in a straightforward manner, will naturally evaluate, along with all subsequent terms, to decimal expressions. **If a decimal environment is used, the term B, which should be zero, may turn out instead to be some small number** (smaller than 10^{-10}). This is the price one pays for using decimal approximations. **This small term B should be discarded in this case.**

The approach we use below and the decimal approach are identical, once the transformation equations are set up. Our point of departure is to evaluate $\cos(\alpha)$ and $\sin(\alpha)$ as algebraic numbers. This can always be done as long as the coefficients of 8.1 are integers or rational numbers. You may recall doing problems like the following in trigonometry. If r is a given rational number, then an expression, for example, like $\sin(\arctan(r))$, can be easily evaluated as an algebraic number, simply by drawing the appropriate right triangle and applying the right triangle definitions of the trigonometric functions.

Mathematica cannot evaluate $\cos(\alpha)$ and $\sin(\alpha)$ directly, as algebraic numbers, for the angle α defined by formula 8.3, but it can evaluate $\cos(2\alpha)$ in this manner. This is not surprising, considering the division by 2 in formula 8.3. We then use the identities

$$\cos^2(\alpha) = \frac{1+\cos(2\alpha)}{2}, \sin^2(\alpha) = \frac{1-\cos(2\alpha)}{2}$$

to evaluate $\cos(\alpha)$ and $\sin(\alpha)$ as algebraic numbers.

These algebraic numbers may not be simple. If they are too complicated, just disregard this algebraic approach, evaluate α as a decimal, and do the entire problem in the straight forward decimal environment mentioned above. If the algebraic numbers are simple enough, however, it is an **interesting exercise to do the calculations in an exact Mathematica environment**

```
In[24]:= eq=2x^2-y^2+4x*y-2x+3y==6
Out[24]= -2 x + 2 x^2 + 3 y + 4 x y - y^2 == 6
```

To avoid a very narrow plot, we used an option that produces a square plot box.

```
In[25]:= ImplicitPlot[eq,{x,-10,10},AspectRatio->1];
```

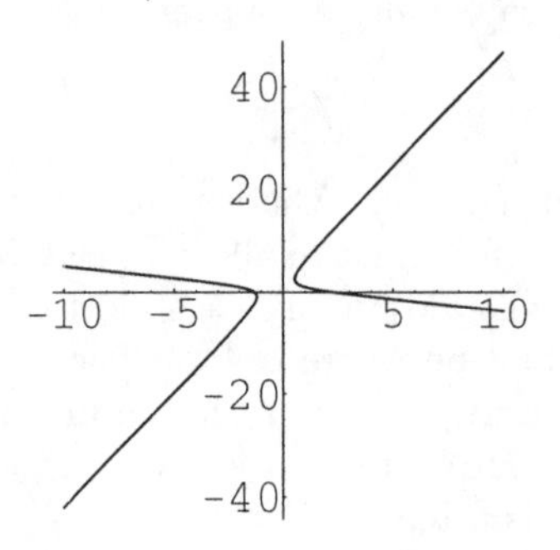

The region where this curve is located was easy to find. Imagine what might have happened, however, if the two branches of this hyperbola were far apart. It would be easy to lock in on only one branch of the hyperbola, miss the other branch, and decide that the curve must be a parabola. **To avoid such a mistake, it would be an easy matter to evaluate the discriminant $b^2 - 4ac$, and decide, in advance, what type of conic section the equation represents.**

In[26]:= α =ArcTan[4/(2+1)]/2

Out[26]= $\frac{1}{2}\ \arctan\left[\frac{4}{3}\right]$

In[27]:= c2=Simplify[Cos[2α]]

Out[27]= $\frac{3}{5}$

The names "cosine" and "sine" that we define next are just names representing $\cos(\alpha)$ and $\sin(\alpha)$, but they were chosen carefully to be meaningful and to avoid conflicting with Mathematica's sine and cosine functions.

```
In[28]:= cosine=Sqrt[(1+c2)/2]
```

Out[28]= $\frac{2}{\sqrt{5}}$

```
In[29]:= sine=Sqrt[(1-c2)/2]
```

Out[29]= $\frac{1}{\sqrt{5}}$

```
In[30]:= {x=cosine*X-sine*Y,y=sine*X+cosine*Y}
```

Out[30]= $\left\{\frac{2\ X}{\sqrt{5}} - \frac{Y}{\sqrt{5}}, \frac{X}{\sqrt{5}} + \frac{2\ Y}{\sqrt{5}}\right\}$

These are permanent assignments, so we already have the equation in the variables X and Y.

```
In[31]:= eq
```

Out[31]= $-2\left(\frac{2X}{\sqrt{5}}-\frac{Y}{\sqrt{5}}\right)+2\left(\frac{2X}{\sqrt{5}}-\frac{Y}{\sqrt{5}}\right)^2+3\left(\frac{X}{\sqrt{5}}+\frac{2Y}{\sqrt{5}}\right)+4\left(\frac{2X}{\sqrt{5}}-\frac{Y}{\sqrt{5}}\right)\left(\frac{X}{\sqrt{5}}+\frac{2Y}{\sqrt{5}}\right)-\left(\frac{X}{\sqrt{5}}+\frac{2Y}{\sqrt{5}}\right)^2 == 6$

```
In[32]:= eq1=Expand[eq[[1]]]==6
```

Out[32]= $-\frac{X}{\sqrt{5}}+3X^2+\frac{8Y}{\sqrt{5}}-2Y^2 == 6$

```
In[33]:= eq2=CompleteSquare[eq1[[1]],X]==6
```

Out[33]= $3\left(-\frac{1}{6\sqrt{5}}+X\right)^2+\frac{1}{12}\left(-\frac{1}{5}+12\left(\frac{8Y}{\sqrt{5}}-2Y^2\right)\right) == 6$

```
In[34]:= eq3=CompleteSquare[eq2[[1]],Y]==6
```

Out[34]= $\frac{1}{8}\left(\frac{64}{5}+8\left(-\frac{1}{60}+3\left(-\frac{1}{6\sqrt{5}}+X\right)^2\right)\right)-2\left(-\frac{2}{\sqrt{5}}+Y\right)^2 == 6$

```
In[35]:= eq4=Collect[eq3[[1]],{-1/(6*Sqrt[5])+X,-2/Sqrt[5]+Y}]==6
```

Out[35]= $\frac{19}{12}+3\left(-\frac{1}{6\sqrt{5}}+X\right)^2-2\left(-\frac{2}{\sqrt{5}}+Y\right)^2 == 6$

```
In[36]:= eq5=eq4[[1]]-19/12==6-19/12
```

Out[36]= $3\left(-\frac{1}{6\sqrt{5}}+X\right)^2-2\left(-\frac{2}{\sqrt{5}}+Y\right)^2 == \frac{53}{12}$

```
In[37]:= eq6=eq5[[1]]*12/53==1
```

Out[37]= $\frac{12}{53}\left(3\left(-\frac{1}{6\sqrt{5}}+X\right)^2-2\left(-\frac{2}{\sqrt{5}}+Y\right)^2\right) == 1$

```
In[38]:= eq7=Collect[eq6[[1]],{-1/(6*Sqrt[5])+X,-2/Sqrt[5]+Y}]==1
```

Out[38]= $\frac{36}{53}\left(-\frac{1}{6\sqrt{5}}+X\right)^2-\frac{24}{53}\left(-\frac{2}{\sqrt{5}}+Y\right)^2 == 1$

We are ready to specify the conic section parameters a, b, and c. The center, foci, and vertices in the rotated system can then be evaluated. Continuing our convention, capital letters are used for coordinates in the rotated system (CT is used for the center). We switch back to lower case letters when we evaluate the coordinates of the center, foci and vertices in the original (unrotated) coordinate system.

```
In[39]:= a=Sqrt[53/36];b=Sqrt[53/24];

In[40]:= c=Sqrt[a^2+b^2]
```

Out[40]= $\frac{\sqrt{\frac{265}{2}}}{6}$

```
In[41]:= CT={1/(6Sqrt[5]),2/Sqrt[5]}
```

Out[41]= $\left\{\frac{1}{6\sqrt{5}}, \frac{2}{\sqrt{5}}\right\}$

```
In[42]:= p=CT[[1]];q=CT[[2]];

In[43]:= {F1,F2}=Simplify[{{p+c,q},{p-c,q}}]
```

Out[43]= $\left\{\left\{\frac{2+5\sqrt{106}}{12\sqrt{5}}, \frac{2}{\sqrt{5}}\right\}, \left\{\frac{2-5\sqrt{106}}{12\sqrt{5}}, \frac{2}{\sqrt{5}}\right\}\right\}$

```
In[44]:= {V1,V2}=Simplify[{{p+a,q},{p-a,q}}]
```

Out[44]= $\left\{\left\{\frac{1}{30}\left(\sqrt{5}+5\sqrt{53}\right), \frac{2}{\sqrt{5}}\right\}, \left\{\frac{1}{30}\left(\sqrt{5}-5\sqrt{53}\right), \frac{2}{\sqrt{5}}\right\}\right\}$

```
In[45]:= ct=Simplify[{x,y}/.{X->CT[[1]],Y->CT[[2]]}]
```

Out[45]= $\left\{-\frac{1}{3}, \frac{5}{6}\right\}$

```
In[46]:= f1=Simplify[{x,y}/.{X->F1[[1]],Y->F1[[2]]}]
```

Out[46]= $\left\{\frac{1}{6}\left(-2+\sqrt{106}\right), \frac{1}{12}\left(10+\sqrt{106}\right)\right\}$

```
In[47]:= f2=Simplify[{x,y}/.{X->F2[[1]],Y->F2[[2]]}]
```

Out[47]= $\left\{\frac{1}{6}\left(-2-\sqrt{106}\right), \frac{5}{6}-\frac{\sqrt{\frac{53}{2}}}{6}\right\}$

```
In[48]:= v1=Simplify[{x,y}/.{X->V1[[1]],Y->V1[[2]]}]
```

Out[48]= $\left\{\frac{1}{15}\left(-5+\sqrt{265}\right), \frac{1}{30}\left(25+\sqrt{265}\right)\right\}$

```
In[49]:= v2=Simplify[{x,y}/.{X->V2[[1]],Y->V2[[2]]}]
```

Out[49]= $\left\{\frac{1}{15}\left(-5-\sqrt{265}\right), \frac{5}{6}-\frac{\sqrt{\frac{53}{5}}}{6}\right\}$

8.2 Parametrically Defined Planar Curves

Suppose that f and g are real valued continuous functions of a real variable t defined on an interval $D = [a, b]$. The set, of all points $(x, y) = (f(t), g(t))$ as t ranges over points in D describes a curve C in the plane. The equations $x = f(t)$, $y = g(t)$ are called parametric equations of the curve C, and C is said to be defined parametrically. The variable t is called the parameter. The point $(f(a), g(a))$ is called the initial point, and $(f(b), g(b))$ is called the terminal point. Intuitively, we think of t as time, so that there is a natural motion along the curve from the initial point to the terminal point.

A parametrically defined curve can be plotted using the ParametricPlot[ ] command. This command acts much like the Plot[ ] command. In particular, it has the same **options, and they should be used generously, if the primary purpose of a plot is to see the shape (perhaps the exact shape) of a curve.** Mathematica will not provide exact shape unless it is demanded. By way of contrast, exact shape may not be important, if a plot is created only as a tool for some other nongraphical purpose. To plot the curve $x = f(t)$, $y = g(t)$ $(a \leq t \leq b)$ we enter:

ParametricPlot[$\{f(t), g(t)\}, \{t, a, b\}$]

To plot a second curve $x = p(t), y = q(t)$ on the same parametric interval, we enter:

ParametricPlot[$\{\{f(t), g(t)\}, \{p(t), q(t)\}\}, \{t, a, b\}$]

Herein lies a problem. Parametric plots frequently have natural parameters associated with them, and **two plots may have different parametric intervals** associated with them, even if **both curves lie in the same region of the plane.** The above input statement **does not permit a different parametric interval** for each curve. When different parametric intervals are used, curves should be **plotted separately and then brought together with the Show[ ] command.** Use the option **(DisplayFunction– >Identity) to suppress the display** of the individual plots until they are brought together. Unfortunately, if this option is used, then **Mathematica must be told to display the plot** when it is desired. The normal value for "DisplayFunction" is **$DisplayFunction.** All this is demonstrated in the next two plots.

```
In[1]:= f=3Cos[2t];g=Sin[4t];
```

```
In[2]:= ParametricPlot[{f,g},{t,0,2*Pi}];
```

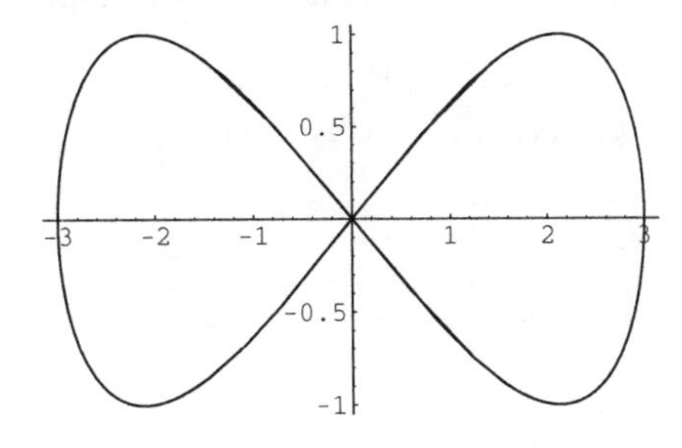


```
In[3]:= f2=Cos[t]/Log[t];g2=Sin[t]/Log[t];
```

```
In[4]:= p1=ParametricPlot[{f,g},{t,0,2*Pi},DisplayFunction->Identity];
```

```
In[5]:= p2=ParametricPlot[{f2,g2},{t,3,20},DisplayFunction->Identity];
```

```
In[6]:= Show[{p1,p2},DisplayFunction->$DisplayFunction];
```

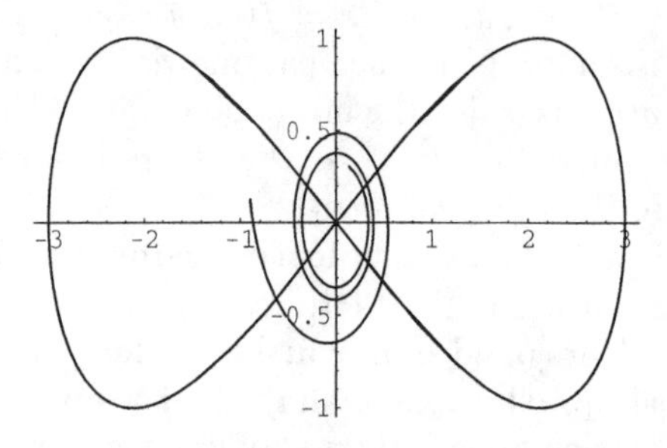

The slope of the line tangent to the graph of the curve C defined parametrically by $x = f(t), y = g(t), (a \le t \le b)$ at the point $(x, y) = (f(t), g(t))$ is

$$\frac{dy}{dx} = \frac{\frac{dy}{dt}}{\frac{dx}{dt}} = \frac{g'(t)}{f'(t)}.$$

The arc length differential is

$$ds = \sqrt{dx^2 + dy^2} = \sqrt{\left(\frac{dx}{dt}\right)^2 + \left(\frac{dy}{dt}\right)^2} = \sqrt{(f'(t))^2 + (g'(t))^2},$$

so that the length of a curve, C, is

$$l(C) = \int_a^b \sqrt{\left(\frac{dx}{dt}\right)^2 + \left(\frac{dy}{dt}\right)^2}\, dt = \int_a^b \sqrt{(f'(t))^2 + (g'(t))^2 dt}.$$

Parametric curves of the form

$$x = a\cos(mt), y = b\sin(nt), 0 \le t \le 2\pi,$$

are known as Lissajous curves. They are well know to provide a wide variety of interesting, complex shapes. In the next example, the subfamily

$$x = a\cos(3t), y = b\sin(2t), 0 \le t \le 2\pi.$$

is considered. We start with a graph of a typical member of this family.

```
In[7]:= ParametricPlot[{2Cos[3t],7Sin[2t]},{t,0,2*Pi},
        PlotRange->{{-7,7},{-7,7}},AspectRatio->1];
```

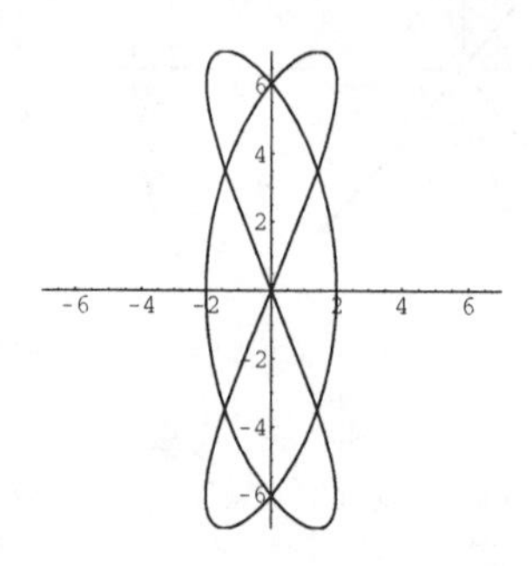

All of the members of this family have a similar shape. As you might expect, the value of a controls the amount of horizontal stretch, and the value of b controls the amount of vertical stretch. We can obtain confirmation of this by using a **new and very exciting Mathematica command called Animate[ ]**. Think of $F(x,t)$ as a function of x for each fixed value of t. Then

Animate[Plot[$F(x,t)$, {x, c, d}], {t, p, q}];

creates (and shows) 24 (a default value) separate plots for a range of values between $t = p$ and $t = q$. **Double click on the first picture to activate the animation (the movie).** The variable t is called the **frame variable**. Remember that Mathematica does not want to plot exact shape. In order to see how the different values of t change the shape of the plot, it is imperative to use a generous mix of options.

Two options which would be suggested include PlotRange– > $\{c, d\}$, where the range $\{c, d\}$ is the same as the one used for x and AspectRatio– > 1,

Mathematica does not list the Animate[ ] command in its Help File. Buried somewhere in its book is a statement that it is available in some package, which does not seem to be supplied. It is not even listed, if

?System`*

is entered in an input cell. **(Recall that this returns all names with a System` context. The asterisk (*) plays the role of a wild card.) Surprisingly enough, the Animate[ ] command works if it is used without loading any package**, so evidently, the command is supplied in some form at start-up.

With this new Mathematica command, we can study the the horizontal and vertical stretch in the Lissajous family of curves under consideration.

```
In[8]:= Animate[ParametricPlot[{2Cos[3t],b*Sin[2t]},
        {t,0,2*Pi},PlotRange->{{-20,20},{-20,20}},
        AspectRatio->1],{b,1,20}];
```

Output is omitted to save space.

```
In[9]:= Animate[ParametricPlot[{aCos[3t],2*Sin[2t]},
        {t,0,2*Pi},PlotRange->{{-20,20},{-20,20}},
        AspectRatio->1],{a,1,20}];
```

Output is omitted to save space.

Don't forget to double click on the first picture to activate the animation. By looking at these animations, each member of the family appears to be symmetric to both the x and y axes, and each of them has several self-intersection points. It seems reasonable to suspect that by choosing a and b appropriately, we can control not only the length of the curve but the angles at some of the self-intersection points. This is the substance of our next example.

Example 8.3 *Consider the family of curves defined parametrically by*

$$x = a\cos(3t), y = b\sin(2t), (0 \leq t \leq 2\pi), \ a, b > 0$$

i) Show that the curves are symmetric with respect to the x-axis and with respect to the y-axis.
ii)Find a member of this family which has an arc length of 20 units, for which self-intersection points off of the coordinate axes are all perpendicular.
iii) Find a point on this curve which is furthest from the origin.

```
In[10]:= x[t_]:=a*Cos[3t];y[t_]:=b*Sin[2*t];P[t_]:={x[t],y[t]}
```

Symmetry with respect to the x-axis is fairly easy to show. The relationship between $[x, y]$ and $[x, -y]$ is established as follows.

```
In[11]:= {P[t],P[-t]}
Out[11]= {{a cos[3 t], b sin[2 t]}, {a cos[3 t], -b sin[2 t]}}
```

Symmetry with respect to the y-axis is less obvious. Plotting one of the curves on different subintervals helps to determine the relationship between $[x, y]$ and $[-x, y]$.

```
In[12]:= ParametricPlot[{2Cos[3t],7Sin[2t]},{t,0,Pi}];
```

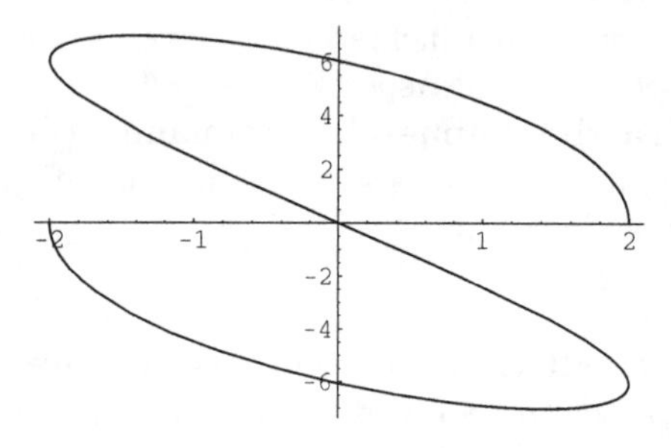


```
In[13]:= {P[t],Simplify[P[t+Pi]]}
Out[13]= {{a cos[3 t], b sin[2 t]}, {-a cos[3 t], b sin[2 t]}}
```

The self-intersection points take the form $P(t1) = P(t2)$ for different t-values $t1, t2$. After equating the x and y coordinates, we get a system of two equations in the two unknowns $t1, t2$. Before we can use Mathematica's FindRoot[ ] command on this system, we must rid the equations of all unassigned letters other than $t1, t2$. Fortunately, this is an easy matter. Since there are so many self-intersection points, we also have to help direct Mathematica to the one we want to find. There are four off of the coordinate axes, and because of symmetry we only need to find the one in the first quadrant. Notice that the x-coordinate of $P(t)$ is 0 for $t = \pi/6, 3\pi/6, 5\pi/6, 7\pi/6, 9\pi/6, 11\pi/6$. This will help to establish that the portion of the curve for $0 < t < \pi/6$ will cross the portion for $7\pi/6 < t < 9\pi/6$.

```
In[14]:= {eq1=P[t1][[1]]/a==P[t2][[1]]/a, eq2=P[t1][[2]]/b==P[t2][[2]]/b}
Out[14]= {cos[3 t1] == cos[3 t2], sin[2 t1] == sin[2 t2]}
```

```
In[15]:= s=FindRoot[{eq1,eq2},{t1,Pi/12},{t2,4Pi/3}]
Out[15]= {t1 → 0.261799, t2 → 4.45059}
```

```
In[16]:= t1=t1/.s[[1]];t2=t2/.s[[2]];
```

```
In[17]:= {P[t1],P[t2]}
Out[17]= {{0.707107 a, 0.5 b},
          {0.707107 a, 0.5 b}}
```

```
In[18]:= m[t_]=y'[t]/x'[t]
```

```
Out[18]= -(2 b cos[2 t] csc[3 t])/(3 a)
```

```
In[19]:= eq3=m[t1]*m[t2]==-1
```

```
Out[19]= -(0.666667 b^2)/(a^2) == -1
```

```
In[20]:= s1=Solve[eq3,b]
Out[20]= {{b -> -1.22474 a},
          {b -> 1.22474 a}}
```

```
In[21]:= b=b/.s1[[2]];
```

```
In[22]:= P[t]
Out[22]= {a cos[3 t], 1.22474 a sin[2 t]}
```

Arc length is next on our agenda.

```
In[23]:= ds=Sqrt[x'[t]^2+y'[t]^2]
Out[23]= Sqrt[6. a^2 cos[2 t]^2 + 9 a^2 sin[3 t]^2]
```

Mathematica cannot evaluate the integral of ds exactly (surely this should be expected), and so numerical approximation techniques must be used. Unfortunately, **We cannot use Mathematica's numerical integration command, NIntegrate[ ], as long as there is an unassigned constant inside the integrand**. Luck is with us, however, because a can be factored out and completely removed from the integral of ds. Mathematica, however, resists any attempt to factor a^2 out of this radical. Several other commands seem promising, but nothing seems to work.

The unassigned letter a can be factored out, however, in a less mathematical, but much simpler way. Substitute $a = 1$ into ds and call the result $ds1$. Then use $ds = a\,ds1$. As the saying goes, "Any port in a storm." We are ready to compute the arc length of the curve, which is just the integral of ds.

```
In[24]:= ds1=(ds/.a->1)
Out[24]= Sqrt[6. cos[2 t]^2 + 9 sin[3 t]^2]
```

```
In[25]:= NIntegrate[a*ds1,{t,0,2Pi}]
```

The output is omitted. The letter a still prevents a numerical approach. It is interesting to observe, however, that **Mathematica can evaluate the following integral**, but it takes an **excessive amount of time**. Evidently, these two forms for numerical integration are **not quite equivalent**.

```
In[26]:= eq4=N[Integrate[a*ds1,{t,0,2Pi}]
```

We omit the output, even though it **slowly** works, and turn our attention to an approach which is certain to work. We want arc length to be 20, so we immediately set this up as an equation.

```
In[27]:= eq4=a*NIntegrate[ds1,{t,0,2Pi}]==20
Out[27]= 16.5044 a == 20
```

```
In[28]:= aa=Solve[eq4,a]
Out[28]= {{a → 1.2118}}
```

```
In[29]:= a=a/.aa[[1]]
Out[29]= 1.2118
```

This establishes the curve we have been looking for, and all that is left is to find the point on the curve which is furthest from the origin. We take advantage of the simple observation that the distance, $\sqrt{x^2+y^2}$, will be a maximum at the same point that the square of this distance, x^2+y^2 is a maximum. This helps to alleviate a needless complication with square roots.

```
In[30]:= f=x[t]^2+y[t]^2
Out[30]= 1.46845 cos[3 t]^2+
           2.20268 sin[2 t]^2
```

```
In[31]:= Plot[f,{t,0,Pi/2}];
```

From the plot (omitted to save space), we can see that the maximum occurs near $t = 1$.

```
In[32]:= eq5=D[f,t]==0
Out[32]= 8.8107 cos[2 t] sin[2 t] - 8.8107 cos[3 t] sin[3 t] == 0
```

```
In[33]:= tt=FindRoot[eq5,{t,1}]
Out[33]= {t → 0.942478}
```

```
In[34]:= t3=t/.tt
Out[34]= 0.942478
```

```
In[35]:= P[t3]
Out[35]= {-1.15249, 1.4115}
```

```
In[36]:= MaxDistance=Sqrt[f/.t->t3]
Out[36]= 1.82224
```

8.3 Polar Coordinates

Mathematical literature is rich with examples of exotic and beautiful curves which can be drawn in polar coordinates. Some of these appear in the exercise set at the end of the chapter. They should be experienced with delight and enjoyment.

Mathematica doesn't seem to have a polar coordinate plotting command. This is not a problem, nor even a disappointment, since it is very easy to create our own command, or simply get along without one. Using the transformation equations

$$x = r\cos(\theta), y = r\sin(\theta)$$

between the polar and rectangular coordinate systems, the polar equation $r = f(\theta)$ can be written in parametric form

$$x = f(\theta)\cos(\theta), y = f(\theta)\sin(\theta),$$

and the curve can then be plotted as follows, using the techniques of the previous section.

ParametricPlot[$\{f(\theta)\cos(\theta), f(\theta)\sin(\theta)\}, \{\theta, 0, 2\pi\}$];

Recall that the Greek letter θ is entered in the form \[Theta], but, to avoid complicated key board activity, any other letter can be used just as well. As we mentioned in the introduction of the previous section, if the primary reason for plotting a polar coordinate curve is to see the exact shape of a curve, then a generous supply of plotting options should be included. Finally, if a large number of polar plots is on the agenda, it may be convenient to create a special polar plot command. We start this new Mathematica notebook with such a command. It includes a plotting option, so that they do not have to be added when the command is used.

```
In[1]:= PolarPlot[r_,s_]:=ParametricPlot[{r*Cos[t],r*Sin[t]},s,AspectRatio->1]
```

Our command is rather primitive. Notice that it requires the use of the letter t for θ in the polar curve $r = f(\theta)$ (no other letter will do). We have no immediate plans to use the command, so we will not bother improving it.

If the equation $r = f(\theta)$ is written parametrically, as we mentioned above, then there is no need to depend on any special forms for the slope of the tangent line or for the arc length differential. We simply use

$$\frac{dy}{dx} = \frac{\frac{dy}{d\theta}}{\frac{dx}{d\theta}}, ds = \sqrt{\left(\frac{dx}{d\theta}\right)^2 + \left(\frac{dy}{d\theta}\right)^2}.$$

However, if desired, these forms readily lead to the formulas

$$\frac{dy}{dx} = \frac{f'(\theta)\sin(\theta) + f(\theta)\cos(\theta)}{f'(\theta)\cos(\theta) - f(\theta)\sin(\theta)},$$

$$ds = \sqrt{r^2 + (\frac{dr}{d\theta})^2 d\theta} = \sqrt{(f(\theta))^2 + (f'(\theta))^2 d\theta}.$$

The arc length of the curve $r = f(\theta)$ on $(\alpha \leq \theta \leq \beta)$ is

$$Length = \int_{\alpha}^{\beta} \sqrt{(f(\theta))^2 + (f'(\theta))^2 d\theta}.$$

If R is a region bounded by $r = f(\theta), \theta = \alpha$, and $\theta = \beta$ with $\alpha < \beta$, then the area of R is

$$Area = \int_{\alpha}^{\beta} \frac{1}{2}(f(\theta))^2\, d\theta.$$

The details of all of these formulas and others, will be found in any standard calculus text.

Example 8.4 *Find the area of the region which lies above the horizontal line $y = 3$, and inside the four-leaved rose $r = 7\cos(2\theta + 1)$.*

The equation $y = 3$ can be expressed in its polar form as $r = 3\csc(\theta)$. Unfortunately, discontinuities are involved in its polar form, and this causes plotting problems. For the sake of plotting this curve in the easiest way, we try the parametric form $x = t$, $y = 3$. After plotting the second curve, it becomes clear that we will want x to range at least between -8 and 8, so we use the parametric interval $[-\pi, \pi]$ instead of $[0, 2\pi]$, and we toss in an extra multiplier, as shown below, to lengthen the horizontal line somewhat.

```
In[2]:= r1=7Cos[2t+1];r2=3Csc[t];
```

```
In[3]:= ParametricPlot[{{r1*Cos[t],r1*Sin[t]},{3t,3}},{t,-Pi,Pi}];
```

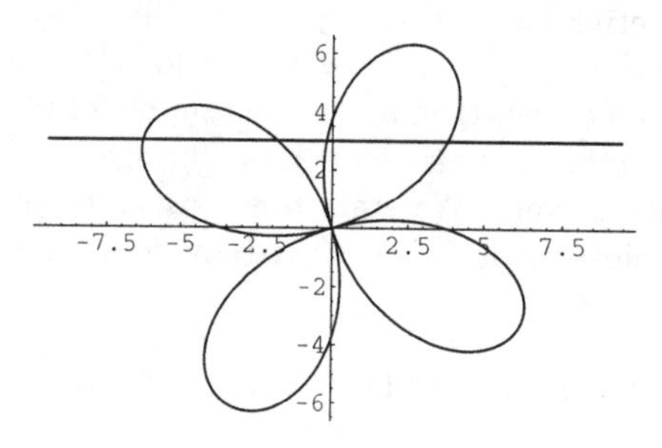

The graph of the region shows that there are four points of intersection. It is too much to ask Mathematica for exact values of the angles at these four points, and so we will use Mathematica's FindRoot[ ] command to get decimal approximations for them instead. Mathematica will need help from us to direct it to the vicinity of each of these four t-angles. By looking at the graph, it appears to be an easy matter to visually approximate these angles. Looks, however, can be deceiving. Remember that negative r-values, are involved in both of the curves seen in the graph. A point (r_0, θ_0) which may appear to be a point of intersection might not satisfy either equation. Instead, the point which satisfies both equations could be $(-r_0, \theta_0 + \pi)$ or some other combination of (r, θ) values. This certainly complicates the matter of guiding Mathematica to the vicinity of each t-value. To help in this matter, we plot, **as an ordinary Cartesian graph**, the difference in y-values as a function of t. By looking at where this graph crosses the horizontal axis (the t-axis), it is now an easy matter to get a range of t-angles in the vicinity of each intersection point. The rest of the problem is straight forward.

```
In[4]:= f=r1*Sin[t]-3
Out[4]= -3 + 7 cos[1 + 2 t] sin[t]
```

```
In[5]:= Plot[f,{t,0,2Pi}];
```

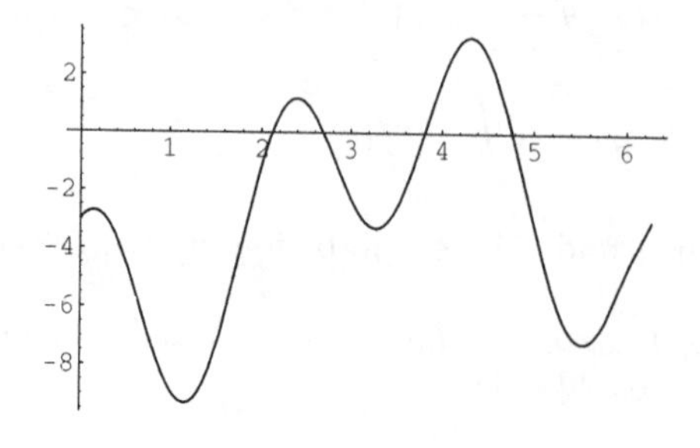

```
In[6]:= t1=FindRoot[f==0,{t,2.1}]
Out[6]= {t → 2.11929}

In[7]:= t2=FindRoot[f==0,{t,2.8}]
Out[7]= {t → 2.69587}

In[8]:= t3=FindRoot[f==0,{t,3.8}]
Out[8]= {t → 3.80925}

In[9]:= t4=FindRoot[f==0,{t,4.8}]
Out[9]= {t → 4.77585}

In[10]:= t1=t/.t1;t2=t/.t2;t3=t/.t3;t4=t/.t4;

In[11]:= A[α_,β_]:=Integrate[r1^2/2-r2^2/2,{t,α, β }]

In[12]:= area=A[t1,t2]+A[t3,t4]
Out[12]= 14.9969
```

In the next example, another way of dealing with the difficulties caused by the nonuniqueness of polar coordinate points is discussed. The example is noteworthy for several other reasons. **Notice the way that the Table[ ] command is used to do some of the computational work**. Also notice, in the computations involving the angle of intersection, the liberal use of functions rather than expressions.

Example 8.5 *Find the polar coordinates of the self-intersection points of the polar coordinate curve*

$$r = 3\sin(7\theta/3).$$

Express the coordinates in a form which satisfies the equation. Find the angle between the intersecting branches of the curve at one of the self-intersecting points (other than the origin).

For the sake of discussion we let $P(r, \theta)$ denote the point in the plane with polar coordinates (r, θ).

What should we choose for a plotting interval? Not the interval $[0, 2\pi]$, because $\sin(7\theta/3)$ is not the same at $\theta = 2\pi$ as it is at $\theta = 0$. **How many circuits about the origin should we make**? Our plotting interval must be of the form $[0, 2m\pi]$, where $7m\pi/3$ is a multiple of 2π. It follows that $[0, 6\pi]$ is the smallest plotting interval. A plot shows an obvious point of intersection at the origin, and so we turn our attention to the seven other points which appear to be (surely they are) equally distributed around a circle. Consequently, it should be enough to find just one of the points, and then use its coordinates, distributed around a circle in an obvious way, to get the coordinates of the other points.

There are several ways in which the curve could self-intersect. To discuss this matter, let f be the function defined by

$$f(\theta) = 3\sin(7\theta/3),$$

so that we can think of r as a variable. We will not use the letter f in our computer notebook, but it will help with this discussion. What if we can find a value θ_0 for which

$$r_0 = f(\theta_0) = f(\theta_0 + 2\pi)?$$

The function f certainly doesn't repeat itself on intervals of length 2π, so the point $P(r_0, \theta_0) = P(r_0, \theta_0 + 2\pi)$ would have to be one of these self-intersecting points. We may not find a point in this way. Then, perhaps we could find a θ_0 such that

$$r_0 = f(\theta_0) = -f(\theta_0 + \pi),$$

or

$$r_0 = f(\theta_0) = f(\theta_0 + 4\pi),$$

or

$$r_0 = f(\theta_0) = -f(\theta_0 + 3\pi).$$

A solution to any one of these equations would produce a self-intersecting point.

To find the angle between the intersecting branches of the curve we will use the trigonometric identity

$$\tan(\alpha - \beta) = \frac{\tan(\alpha) - \tan(\beta)}{1 + \tan(\alpha)\tan(\beta)}.$$

The slope of the tangent line to a curve at a point is the tangent of the angle of inclination of the line.

```
In[13]:= Clear["Global`*"]
```

```
In[14]:= r[t_]:=3Sin[7t/3]
```

```
In[15]:= ParametricPlot[{r[t]*Cos[t],r[t]*Sin[t]},{t,0,6Pi},AspectRatio->1];
```

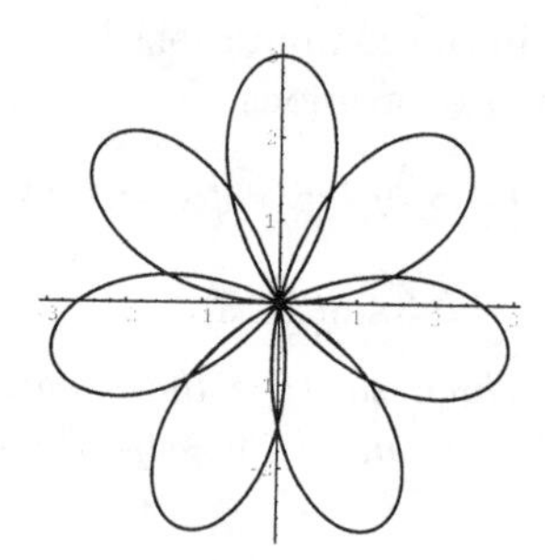

```
In[16]:= eq=r[t]==r[+t+2Pi]
```

Out[16]= $3 \sin\left[\frac{7\,t}{3}\right] == 3 \sin\left[\frac{7}{3}(2\pi + t)\right]$

Next, we replace $7t/3$ by θ, so that we can expand the trigonometric expressions without expanding relative to $7t/3$.

```
In[17]:= eq1=eq/.t->3θ/7
```

Out[17]= $3 \sin[\theta] == 3 \sin\left[\frac{7}{3}\left(2\pi + \frac{3\theta}{7}\right)\right]$

```
In[18]:= eq2=Simplify[eq1]
```

Out[18]= $3 \sin[\theta] == 3 \sin\left[\frac{14\pi}{3} + \theta\right]$

```
In[19]:= eq3=eq2[[1]]==TrigExpand[eq2[[2]]]
```

We omit the output. Let's call it Mathematica nonsense. The baffling expression contains several complex values terms such as $(-1)^{1/6}$. This should be real valued. If so, the imaginary term should disappear after an application of the next command.

```
In[20]:= eq4=ComplexExpand[eq3]
```

Out[20]= $3\ \sin[\theta] == \frac{3}{2}\ \sqrt{3}\ \cos[\theta] - \frac{3\ \sin[\theta]}{2}$

```
In[21]:= s=Solve[eq4,θ ]
Solve::ifun : Inverse functions are being used
    by Solve, so some solutions may not be found.
```

Out[21]= $\{\{\theta \to -\frac{5\ \pi}{6}\}, \{\theta \to \frac{\pi}{6}\}\}$

```
In[22]:= {t1=(3θ /7)/.θ ->Pi/6,r1=r[t1]}
```

Out[22]= $\{\frac{\pi}{14}, \frac{3}{2}\}$

Using the polar graph of the curve and the basic self-intersecting point we just found, it is clear that the seven points of self-intersection have coordinates (r, θ), where $r = 3/2$ and $\theta = T[[j]]$ $(j = 1, 2, \ldots, 7)$, determined in the following way.

```
In[23]:= T=Table[t1+2Pi/7*j,{j,0,6}]
```

Out[23]= $\{\frac{\pi}{14}, \frac{5\ \pi}{14}, \frac{9\ \pi}{14}, \frac{13\ \pi}{14}, \frac{17\ \pi}{14}, \frac{3\ \pi}{2}, \frac{25\ \pi}{14}\}$

There is little doubt that these are, indeed, the coordinates of the points, but only some of them satisfy the equation $r = f(\theta)$. This is a classic polar coordinate problem caused by the nonunique representation of points in polar space. A point (r_0, θ_0) can be on a curve $r = f(\theta)$, even though $r = r_0$, $\theta = \theta_0$ fails to satisfy the equation of the curve. The points (r_0, θ_0), $(-r_0, \theta_0 + \pi)$, and $(r_0, \theta_0 + 2\pi)$, to list a few, are all polar coordinates of the same point. It is possible for one of these ordered pairs to satisfy a polar equation, and for another ordered pair to fail to satisfy the equation.

With this in mind, let us return to the self-intersection points we determined above. In order to express the coordinates of these points in another form which will satisfy the equation $r = f(\theta)$ of the curve, we compare the value of $r = f(\theta)$ at $\theta = t[j], \theta = t[j]+\pi$, and $\theta = t[j] + 2\pi$ for each $j = 1, 2, \ldots, 7$. This comparison will shed light on what coordinates we should use, and also what the "return" coordinates should be. **Notice how easily this is done with the Table[ ] command.**

```
In[24]:= Table[{r[T[[j]]],r[T[[j]]+Pi],r[T[[j]]+2*Pi]},{j,1,7}]
```

Out[24]= $\{\{\frac{3}{2}, 3, \frac{3}{2}\}, \{\frac{3}{2}, -\frac{3}{2}, -3\}, \{-3, -\frac{3}{2}, \frac{3}{2}\}, \{\frac{3}{2}, 3, \frac{3}{2}\}, \{\frac{3}{2}, -\frac{3}{2}, -3\}, \{-3, -\frac{3}{2}, \frac{3}{2}\}, \{\frac{3}{2}, 3, \frac{3}{2}\}\}$

Using these comparisons, we can list the coordinates and the "return" coordinates, of each self-intersection point, each of which, by design, satisfies the equation of the curve. Later on, our angle calculations will put to rest any lingering doubt as to whether the coordinates and "return" coordinates of any point correspond to different or overlapping subarcs of the curve.

Look, for example, at the third and forth triplets in this list. We use this to specify the coordinates of the third point, $p3$, its return point $p3r$, and likewise the forth point, $p4$, and its return point $p4r$.

```
In[25]:= r0=3/2;p3={-r0,T[3]+Pi};p3r={r0,T[3]+2Pi};

         p4={r0,T[4]};p4r={r0,T[4]+2Pi};
```

```
In[26]:= {p4,p4r}
```

Out[26]= $\{\{\frac{3}{2}, \frac{13\,\pi}{14}\}, \{\frac{3}{2}, \frac{41\,\pi}{14}\}\}$

Using ideas given in the introduction of this section, we compute next a formula for the slope of the tangent line at a point on the curve. What we really want is the angle of inclination that a tangent line makes with the positive x-axis. This is easy to get from the slope, since slope (rise over run) is just the tangent of the angle of inclination.

```
In[27]:= x[t_]:=r[t]*Cos[t];y[t_]:=r[t]*Sin[t]
```

```
In[28]:= m[t_]:=y'[t]/x'[t]
```

```
In[29]:= TanAngle[t_,u_]:=Abs[(m[t]-m[u])/(1+m[t]*m[u])]
```

```
In[30]:= Angle[t_,u_]:=N[ArcTan[TanAngle[t,u]]]
```

```
In[31]:= Angle[T[1],T[1]+2Pi]
Out[31]= 0.485128
```

```
In[32]:= Angle[T[2],T[2]+Pi]
Out[32]= 0.485128
```

```
In[33]:= DegreeAngle=%*180/Pi
Out[33]= 27.7958
```

Example 8.6 *Find the arc length of the ellipse defined by the polar coordinate equation*

$$r = \frac{30}{13 - 12\sin(\theta + \frac{\pi}{5})}.$$

```
In[34]:= Clear["Global`*"]
```

```
In[35]:= r[t_]:=30/(13-12Sin[t+Pi/5])
```

```
In[36]:= ParametricPlot[{r[t]*Cos[t],r[t]*Sin[t]},{t,0,2Pi},AspectRatio->1];
```

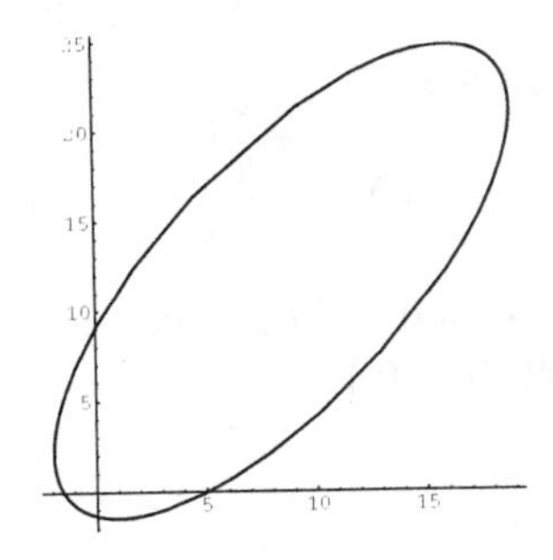


```
In[37]:= ds=Simplify[Sqrt[r[t]^2+r'[t]^2]]
```

Out[37]= $30\sqrt{\dfrac{144\cos\left[\frac{\pi}{5}+t\right]^2+\left(13-12\sin\left[\frac{\pi}{5}+t\right]\right)^2}{\left(13-12\sin\left[\frac{\pi}{5}+t\right]\right)^4}}$

```
In[38]:= L=NIntegrate[ds,{t,0,2Pi}]
Out[38]= 71.253
```

8.4 Exercise Set

Some New and Some Old Mathematica features and Commands

Animate[ ]	Apart[ ]	Clear[ ]
Coefficient[ ]	Collect[ ]	Expand[ ]
FindRoot[ ]	FullSimplify[ ]	Get[ ] or <<
<<Graphics`ImplicitPlot`	ImplicitPlot[ ]	Integrate[ ]
Limit[ ]	NIntegrate[ ]	N[Integrate[ ]]
ParametricPlot[ ]	Part[ ]	PolarPlot[ ]
PowerExpand[ ]	Show[ ]	Table[ ]
Together[ ]	TrigExpand[ ]	

Plot options

AspectRatio– > 1	DisplayFunction– >Identity
DisplayFunction– >$DisplayFunction	PlotRange– > { }

The PolarPlot[ ] command listed above is a user defined command.

1. Graph the following second degree equations. If they exist, find the coordinates of the foci, vertices, and the slopes of the asymptotes. If the graph happens to be a degenerate conic, is it necessary to go through a formal rotation to confirm this fact?
   a) $4x^2 + 15xy + 12y^2 - 20x + 44y + 80 = 0$
   b) $9x^2 + 6xy + y^2 + 7x + 3y + 2 = 0$
   c) $2x^2 - 5y^2 + 3x + 8y + 9 = 0$
   d) $3x^2 - 13xy - 10y^2 + 11x + 13y - 4 = 0$
   e) $2x^2 + 5xy + 14y^2 + 3x - 4y - 7 = 0$

2. Use Mathematica to show that the discriminant of a second degree equation is unchanged by a rotation, regardless of the angle of the rotation.

3. Use Mathematica to derive the formula

$$\tan(2\alpha) = \frac{b}{a-c}$$

that is used to analyze the graph of a second degree equation with a nonzero xy-term. (Hint: Perform an arbitrary rotation and set $B = 0$.)

4. Parametric equations of the form

$$x = x_0 + a\cos(mt), y = y_0 + b\sin(mt), 0 \leq t \leq 2\pi$$

are effective ways of representing a circle or an ellipse having a horizontal or vertical major axes. Ultimately these parametric equations should become so familiar to a student of mathematics that Mathematica's plotting tools are not needed. Use pencil and paper techniques and Mathematica's plotting tools to help gain some insight into these forms. Determine the role played by x_0, y_0, by a, b, and by m. In particular, how can Mathematica's plot command be used to determine the role played by m?

5.A straight line in the plane is represented parametrically by

$$x = x_0 + pt, y = y_0 + qt, -\infty < t < \infty,$$

but parametric equations of straight lines do not have to be linear. Use Mathematica to plot

$$x = 3 + 2f(t), y = 5 - 3f(t), t_0 < t < t_1.$$

for the following sample of functions f and intervals $[t_0, t_1]$.
   a) $f(t) = t$ on $(-\infty, \infty)$ b) $f(t) = t^3$ on $(-\infty, \infty)$
   c) $f(t) = t^2$ on $(-\infty, \infty)$ d) $f(t) = \tan(t)$ on $(-\frac{\pi}{2}, \frac{\pi}{2})$
What role does the function f play in these representations? Experiment further, if necessary, to answer this question.

6. Plot the Hypotrochoid

$$x = \cos(t) + 5\cos(3t), y = 6\cos(t) - 5\sin(3t), 0 \leq t \leq 2\pi.$$

7. A circle of radius 1, with a fixed point P marked on the circle, lies inside a circle of radius 10 centered at the origin, and is allowed to roll along the circumference of the larger circle. Initially, the point P is at the point (10,0) and the small circle rolls in a counterclockwise direction along the inside of the larger circle. The point P traces out a curve defined parametrically by

$$x = 9\cos(t) + \cos(9t), y = 9\sin(t) - \sin(9t), 0 \leq t \leq 2\pi.$$

Use Mathematica to plot this curve.

8. Parametric curves of the form

$$x = a\cos(mt), y = b\sin(nt), 0 \leq t \leq 2\pi,$$

are known as Lissajous curves. Use the ParametricPlot[ ] command to graph a sample of them and to determine the role played by a, b, m, n.
   a) Fix $n = 2, m = 3$, and plot over a range of a and b values. A nice way to do this is with Mathematica's animate command. With this command, you could, for example, set $a = s, b = 31 - s$ and $s = 1..30$, where s is the "frame variable."
   b) Plot with $a = 2, b = 5, n = 1$ fixed and $m = 2, 3, 6$.

c) Plot with $a = 2, b = 5, n = 2$ fixed and $m = 3, 5, 7$.

d) Plot with $a = 2, b = 5$ fixed and $n = 2s,\ m = 3s$ for $s = 2, 3, 4$. Notice that a common factor, j, between $n = n_1 j$ and $m = m_1 j$ has no effect on the graph. Why? If we happen to know the basic shapes of these curves for all integers m and n, is there anything gained by considering rational numbers m and n, which are not integers?

9. a) Plot the rose petal curve defined by the polar coordinate equation $r = 3\cos(2\theta)$. b) Determine the role played by the fixed angle s in $r = 3\cos(2\theta + s)$ Use Mathematica's animation command with s as the frame variable.

10. Plot the following conic sections in polar coordinates and find the vertices and the second focus in polar and rectangular coordinates.

$a)\ r = \frac{4}{3+2\cos(\theta-\frac{3\pi}{4})}$ $\quad b)\ r = \frac{8}{3-3\cos(\theta+\frac{\pi}{6})}$ $\quad c)\ r = \frac{21}{5+7\cos(\theta+1)}$

11. Consider the family of curves defined by $r = \sin(m\theta)$. The value of m has a dramatic effect on the graph.

a) When m is an positive integer the curves are called "rose petal" curves. Plot these curves for $n = 2, 3, 4, 5, 6, 7$. Make a conjecture about value of the integer m and the shape of the curve. Explain, analytically, why the patterns are different for m even or odd.

b) Plot the curves for $m = 5/3$, and then for $m = 7/2, 7/3, 7/4, 7/6, 7/11$. Any conjectures?

c) A larger denominator seems to complicate the curve. What would be the result be if m were irrational? Try to answer this question first, then plot the curve for $m = \sqrt{2}$. Use a large θ-range. Why? Then sit back and relax. This will take some time.

d) What about curves of the form $r = \cos(m\theta)$? Do we have to go through all this work again? Think about Problem 9 and an elementary trigonometric identity.

12. Plot the beautiful **Fay Butterfly**

$$r = e^{\cos(\theta)} - 2\cos(4\theta) + \sin^5(\frac{\theta}{12}).$$

Notice that a θ-interval larger than 2π must be used. How big must the plot range be in order to display the whole curve? This curve appears in "The Butterfly Curve" by Temple H. Fay, Amer.Math.Monthly **96** No.5, May 1989, pp.442-443.

This curve will take some time for Mathematica to draw, but be patient. The result is worth the wait—enjoy!

13. Plot the curve $x = t^2 - t; y = 2t^2 - t^3 + 7$ in a window which shows its self-intersection point. Find the coordinates of the point of intersection expressed as decimal numbers. Find the angle (acute angle) between the two intersecting subarcs at the intersection point. express your answer in degree measure. Find the length of the arc (loop) beginning and ending at this intersection point.

14. Consider the torus generated by rotating, about the x-axis, the circle of radius r centered at $(x_0, y_0) = (0, R)$, where r, and R are fixed positive numbers with $R > r$. Find a formula for the surface area of the torus in terms of r and R. Use a parametric form for the equation of the circle.

15. Find the area of the region inside the graph of $r = -4\sin(t)$, and outside the graph of $r = 6\sin(3t)$.

16. Find the area of the region outside of the graph of the cardioid $r = 2 + 2\cos(\theta)$ and inside the polar coordinate curve $r^2 = 25\cos(\theta)$. plotting the second curve will require some special handling. The restricted domain of the second curve may also cause some complications. A full plot of the region involved should be included. It may help to use the Show[ ] command.

17. Use a parametric form for the equation of the ellipse

$$\frac{x^2}{a^2} + \frac{y^2}{b^2} = 1, (\text{ assume } a > b)$$

to show the following classic reflective property of the ellipse: Let P be a point on the ellipse, let L be the tangent line to the ellipse at P, and let C_1, C_2 be the foci of the ellipse. Then the The acute angle between L and the line from C_1 to P, is the same as the acute angle between the L and the line from P to C_2. This result is often phrased as, "The angle of incidence equals the angle of reflection."

How does one show that $expr_1 = expr_2$ when $expr_1$ and $expr_2$ are complicated expressions? One way is to show that $expr_1 - expr_2 = 0$, but there is another way, somewhat similar to this, that might work better. Describe an approach of this sort. It may help in this or some future problem.

Incidentally, in this problem it is enough to show that the tangents of the two angles involved are equal.

18. The reflective property for a hyperbola is shown below, where $F1$ and $F2$ denote the foci of the hyperbola

$$\frac{x^2}{a^2} - \frac{y^2}{b^2} = 1.$$

For any point P, the line from P to Q (towards $F1$) and then to $F2$ produces angles α and β of equal measure at the point Q on the hyperbola.

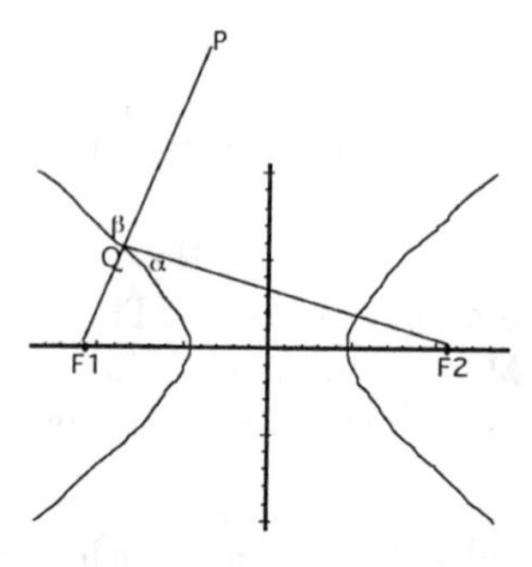

Use the parametric form for the equation of the hyperbola to prove this reflective property. See the comment following the preceding exercise for some advise on showing equality.

Project: Half of a Rose

Find a horizontal line $y = a$ above the x-axis, such that the area of the region above this line and inside the three-leaved rose $r = 5\sin(3\theta)$ is half the area of the entire rose.

This problem, which resembles Example 8.4 seems simple enough. The area of the entire rose is easy to compute. Then, all one has to do is find the points of intersection of the line and the rose and the rest is fairly straight forward. An exact solution to the equation involved in this point of intersection, however, is too complicated to be useful, and so it is natural to turn to approximate solutions using Mathematica's FindRoot[] command. Unfortunately, the value of a that controls the horizontal line is unknown, and in order to use the FindRoot[] command in an equation, every name in the equation must evaluate to a numeric except for the unknown that the equation is being solved for.

The easiest way to proceed is by guess work. Guess at an answer, compute the area involved, use this to adjust your guess. This can be repeated again and again, each step leading to a better answer.

A better way is to write a program which mimics your guess work. Perhaps, you can think of an approach that works better than the one discussed below. Notice, however, how "natural" the idea is, how methods of this sort can be created without any background in numerical approximation techniques.

Start with a value for a in $y = a$, call it *hi*, which is too big and another value for a, call it *low*, which is too small. Clearly a =*hi*= 4 and a =*low*= 0 will work. Let *mid* be the average of *low* and *hi*. One can then check the area corresponding to a=*mid* to see if it is larger or smaller than half the area of the rose. Use this to reassign either *low* or *hi* to be the value of *mid*. This would complete one pass through the program loop, and the steps are then repeated as often as necessary until a =*mid* is close enough.

Write a program which solves this problem. To avoid integrating each pass through the program loop, **integrate just once, and generate a formula** that can be used instead to evaluate area in each pass through the program loop.

Mathematica will have to be told to stop the program when the answer is close enough. To do this we use the While[ ] command.

Here is a sample program, which uses similar techniques, that could be used to find a positive solution to the equation $x^2 = 3$. Start with a guess ($x = hi$), which is too big and a guess ($x = low$), which is too small.

```
low=0.0; hi=2.0; While[hi-low>= 10^(-8),
mid=(low+hi)/2; If[mid^2 < 3,lo=mid,hi=mid]];mid
```

We started with decimal numbers, in order to maintain a decimal profile. Results of programs are normally not printed (semicolons are used or implied throughout), so the input statement ended with a separate call to evaluate *mid*. (We could have used *hi* or *low* just as well.)

Project: Working on the Railroads

A traffic control engineer working for a railroad corporation watches the progress of two trains on a computer console. The trains are traveling on different tracks which intersect at a point ahead of both trains. The coal train is 1.57 miles long and the position of its head (its front most point) at time t is

$$x = 17t - 4\cos(2t); y = 2t + 3\sin(5t).$$

The freight train is 1.14 miles long and the position of its head at time t is

$$x = 30 + 11t, y = 117 + 18t - 7t^2.$$

In both cases t represents time, in hours, and x and y are in miles. The engineer makes a quick calculation and decides that the trains will safely pass through the intersection point. Unfortunately, this is not a good decision.

a) Show conclusively that the trains collide, and determine which train crashes into the other.

When the mess is cleaned up, management decides that it better send all of its traffic control engineers to school to learn some calculus. When the course is completed each engineer will be given a test. Failure will mean immediate job dismissal!

Let us tune back in to this drama, then, after some time has past, and suppose that the moment has arrived. As one of the traffic control engineers, here is the test that you must now take.

b) Given the exact circumstances surrounding the accident, what is the minimal amount of extra time that the coal train must have, in order for it to pass the intersection point (crash free) just ahead of the freight train? What is the minimal amount of extra time that the freight train must have, in order for it to pass the intersection point (crash free) just ahead of the coal train?

If one of the trains had exactly this amount of extra time, then the trains would safely pass through the intersection point of the two railroad tracks, but the trains would come so close to each other, that the experience would clearly scare everyone involved. Consequently, you are now asked to add a margin of safety to your calculations, so that the trains will "safely" pass each other. Think of two distinctly different ways of adding a safety margin, describe them carefully, and recalculate answers with a built-in safety margin.

Part (b) (the test) is not easy. Each answer is the solution to an arc length equation, but Mathematica may be unwilling to solve the equation, even with the FindRoot[ ] command. Nevertheless, setting up the appropriate arc length equation is a good way to begin the solution.

The easiest way to proceed is by guess work. Using the results from part a), you know a t-value which is too small (the trains crash). By guess-work, find a t-value which is too big (the trains do not crash). It follows that the correct answer is some t-value in between. Guess again, and decide whether your guess is too big or too small. Now what interval is the correct answer in? Guess again and again. With each guess, the correct answer is narrowed down to a smaller interval.

A better approach would be to write a program which mimics all of this guess work. Start with a t-value (call it *low*) which is too small and a t-value (call it *hi*) which to too big. and let *mid* be the average of *low* and *hi*. Decide whether $t = mid$ is too big or too small, and use this to reassign either *low* or *hi* to be the value of *mid*. This would complete one pass through the program loop, and the steps are then repeated as often as necessary until *mid* is close enough.

Mathematica will have to be told to stop the program when the answer is close enough. To do this we use the While[ ] command.

A sample program is included at the end of the "Half of a Rose" project preceding this project.

Chapter 9

Vectors and Analytic Geometry

The concept of a vector is one of the most central ideas in all of mathematics. We think of a vector as a directed line segment, having both a length (size) and a direction. Force, which has both a magnitude and a direction is a classic example. Vectors, however, play a much larger and much more central role in mathematics. A vector in the plane is determined by an ordered pair of numbers. A point in the plane is determined in the same way, by an ordered pair of numbers. Since an ordered pair of numbers can be a point or a vector, it follows that a point can also be regarded as a vector and vice versa. In the same way, points in (3-dimensional) space can be regarded as vectors in space, and, more generally, points in any higher dimensional space can be regarded as vectors in the same space. When we study vectors, we are studying more than just physical concepts like force. We are studying points themselves. What could be more general or basic?

Why is it important to think of a point as a vector? It turns out that we are able to define a very useful vector algebra and associated vector geometry. Since vectors and points are essentially one and the same thing, this rich system of algebra and geometry can be used to study a diverse range of topics dealing with points in the plane or points in space.

9.1 Vectors, Addition, and Scalar Multiplication

A vector is usually defined to be a directed line segment, or more precisely, a class of all directed line segments having the same length and direction. Soon after addition and scalar multiplication are defined geometrically, we arrive at the familiar form $\vec{v} = a\vec{i} + b\vec{j}$ for a vector in two-dimensional space and $\vec{w} = a\vec{i}+b\vec{j}+c\vec{k}$ for a vector in three-dimensional space. These forms can be reduced further to the ordered pair $\vec{v} = (a, b)$ and the ordered triplet $\vec{w} = (a, b, c)$. We take up the subject with this symbolic vector form.

Since a vector in the plane or in space is determined by an ordered pair or ordered triplet, it will be no surprise to discover that a **vector in Mathematica** is just a **list** of the form

$v = \{v_1, v_2\};\ \ w = \{w_1, w_2, w_3\};$

We have used lists in Mathematica from the start, so this should be a very familiar topic. The components of a vector can be extracted in a familiar way.

```
In[1]:= v={5,-11,-8};v[[2]]
Out[1]= -11
```

An unassigned name in Mathematica is treated as an arbitrary real (actually complex valued) number, also referred to as a *variable*. A very simple assignment can be used to construct an arbitrary vector, or a vector valued variable. **If $v1, v2$, and $v3$ are**

unassigned names, then the assignment $v = \{v1, v2, v3\}$ creates an arbitrary 3-dimensional vector. The command Array[ ] can also be used to create an arbitrary vector. If v is an unassigned letter and n is a positive integer, then Array$[v, n]$ creates the vector $\{v[1], v[2], \ldots, v[n]\}$. Curiously, the **name v remains unassigned, and the assignment v =Array$[v, n]$ cannot be used, because it would create a deadly infinite loop**. The Array[ ] command is a useful way to create an arbitrary vector with a large number of components.

```
In[2]:= W=Array[w,8]
Out[2]= {w[1],w[2],w[3],w[4],w[5],w[6],w[7],w[8]}
```

Notice that the letter W in the above assignment, is different from the letter w. As we mentioned, this is important. Actually, our vectors will only have two or three components, so we will have little occasion to use this command. An assignment like $w = \{w1, w2, w3\}$ is preferable.

The ability to create arbitrary vectors gives us a very effectively way to show **how to perform vector addition and scalar multiplication** on a Mathematica notebook.

```
In[3]:= v={v1,v2,v3};w={w1,w2,w3};
```

```
In[4]:= v+w
Out[4]= {v1 + w1, v2 + w2, v3 + w3}
```

```
In[5]:= a*v
Out[5]= {a  v1, a  v2, a  v3}
```

```
In[6]:= 3w
Out[6]= {3  w1, 3  w2, 3  w3}
```

The notational rules for scalar multiplication, are the same as they are for ordinary multiplication. We had to use the multiplication symbol(*), or a blank space, between the two letters a and v above to create the scalar multiple $a\vec{v}$. In contrast, $3w$ was used to create the scalar multiple $3\vec{w}$. **The symbol av in an input cell would be interpreted as another unassigned name representing an arbitrary real (or complex) number**.

```
In[7]:= p={5,-11,-8};q={6,2,-1};
```

```
In[8]:= 3p-2q
Out[8]= {3, -37, -22}
```

The familiar notation $\vec{v} = a\vec{i} + b\vec{j} + c\vec{k}$ used to represent a vector on paper could be created on a notebook, but it is probably not a good idea to use it. Defining the vectors $\vec{i}$, $\vec{j}$, $\vec{k}$, with input statements like $i = \{1, 0, 0\}$, $j = \{0, 1, 0\}$, $k = \{0, 0, 1\}$ would be valid, mathematically, but it would not be be very useful. Mathematica would simply return $\{a, b, c\}$ for $v = ai + bj + ck$. To get the desired output, we must let i, j, k, be unassigned names. Such a strategy, however, is somewhat dishonest mathematically, is awkward to use in practice, and will not be used in this manual.

Two dimensional vectors can also be displayed. This may have limited use as a problem solving tool, but, occasionally, a display can help to provide insight, especially during our initial work. The command **Arrow[ ] resides in the Graphics`Arrow` package**. The output of this command is **not a display**, but is a new basic Mathematica object called a **graphics primitive**. Pictures frequently originate as objects of this sort. A **graphics primitive** is, what the name suggests, a very primitive set of instructions for creating a certain display. Before a graphics primitive can be displayed, it must be **turned into a graphics object, by using the command Graphics[ ]**. The output of this command is **also not a display**, but is instead a higher level set of instructions for creating a display. **To actually display a picture of a "graphics object," we use the Show[ ] command.**

The input expression Arrow[{a, b}, {c, d}] creates a graphics primitive for the vector from the point (a, b) to the point (c, d). We use this below to demonstrate the parallelogram law for vector addition. If z is a decimal number ($0 \leq z \leq 1$) the **graphics directive Hue[z]** in the command Graphics[{...,Hue[z],...}], **changes the color of everything to the right of the position where Hue[z] appears**. On a color monitor, the vector sum, $r = p + q$ would appear in red.

```
In[9]:= <<Graphics'Arrow'
```

This loads the package.

```
In[10]:= p={2,3};q={-1,2};r=p+q;s=2p;
```

```
In[11]:= Show[Graphics[{Arrow[{0,0},p],
         Arrow[{0,0},q],Hue[0],Arrow[{0,0},r]}]];
```

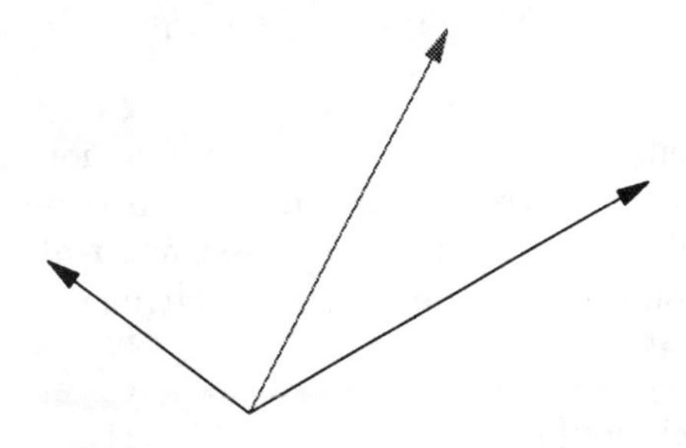

Example 9.1 *Find a unit vector in direction* $\vec{v} = 13\vec{\imath} + 4\vec{\jmath} - 9\vec{k}$. *Find a vector of length 18 in direction* $\vec{v}$.

Often, the first step in solving a problem involving vectors is to turn a vector into a unit vector. This frequently makes the vector much easier to deal with. Scalar multiplication does not change the direction of a vector, and so **scalar multiplying a vector by the reciprocal of its length will turn it into a unit vector** with the same direction. .

Mathematica does not have a command for computing the length or norm of a vector. It is so easy to compute length directly, that there is no need for such a command. Actually, in the next section, we will mention a more convenient way to compute length, but for now we will use a direct approach. Incidentally, Mathematica uses the word *length* of a vector to mean the number of components in the vector. We will not use the word *length* in this way.

```
In[12]:= v={13,4,-9}
Out[12]= {13, 4, -9}
```

```
In[13]:= l=Sqrt[13^2+4^2+(-9)^2]
Out[13]= √266
```

```
In[14]:= u=(1/l)*v
```

Out[14]= $\left\{\frac{13}{\sqrt{266}}, 2\sqrt{\frac{2}{133}}, -\frac{9}{\sqrt{266}}\right\}$

```
In[15]:= w=(18/l)*v
```

Out[15]= $\left\{117\sqrt{\frac{2}{133}}, 36\sqrt{\frac{2}{133}}, -81\sqrt{\frac{2}{133}}\right\}$

```
In[16]:= check1=Sqrt[u[[1]]^2+u[[2]]^2+u[[3]]^2]
Out[16]= 1
```

```
In[17]:= check2=Sqrt[w[[1]]^2+w[[2]]^2+w[[3]]^2]
Out[17]= 18
```

The next example may seem pointless, but problems like this frequently appear as steps in more practical problems.

Example 9.2 *Let $\vec{f}$ be the vector $\vec{f} = x^2\vec{i}+(x-4)^2\vec{j}+(9-2x)\vec{k}$ and suppose that $\vec{p} = \vec{f}+\vec{K}$ for some unknown constant vector $\vec{K}$. Determine a $\vec{K}$ such that $\vec{p} = 18\vec{i} - 33\vec{j} + 4\vec{k}$ when $x = 7$.*

It would probably be more efficient to introduce $\vec{f}$ as a function, but we avoid doing this just to offer a slightly different experience. **The unknown constant $\vec{K}$ cannot be left unassigned.** If it is, it will be regarded as an unassigned real or complex number. This issue is critical. Try the following solution, with $\vec{K}$ left unassigned, and notice the overwhelming nature of the mistake.

```
In[18]:= f={x^2,(x-4)^2,9-2x};pp={18,-33,4};K={k1,k2,k3};
```

```
In[19]:= p=f+K
```

Out[19]= $\{k1 + x^2, k2 + (-4 + x)^2, 9 + k3 - 2\ x\}$

```
In[20]:= eq7=pp==p/.x->7
Out[20]= {18, -33, 4} == {49 + k1, 9 + k2, -5 + k3}
```

The input statements look better than the output statements, but this awkward looking output equation is easily solved. Take note of this.

```
In[21]:= s=Solve[eq7,K]
Out[21]= {{k1 → -31, k2 → -42, k3 → 9}}
```

```
In[22]:= K=(K/.s[[1]])
Out[22]= {-31, -42, 9}
```

```
In[23]:= p
```

Out[23]= $\{k1 + x^2, k2 + (-4 + x)^2, 9 + k3 - 2\ x\}$

Notice that $\vec{p}$ doesn't seem to recognize the value for $\vec{K}$. To solve this awkward problem, we just redefine $\vec{p}$, now that $\vec{K}$ has a value.

```
In[24]:= p=f+K
Out[24]= {- 31 + x^2, -42 + (-4 + x)^2, 18 - 2 x}
```

A better approach involves a short "pencil and paper" strategy session prior to a Mathematica solution. Actually, just look at the input cell (not the output cell), where p is assigned to see a quick solution. Since $\vec{p} = \vec{f} + \vec{K}$ for every x, and since $\vec{p}$ and $\vec{f}$ are known when $x = 7$, it follows that $\vec{K} = \vec{pp} - \vec{f}$, where f is evaluated at $x = 7$. Certainly, the following approach is easier.

```
In[25]:= K=pp-(f/.x->7)
Out[25]= {-31, -42, 9}
```

9.2 The Dot and Cross Products

The two most interesting vector operations are the dot product and cross product. **The Mathematica commands, Dot[ ] and Cross[ ],** are very straightforward to use. These commands also have very nice infix operator forms, which mimic the notation used on paper. **A period (.) between vectors determines a dot product, and a × between vectors determines a cross product**. These infix operator forms are so natural to use, that they will probably be preferred over their equivalent command based forms. **To enter the symbol × in an input cell, type \[Cross]**. This is similar to the way we enter Greek letters in input cells. As soon as we type the closing bracket, the symbol \[Cross] will disappear and be replaced by the × symbol. For the record, we use arbitrary vectors to nail down the dot and cross product computations. As you can see, these results agree with the standard algebraic definitions of the dot and cross products.

```
In[1]:= v={v1,v2,v3};w={w1,w2,w3};

In[2]:= v.w
Out[2]= v1 w1 + v2 w2 + v3 w3

In[3]:= v×w
Out[3]= {- v3 w2 + v2 w3, v3 w1 - v1 w3, - v2 w1 + v1 w2}
```

The important geometric rule,

$$\vec{v} \cdot \vec{w} = |\vec{v}||\vec{w}| \cos(\theta),$$

can be used to determine the angle, θ, between $\vec{v}$, and $\vec{w}$. . There is a similar (equally important) rule,

$$|\vec{v} \times \vec{w}| = |\vec{v}||\vec{w}| \sin(\theta),$$

for cross products, but it is generally not a good idea to use this rule to determine the angle, θ, between $\vec{v}$, and $\vec{w}$. Finding the angle, θ, in this manner, means writing this equation in the form $\sin(\theta) = p$ for θ, where $p \geq 0$ is the obvious product and quotient of lengths of vectors. **For every such p, however, there are two angles, θ, in the interval $0 \leq \theta \leq \pi$, which satisfy this equation**. Only one of these angles is an appropriate answer, and this is the root of the problem.

The dot product also gives us a very convenient way to determine the length of a vector. It is easy to see that $\vec{v} \cdot \vec{v} = |\vec{v}|^2$ for every vector $\vec{v}$, where $|\vec{v}|$ denotes the geometric length of $\vec{v}$. **The resulting formula $|\vec{v}| = \sqrt{\vec{v} \cdot \vec{v}}$ is a convenient way to evaluate the length of $\vec{v}$.**

A few sample calculations, using these commands are shown below.

```
In[4]:= v={2,-7,3};w={8,1,-6};
```

```
In[5]:= v.w
Out[5]= -9
```

```
In[6]:= v×w
Out[6]= {39, 36, 58}
```

```
In[7]:= L=Sqrt[v.v]
Out[7]= √62
```

```
In[8]:= θ=N[ArcCos[v.w/Sqrt[(v.v)*(w.w)]]]
Out[8]= 1.68478
```

The value of L above is the geometric length of $\vec{v}$. This angle θ between $\vec{v}$ and $\vec{w}$ is in radian measure. We convert it to degree measure in the next calculation.

```
In[9]:= θ1=θ *180/Pi
Out[9]= 96.5305
```

Using Mathematica, it is easy to show that the cross product of two vectors in three-dimensional space is just the determinant of a certain 3×3 matrix. This is the familiar rule that we use to compute cross products by hand. We computed $\vec{v} \times \vec{w}$ earlier in this section, so we can compare that output with the following computation.

A matrix in Mathematica is just a list, where the entries are themselves other lists representing the rows of the matrix. **The command for computing the determinant of a matrix is Det[ ].** Neither matrices nor determinants will be used much in this manual, but this is a worthwhile exercise. Notice below, in the verification of a cross product as a determinant, that the vectors $\vec{i}, \vec{j}, \vec{k}$ are represented by unassigned names i, j, k. Mathematica has no idea that they represent the three unit coordinate vectors. Remember that unassigned names are regarded as real or complex numbers by Mathematica. Let's not tell Mathematica what we mean by i, j, k. It is enough to just hold on to this meaning in our minds.

```
In[10]:= v={v1,v2,v3};w={w1,w2,w3};
```

```
In[11]:= A={{i,j,k},v,w}
Out[11]= {{i, j, k}, {v1, v2, v3}, {w1, w2, w3}}
```

```
In[12]:= MatrixForm[A]
```

$$\text{Out[12]=} \begin{pmatrix} i & j & k \\ v1 & v2 & v3 \\ w1 & w2 & w3 \end{pmatrix}$$

The command MatrixForm[] serves no real computational purpose. It merely displays the matrix A in its familiar form.

```
In[13]:= M=Det[A]
Out[13]= -k v2 w1+j v3 w1+k v1 w2-i v3 w2-j v1 w3+i v2 w3
```

```
In[14]:= Collect[M,{i,j,k}]
Out[14]= k (-v2 w1+v1 w2)+j (v3 w1-v1 w3)+i (-v3 w2+v2 w3)
```

We used **the new command, Collect[ ]** in the above work. Its purpose is straight forward, so we will not comment further, but **it should be noted, so that it can be used in the future** when it is needed.

The algebraic rules governing the behavior of dot products and cross products are easy to justify. The dot product, for example, is obviously a commutative operation. That is to say, $\vec{u} \cdot \vec{v} = \vec{v} \cdot \vec{v}$, for any vectors $\vec{u}$ and $\vec{v}$. The less obvious rules can be proved quite easily using Mathematica. One such rule is the *Distributive Law of Cross Product over Addition.* Its proof and those of several other algebraic rules, are interesting problems that appear in the exercise set at the end of the chapter.

In the next Mathematica work session, we explore the proof of the algebraic rule $a(\vec{v} \times \vec{w}) = (a\vec{v}) \times \vec{w}$. We have already entered $\vec{v}$ and $\vec{w}$ as arbitrary vectors, so we proceed with a computation of the left and right hand sides.

```
In[15]:= L=a*(v×w)
Out[15]= {a (-v3 w2+v2 w3), a (v3 w1-v1 w3), a (-v2 w1+v1 w2)}
```

```
In[16]:= R=(a*v)×w
Out[16]= {-a v3 w2+a v2 w3, a v3 w1-a v1 w3, -a v2 w1+a v1 w2}
```

```
In[17]:= z=L-R
Out[17]= {a v3 w2-a v2 w3+a (-v3 w2+v2 w3),
           -a v3 w1+a v1 w3+a (v3 w1-v1 w3),
           a v2 w1-a v1 w2+a (-v2 w1+v1 w2)}
```

We named this vector z because it should be the zero vector.

```
In[18]:= z1=Simplify[z]
Out[18]= {0, 0, 0}
```

This means that $\vec{L} = \vec{R}$, which proves the algebraic rule. We used a simple strategy in the above work session, which deserves more attention. It can be visually taxing to see that two complicated vectors are equal to each other, by just looking at them. We used an approach which was easier on our eyes. **To show that $\vec{L} = \vec{R}$, just show instead, that $\vec{R} - \vec{L} = \vec{0}$. This approach can be used on almost any two Mathematica objects (vectors or otherwise) to show their equivalence. This is worth remembering!**

Some of the familiar algebraic rules that hold for addition and multiplication of real numbers do not hold in vector algebra. **Cross product, for example, is not an associative operation.** In the next Mathematica work session we investigate this curious behavior in the cross product. Almost any combination of three vectors $\vec{u}$, $\vec{v}$, and $\vec{w}$, will show that $\vec{u} \times (\vec{v} \times \vec{w})$ is different from $(\vec{u} \times \vec{v}) \times \vec{w}$. (Some combinations satisfy associativity, but the chances are remote that a random choice would produce such a combination.)

```
In[19]:= u={2,-5,1};v={-1,8,2};w={6,1,3};
```

```
In[20]:= {u×(v×w),(u×v)×w}
Out[20]= {{230, 120, 140}, {-26, 120, 12}}
```

This proves that cross product is not associative. What, then, can we say about these two triple cross products? Among other things, the vector $(\vec{u} \times \vec{v}) \times \vec{w}$ is perpendicular to $\vec{u} \times \vec{v}$, and so it is in the plane formed by $\vec{u}$ and $\vec{v}$ (if all of the vectors are based at the same point). This means that it is a linear combination of $\vec{u}$ and $\vec{v}$. Given arbitrary vectors, $\vec{u}$, $\vec{v}$, and $\vec{w}$, Mathematica can be used to find scalars a and b such that $(\vec{u} \times \vec{v}) \times \vec{w} = a\vec{u} + b\vec{v}$. This and other interesting properties of this triple cross product will be explored in the exercises.

The operations we have been discussing in the last two sections can be used effectively in a large variety of mathematical problems. A sample of these standard mathematical problems are considered in the next section. The operations can also be used to solve very interesting and impressive applied problems. Some of these appear in the exercises at the end of this chapter. In particular, Captain Ralph reappears after a long vacation. Now that we are working in three dimensional space we will have the opportunity to accompany Captain Ralph more frequently on his adventures.

We end this section with one standard mathematical application. Frequently, vectors need to be decomposed into a sum of two vectors, one which is parallel to a given direction and the other which is perpendicular. For example, if $\vec{f}$ is a force vector being applied to an object traveling along a curve C, then it would make sense to express $\vec{f}$ in the form

$$\vec{f} = \vec{f_t} + \vec{f_n},$$

where $\vec{f_t}$ is tangent to the curve C, and $\vec{f_n}$ is normal to the curve C.

Example 9.3 *Express the vector $\vec{v} = 13\vec{i} - 32\vec{j} + 7\vec{k}$ as the sum of a vector $\vec{w}_1$ parallel to $\vec{w} = -8\vec{i} + 9\vec{j} + 2\vec{k}$, and a vector $\vec{w}_2$ which is perpendicular to $\vec{w}$.*

We project $\vec{v}$ onto $\vec{w}$ to get $\vec{w}_1$. Once we have $\vec{w}_1$, the vector $\vec{w}_2$ is easy to get from $\vec{v} = \vec{w}_1 + \vec{w}_2$ by subtraction.

Calculus texts always have a formula for the projection vector, but there is no need to memorize this formula. Draw a picture on paper of any two vectors $\vec{v}$ and $\vec{w}$ based at the same point and proceed as follows. First turn $\vec{w}$ into a unit vector $\vec{u}$ (this is the direction of $\vec{w}_1$ or its opposite direction). Then, $\vec{w}_1 = L\,\vec{u}$, where L is either the length of the projection vector or its negative, depending on whether $\vec{w}_1$ and $\vec{u}$ have the same or opposite directions. Elementary trigonometry easily implies that $L = |\vec{v}|\cos(\theta)$, where θ is the angle between $\vec{v}$ and $\vec{u}$. This number L automatically has the right sign as well as the right size.

```
In[21]:= v={13,-32,7};w={-8,9,2};
```

```
In[22]:= u=(1/Sqrt[w.w])*w
```

Out[22]= $\left\{-\frac{8}{\sqrt{149}}, \frac{9}{\sqrt{149}}, \frac{2}{\sqrt{149}}\right\}$

The next computation represents the cosine of the angle between $\vec{v}$ and $\vec{u}$. Mathematica will not let us name it in such a way, so to create a suggestive name, that was acceptable to Mathematica, we had to really misspell the word. We can apologize to all our English teachers, but suggestive names help.

```
In[23]:= kosine=(u.v)/Sqrt[v.v]
```

Out[23]= $-21\sqrt{\frac{6}{3427}}$

```
In[24]:= w1=Sqrt[v.v]*kosine*u
Out[24]= {3024/149, -3402/149, -756/149}
```

```
In[25]:= w2=v-w1
Out[25]= {-1087/149, -1366/149, 1799/149}
```

Just to be sure, we check to see if $\vec{v}$ is the sum of $\vec{w1}$ and $\vec{w2}$. Surely $\vec{w1}$ will be parallel to $\vec{w}$, but we check to see if $\vec{w2}$ is perpendicular to $\vec{w}$.

```
In[26]:= {w1+w2,w2.w}
Out[26]= {{13, -32, 7}, 0}
```

9.3 Lines and Planes

Example 9.4 *Find the equation of the plane through the points P(2,-5,-4), Q(-1,6,7), R(9,5,-6).*

Mathematically, a point and a vector in three dimensional space are equivalent. They are both just ordered triplets of numbers. The only difference is a point of view. While this is frequently an advantage, it can also cause confusion. In this one example, we will try to separate the two ideas, by using capital letters, when we are thinking of points and small case letters when we are thinking of vectors. You might want to use this practice in your own work, if it helps. After this example, we will simply treat both points and vectors, right from the start, as vectors.

```
In[1]:= P={2,-5,-4};Q={-1,6,7};R={9,5,-6};
```

```
In[2]:= n=(Q-P)×(R-P)
Out[2]= {-132, 71, -107}
```

The vector $\vec{n}$ is normal to the plane.

```
In[3]:= X={x,y,z};
```

The point $X(x, y, z)$ is in the plane if and only if the vector from P to X is orthogonal to the normal vector $\vec{n}$.

```
In[4]:= eq=n.(X-P)==0
Out[4]= -132 (-2 + x) + 71 (5 + y) - 107 (4 + z) == 0
```

```
In[5]:= plane=Simplify[eq]
Out[5]= 191 - 132 x + 71 y - 107 z == 0
```

Example 9.5 *Find the distance between the line through* $A(4,-3,8)$, $B(-1,9,5)$ *and the line through* $C(5,7,-2)$ *and* $D(13,2,-6)$.

Imagine a plane which contains one of the lines and is parallel to the second line. Then the distance in question is just the distance from a point (any point on the second line) to the plane. Most calculus books provide a formula for the distance from a point Q to a plane, but it is fairly easy to see geometrically. Just take any vector from a point in the plane to Q, project it onto the plane's normal vector, and compute the length of the projection.

This problem can also be done by minimizing a certain function of two variables, namely the function giving the distance between a variable point on one line and a variable point on the other. We will return to this example in our study of functions of several variables.

```
In[6]:= a={4,-3,8};b={-1,9,5};c={5,7,-2};d={13,2,-6};
```

```
In[7]:= {u=a-b,v=c-d}
Out[7]= {{5, -12, 3}, {-8, 5, 4}}
```

A plane parallel to one line and containing the other has the following normal vector $\vec{n}$.

```
In[8]:= n=u×v
Out[8]= {-63, -44, -71}
```

To find the distance, we project the vector from A to C onto $\vec{n}$ and compute its length.

```
In[9]:= {d=Abs[n.(c-a)]/Sqrt[n.n],N[d]}
```

Out[9]= $\left\{\frac{207}{\sqrt{10946}}, 1.97853\right\}$

Example 9.6 *Find an equation for the line of intersection between the plane through* $A(5,1,-8)$, $B(-6,2,7)$, $C(3,9,1)$ *and the plane through* $P(-8,1,1)$, $Q(4,-7,-3)$ *and* $R(-5,-7,-1)$.

We will discuss Mathematica's three dimensional plotting devises in the next section. They could be used to see the line of intersection, but they contribute little to this problem.

There are two basic ways of proceeding. In one approach, the equation of the line is found by solving a system of two equations in three unknowns. This problem is a part of a more general problem involving m equations in n unknowns, and it is one of the main topics of discussion in a linear algebra course usually taken soon after a calculus sequence.

The approach we take is more geometrical. Let $\vec{n_1}$ and $\vec{n_2}$ be normal vectors for each of the two planes. The line of intersection lies in both planes and so it is orthogonal to both $\vec{n_1}$ and $\vec{n_2}$. It follows that the vector $\vec{v} = \vec{n_1} \times \vec{n_2}$ is a direction vector for the line of intersection. All we need, then, is a point of intersection of the two lines.

```
In[10]:= a={5,1,-8};b={-6,2,7};c={3,9,1};
         p={-8,1,1};q={4,-7,-3};r={-5,-7,-1};
```

```
In[11]:= {ab=b-a,ac=c-a}
Out[11]= {{-11, 1, 15}, {-2, 8, 9}}
```

```
In[12]:= {pq=q-p,pr=r-p}
Out[12]= {{12, -8, -4}, {3, -8, -2}}
```

```
In[13]:= {n1=ab×ac,n2=pq×pr}
Out[13]= {{-111, 69, -86}, {-16, 12, -72}}
```

```
In[14]:= v=n1×n2
Out[14]= {-3936, -6616, -228}
```

The vector $\vec{v}$ is the direction vector for the line. We discussed this in the introduction to this solution. The equations of the two planes are computed next. We let X denote an arbitrary point on either plane.

```
In[15]:= X={x,y,z}
Out[15]= {x, y, z}

In[16]:= ABC=n1.(X-a)==0
Out[16]= -111 (-5 + x) + 69 (-1 + y) - 86 (8 + z) == 0

In[17]:= ABC=Simplify[ABC]
Out[17]= -202 - 111 x + 69 y - 86 z == 0

In[18]:= PQR=Simplify[n2.(X-p)==0]
Out[18]= -4 (17 + 4 x - 3 y + 18 z) == 0

In[19]:= PQR=PQR[[1]]/4==0
Out[19]= -17 - 4 x + 3 y - 18 z == 0
```

All we need is to find one point on the line of intersection. Since we have two equations and three unknowns, we can assign any number to one of the variables, and solve for the remaining two variables.

```
In[20]:= x=-2;

In[21]:= s=Solve[{ABC,PQR},{y,z}]
```

Out[21]= $\left\{\left\{y \to -\frac{189}{164}, z \to -\frac{227}{328}\right\}\right\}$

```
In[22]:= {y,z}=({y,z}/.s[[1]])
```

Out[22]= $\left\{-\frac{189}{164}, -\frac{227}{328}\right\}$

```
In[23]:= X
```

Out[23]= $\left\{-2, -\frac{189}{164}, -\frac{227}{328}\right\}$

```
In[24]:= L=X+tv
```

Out[24]= $\left\{-2 + tv, -\frac{189}{164} + tv, -\frac{227}{328} + tv\right\}$

The vector (or point) $\vec{L}$ represents an arbitrary point on the line of intersection. We can express this in a more familiar parametric form as follows. The punctuation marks around the letters x, y, and z are needed in this input statement. Why?

```
In[25]:= {"x"==L[[1]],"y"==L[[2]],"z"==L[[3]]}
```

Out[25]= $\left\{\text{"x"} == -2 + tv, \text{"y"} == -\frac{189}{164} + tv, \text{"z"} == -\frac{227}{328} + tv\right\}$

Projection vectors were used frequently throughout this section, and we always went back to basic vector work to construct the projections. This may be a good learning devise, but eventually, it is more efficient to **simply create a special projection command**. This is a problem in the exercise set. Once such a command is created, it can be saved on the "mylib" file, where we have placed several other personal commands from previous chapters. It can then be used to easily compute projection vectors in the future.

If you create your own personal projection command (or any other personal command) and wish to use it on another notebook, it would, of course, have to be entered as an input line on the new notebook. Our personal library file could also be **saved as a package**. A personal "package" would be used like any other package in Mathematica. If you are interested in creating a package, look in the Mathematica Book in the Help File for guidance. Section 2.6.10, *Setting Up Mathematica Packages* would be a good place to start. The creation of special packages is presented only as a suggestion to someone who has an extraordinary interest in Mathematica and the time to experiment with this idea. It is not at all a necessary part of this manual, nor is it important for a better understanding of mathematics.

9.4 Surfaces in Space

The graph of an equation in three variables is a surface in three dimensional space. The graph of a function f of two variables x and y, is the graph of the equation $z = f(x, y)$. Traditionally, the letters x and y are used as independent variables and z as the dependent variable, but Mathematica has no particular preference for one letter choice over another. There are a variety of 3-dimensional graphical tools in mathematica, and only the rudiments are presented in this manual. You are encouraged to explore the Help File, both the main kernel and the packages. Some of our graphs will not be presented in the best way, because the effort required to produce a better graph takes us too far away from the main theme of calculus.

Mathematica's main three-dimensional plotting tool, is the **command Plot3D[ ]**. To plot the equation $z = f(x, y)$, we would type the following in an input cell.

Plot3D[$f(x, y), \{x, a, b\}, \{y, c, d\}$, *options*];

The term *options* represents one or more plotting options, separated by commas. **Plotting options are "Rules," so they all have the form, *option*− > *value***. There is a **large number of options** that can be used along with any graphical tool. These options are not always easy to use, but they are particularly important in 3-dimensional graphing. Make it a habit to us as many options as necessary to get the desired output.

Finally, before we attempt any Mathematica activity, realize that **plotting is a memory intensive activity**. It is easy to run out of memory before a plot request is finished. **Mathematica may let you save before it forces you to quit, or it may force you to quit without saving**! It is a good practice, especially when using plotting commands to **save your file before you enter a request for a memory intensive activity**.

The amount of memory, which has been used can be seen in a small bar chart. You can arrange to have this bar chart appear automatically on your monitor each time the Kernel is opened up. To do this, open up the Kernel and from the Kernel (not the Front End), pull down the File Menu, and click on **Show Memory Usage**. Drag the chart to a quiet corner of your computer screen, and it will appear there automatically whenever you run Mathmatica.

Mathematica's plotting commands are effective tools for understanding mathematics and solving problems. Now, let us experiment with these commands. The plots given below should be experienced first hand. **Try to predict the basic shape of the surface before you enter the plot request—This is always a good experience**.

```
In[1]:= z1=(9-x^2-y^2)*Cos[Abs[x]+Abs[y]]
```

Out[1]= $(9 - x^2 - y^2)$ cos[Abs[x] + Abs[y]]

```
In[2]:= Plot3D[z1,{x,-5,5},{y,-5,5}];
```

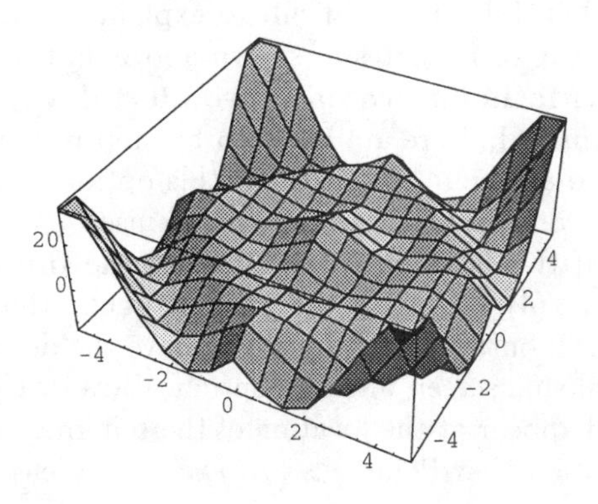

To get a cut-away view, enter a plot request with a different range. A few options were used just to start using them. This particular combination of options **creates a really beautiful picture**, but increasing the value of *PlotPoints* will use up more memory and take more time to compute.

```
In[3]:= Plot3D[z1,{x,-5,5},{y,0,5},PlotPoints->40,Mesh->False];
```

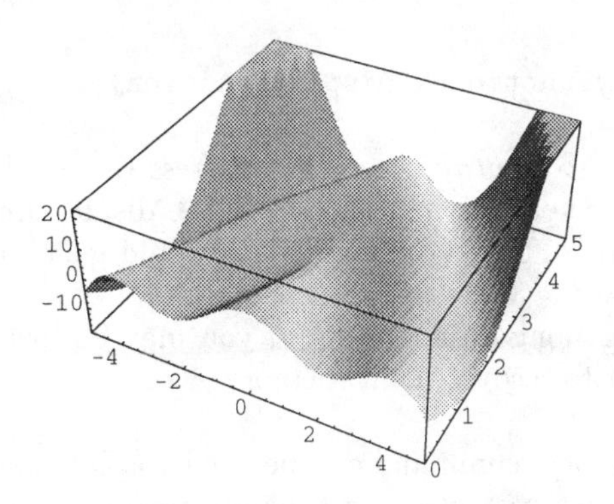

For important mathematical reasons, we will frequently want to **turn graphs around, to look at then at a different angle or a different point of view.**

```
In[4]:= Plot3D[z1,{x,-5,5},{y,0,5},ViewPoint->{0,4,-1}];
```

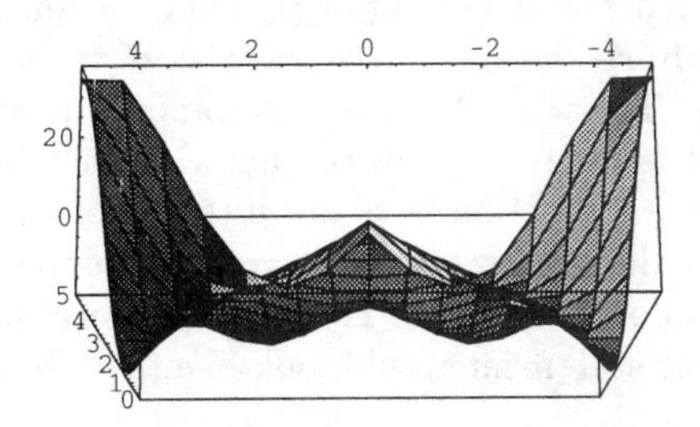

The list $\{0, 4, -1\}$ is a point where the viewing takes place. The coordinates of this point depend on the "box" which contains the plot. If you are interested, look up this topic on the Mathematica Book on the Help File. It is difficult to explain and more difficult yet to use this option by directly typing in coordinates as shown above, but fortunately, we don't have to bother with any of this. **Mathematica has a wonderful way of inserting this option into a Plot3D[ ] command**. Here are the step by step instructions. (1) Type a 3-dimensional plot command into an input cell (without this option). (2) Move the mouse pointer to the place inside of the command where you wish to insert the option and click, so that a blinking cursor appears at this location. (3) Pull down the **Input Menu** and click on **3D ViewPoint Selector...** (4) Play with the graphical cube that appears from the menu bar. Drag it. Move the scroll buttons, etc., until you have the desired view point. (5) Click the **Paste** tab on this dialogue screen and the option **ViewPoint− > $\{j_1, j_2, j_3\}$** with the desired coordinates will appear at the location of the blinking cursor (back on step 1). This technique makes the option ViewPoint− > $\{j_1, j_2, j_3\}$ very easy to use.

The 2-dimensional plot command can plot several curves at the same time, but the **Plot3D[ ] cannot do this**. The Show[ ] command must be used to plot two or more 3-dimensional plots.

```
In[5]:= g=x^2+y^2-400;
```

```
In[6]:= p1=Plot3D[f,{x,-10,10},{y,-10,10},DisplayFunction->Identity];
        p2=Plot3D[g,{x,-10,10},{y,-10,10},DisplayFunction->Identity];
```

```
In[7]:= Show[{p1,p2},DisplayFunction->$DisplayFunction];
```

The option *DisplayFunction− >Identity* is used to suppress the display. You don't need to use it, unless you wish to save space, but if it is used, the display must be turned back on again in the Show[ ] command. The statement above would generate a display, but it was deleted to save space.

By now, if you are following along on a computer, you may be low on memory. If so, you will need to quit and restart to recover spent memory. Keep an eye on the memory bar chart on you computer screen.

The 2-dimensional ImplicitPlot[ ] command can be used to plot an equation (a planar curve) (or several equations) in two variables. An equation involves the use of the equality symbol (== in Mathematica), and **this symbol cannot appear** inside the Plot[ ] or Plot3D[ ] commands. This is why another command is needed. Mathematica **does not have** an ImplicitPlot3D[ ]command, a command for plotting equations (surfaces) in three variables. It does, however, have a command which will do the job for us.

We will use the ContourPlot3D[ ] command. To fully explain this command would require a preemptive attack on Chapter 11, and rather than say too much, we choose instead to say as little as possible. If $f(x, y, z)$ is a function of x, y, and z, then the **surfaces defined by $f(x, y, z) = k$, where k is an arbitrary constant**, are called the **level sets or contours of $f(x, y, z)$**. The word "contour" is somewhat inappropriate to use in a 3-dimensional situation, but it helps to explain the name chosen for this command. In the 2-dimensional case, if $g(x, y)$ is a function of x and y (think of $g(x, y)$ as elevation at a point (x, y) on a flat map), then the level sets $g(x, y) = k$ are curves or **contours** on the map. This is why the word "contour" is used. **To plot the surfaces defined by the equations $f(x, y) = k$**, an input statement would look like the following:

ContourPlot3D[$f(x, y, z), \{x, x_{min}, x_{max}\}, \{y, y_{min}, y_{max}\}, \{z, z_{min}, z_{max}\}$];

Notice that the first entry is a **function, not an equation**. The default values for the **range of k values is just $k = 0$**, so this command plots the **equation $f(x, y, z) = 0$**.

Finally, this command resides in the **Graphics`ContourPlot3D` package**, so this package must be loaded before the command can be used.

```
In[8]:= eq=x^2/4+y^2/16+z^2/81==1
```

$$\text{Out[8]=}\ \frac{x^2}{4}+\frac{y^2}{16}+\frac{z^2}{81}==1$$

```
In[9]:= <<Graphics`ContourPlot3D`
```

```
In[10]:= ContourPlot3D[eq[[1]]-1,{x,-2,2},{y,-4,4},{z,-9,9}];
```

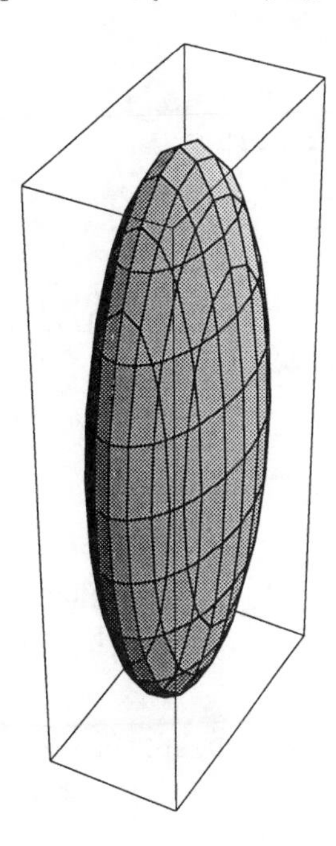

The equation *eq* must be expressed in the form $f(x, y, z) = 0$ and then $f(x, y, z)$ is entered as the first argument. Recall that *eq*[[1]], the first part of *eq* is the left hand side of the equation.

Example 9.7 *Draw the region bounded by the graphs of*

$$f(x, y) = 9 - \frac{y^2}{16} - \frac{x^2}{84} \text{ and } g(x, y) = y^2 + \frac{x^2}{64} + 5.$$

What should we use for ranges for x and y? If a standard range of $-10..10$ is used for each variable, the region bounded by the two surfaces will not quite fit in one direction, but it will still be **dwarfed by the large surfaces involved**. By a "hit and miss" procedure these ranges could be adjusted to eventually produce a fairly large x range and a fairly small y range. However, each attempt to plot the region will require a long wait for the plot, so this should really be called a *hit, wait, and miss* procedure.

A better approach would be to look at the curve $f(x,y) = g(x,y)$, and choose a range on x and y so that the corresponding rectangle in the x,y-plane just covers this curve and its interior. After all, the region under consideration surely consists of all points $P(x,y,z)$ with $g(x,y) \leq z \leq f(x,y)$.

```
In[11]:= Clear["Global`*"]
```

```
In[12]:= z1=9-y^2/16-x^2/84;z2=y^2+x^2/64+5;
```

```
In[13]:= eq=z1-z2==0
```

Out[13]= $4 - \frac{37\ x^2}{1344} - \frac{17\ y^2}{16} == 0$

```
In[14]:= {eqx=(eq/.y->0),eqy=(eq/.x->0)}
```

Out[14]= $\left\{4 - \frac{37\ x^2}{1344} == 0,\ 4 - \frac{17\ y^2}{16} == 0\right\}$

```
In[15]:= {sx=FindRoot[eqx,{x,10}],sy=FindRoot[eqy,{y,2}]}
Out[15]= {{x → 12.0539}, {y → 1.94029}}
```

```
In[16]:= a=(x/.sx);b=(y/.sy);
```

```
In[17]:= p1=Plot3D[z1,{x,-a,a},{y,-b,b},DisplayFunction->Identity];
         p2=Plot3D[z2,{x,-a,a},{y,-b,b},DisplayFunction->Identity];
```

```
In[18]:= Show[{p1,p2},DisplayFunction->$DisplayFunction]
```

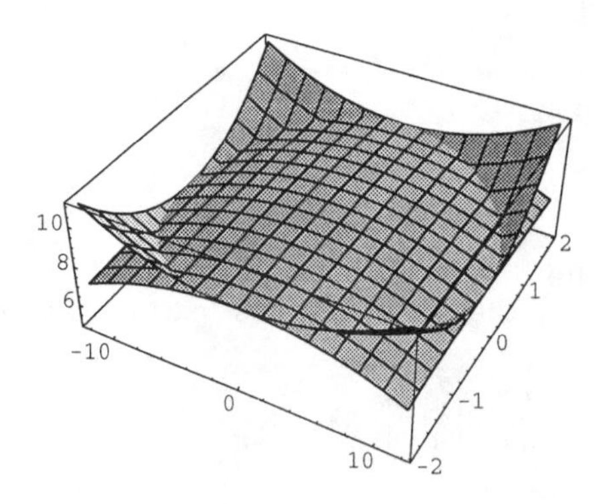

The above plot adequately shows the region, especially if suitable options are used to enhance the plot, but **we can do much better**! As you can see by looking at the numbers on each axis, it is far from true scale.

```
In[19]:= Show[{p1,p2},DisplayFunction->$DisplayFunction,
         BoxRatios->{20,4,6}];
```

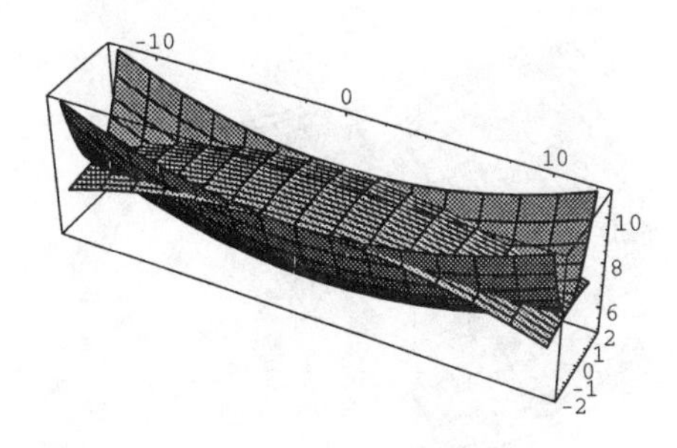

Example 9.8 *Plot the region bounded by the surfaces $y = 9 - x^2$, $99z - 10y = 900$, and the xy-plane.*

The xy-plane (the equation $z = 0$), could be viewed as one side of the plotting box. The second equation could be expressed in the form $z = f(x, y)$, and it could be plotted with the Plot3D[ ] command. The first equation, however, cannot be expressed in the required form $z = g(x, y)$.

We can still use the Plot3D[ ] command, but we must use a very different approach. Both equations can be expressed in the form $y = f(x, z)$. The first is constant with respect to z, but it still has that form. We use the Plot3D[ ] command to plot these two functions of x and z. This will force the y-axis to be vertical, but we can live with that, **or change it with a plotting option**.

```
In[1]:= y1=9-x^2;eq=99z-10y==900;
```

```
In[2]:= s=Solve[eq,y]
```

Out[2]= $\{\{y \to \frac{9}{10}\ (-100 + 11\ z)\}\}$

```
In[3]:= y2=(y/.s[[1]])
```

Out[3]= $\frac{9}{10}\ (-100 + 11\ z)$

```
In[4]:= p1=Plot3D[y1,{x,-10,10},{z,0,10},DisplayFunction->Identity];
        p2=Plot3D[y2,{x,-10,10},{z,0,10},DisplayFunction->Identity];
```

```
In[5]:= Show[{p1,p2},DisplayFunction->$DisplayFunction,
        AxesLabel->{"x","z","y"}]
```

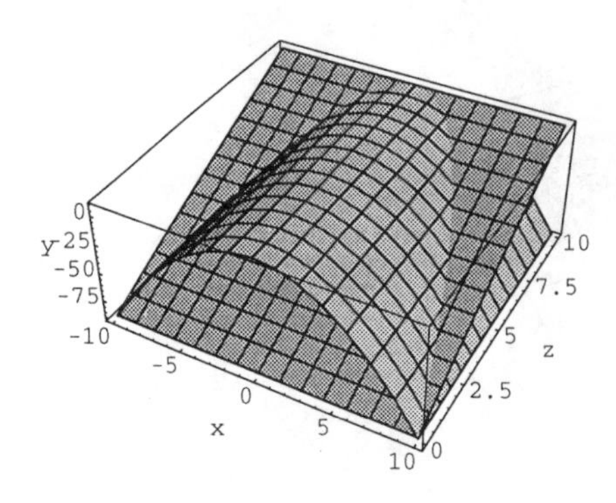

The y-axis is vertical—**notice the ordering, $\{x, z, y\}$ of the letters above**. There is no mathematical disadvantage to having a vertical y-axis, but we can change it with an option.

```
In[6]:= Show[{p1,p2},DisplayFunction->$DisplayFunction,
        AxesLabel->{"x","z","y"},ViewVertical->{0,1,0}];
```

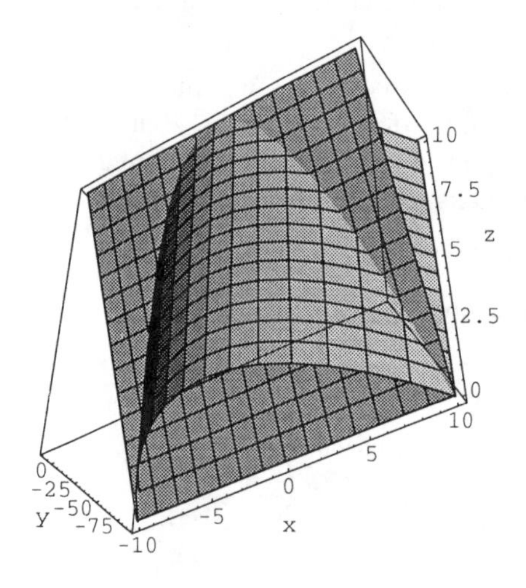

With a value of $\{0, 1, 0\}$ for *ViewVertical*, the second letter (wherever the 1 is located) becomes the vertical axis.

Example 9.9 *Plot, in the same coordinate system, the surface whose equation in cylindrical coordinates is*

$$r = 2\cos(2\theta), 0 \leq z \leq 4,$$

and the surface whose equation in rectangular coordinates is

$$z = \frac{4 + (4 - x^2)(4 - y^2)}{5}, -2 \leq x, y \leq 2.$$

One of the reasons for selecting this particular example is to show that very different plotting environments can be brought together with the Show[ ] command. This is also an opportunity to introduce **another major plotting command.**

The first equation is given in cylindrical coordinates, which are polar coordinates in x and y and rectangular coordinates in z. We used the ParametricPlot[ ] command before to plot polar coordinate curves. The **ParametricPlot3D[ ] command** can be used to plot a 3-dimensional surface of the form

$$x = f(s,t),\ \ y = g(s,t),\ \ z = h(s,t).$$

The variables s and t are the parameters. The plot input statement would be entered as follows:

ParametricPlot[{$f(s,t), g(s,t), h(s,t)$}, {s, a, b}, {t, c, d}];

We use the polar coordinate equations, $x = r\cos(\theta)$, $y = r\sin(\theta)$ and replace r by the given expression. This gives us something of the form $x = f(\theta, z)$, $y = g(\theta, z)$, and $z = h(\theta, z)$. Actually, $f(\theta, z)$ and $g(\theta, z)$ are independent of z, and $h(\theta, z) = z$ is independent of θ. See Section 8.3 for further insight.

```
In[7]:= z1=4+(4-x^2)*(4-y^2)/5;r=2Cos[2t];
```

```
In[8]:= p1=ParametricPlot3D[{r*Cos[t],r*Sin[t],z},{t,0,2Pi},{z,0,4},
        DisplayFunction->Identity];
        p2=Plot3D[z1,{x,-2,2},{y,-2,2},DisplayFunction->Identity];
```

```
In[9]:= Show[{p1,p2},DisplayFunction->$DisplayFunction]
```

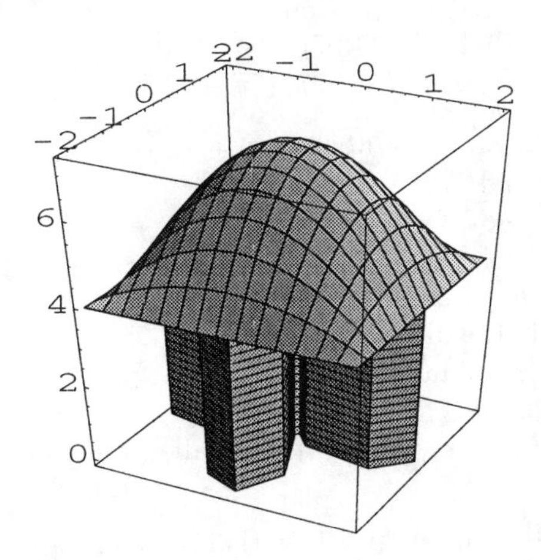

9.5 Exercise Set

Some New and Some Old Mathematica features and Commands

Animate[ ]	Array[ ]	Arrow[ ]
ContourPlot3D[ ]	Cross[ ] §	Dot[ ] §
Det[ ]	FindRoot[ ]	FullSimplify[ ]
Get[ ] or <<	Graphics[ ]	<<Graphics`Arrow`
Hue[ ]	If[ ]	ImplicitPlot[ ]
<<Graphics`ImplicitPlot`	MatrixForm[ ]	ParametricPlot3D[ ]
Part[ ] or [[ ]]	Plot3D[ ]	Show[ ]
Table[ ]		

Plot options

AspectRatio− > 1	AxesLabel− > {"x","z","y"}
BoxRatios− > { }	DisplayFunction− >Identity
DisplayFunction− >$DisplayFunction	Mesh− >False
PlotPoints− > n (a positive integer)†	PlotRange− > { }
ViewPoint− > { } ‡	ViewVertical− > { }

§ Use the infix forms for scalar, dot and cross products: $a\vec{v}$, $\vec{v}\cdot\vec{w}$, and $\vec{v}\times\vec{w}$ can be computed using input statements $a * v$, $v.w$, and $v \times w$ (enter \[Cross] for $\times$).

† the default value for PlotPoints is $n = 25$. Don't make it too big.

‡ Use the **3D ViewPoint Selector...** in the **Input Menu** to insert the ViewPoint option. See page 184 for details.

Use the identity $|\vec{v}|^2 = \vec{v}\cdot\vec{v}$ to compute the length of a vector $\vec{v}$.

1. Let $\vec{u}$, $\vec{v}$ $\vec{w}$ be the vectors defined by $\vec{u} = 7\vec{\imath} - 11\vec{\jmath} - 9\vec{k}$, $\vec{v} = -13\vec{\imath} + 5\vec{\jmath} - \vec{k}$, and $\vec{w} = 2\vec{\imath} + 3\vec{\jmath} + 6\vec{k}$. Use Mathematica's vector commands to compute the following. Some of them have obvious answers. Try to catch them in advance, and state the answer before you use Mathematica. Why are parentheses not used in parts (m) and (n)?

a) $\vec{u}+\vec{v}$	b) $\vec{w}-\vec{v}$	c) $3\vec{u}+7\vec{w}$
d) $7\vec{v}-12\vec{w}$	e) $\vec{u}+\vec{v}+\vec{w}$	f) $5\vec{u}+8\vec{v}-3\vec{w}$
g) $\vec{u}\cdot\vec{u}$	h) $\vec{u}\times\vec{v}$	i) $\vec{u}\times(\vec{v}\times\vec{w})$
j) $(\vec{u}\times\vec{v})\times\vec{w}$	k) $(\vec{u}\times\vec{v})\times\vec{v}$	l) $\vec{w}\times(13\vec{w})$
m) $\vec{u}\cdot\vec{v}\times\vec{w}$	n) $\vec{v}\cdot\vec{v}\times\vec{w}$	o) $(3\vec{u}+7\vec{w})\cdot(7\vec{v}-12\vec{w})$

2. Let $\vec{u}$ and $\vec{v}$ be the vectors $\vec{u} = 8\vec{\imath} - 3\vec{\jmath} - 6\vec{k}$, $\vec{v} = 2\vec{\imath} + 5\vec{\jmath} + 7\vec{k}$.
   a) Compute the lengths of $\vec{u}$ and $\vec{v}$.
   b) Compute the length of $5\vec{u} - 8\vec{v}$.
   c) Compute the length of $\vec{u}\times\vec{v}$.
   d) Find a unit vector in the direction of $2\vec{u} - 3\vec{v}$.
   e) Find a vector of length 7 in the direction of $\vec{u}\times\vec{v}$.
   f) Find the angle between $\vec{u}$ and $\vec{v}$ in degree measure.
   g) Find the projection of $\vec{v}$ onto $\vec{u}$.
   h) Express $\vec{v}$ in the form $\vec{v} = \vec{v_1} + \vec{v_2}$, where $\vec{v_1}$ is parallel to $\vec{u}$ and $\vec{v_2}$ is orthogonal to $\vec{u}$.

3. Create a projection command—name it "proj," so that if u and v are Mathematica names for the vectors $\vec{u}$ and $\vec{v}$, then proj[u, v] is the projection of $\vec{u}$ onto $\vec{v}$. You may want to save it in your own library, so that you can use it in the future. Use it to compute the projection of $\vec{a} = 6\vec{\imath} - \vec{\jmath} - 613\vec{k}$, onto $\vec{b} = 2\vec{\imath} + 11\vec{\jmath} - 5\vec{k}$ and the projection of $\vec{b}$ onto $\vec{a}$.

4. Use Mathematica to prove the identity

$$(\vec{u}-\vec{v})\cdot(\vec{u}+\vec{v}) = |\vec{u}|^2 - |\vec{v}|^2.$$

5. Use Mathematica to prove the following rule, known as the *Distributive Law of Dot Product Over Addition.*

$$\vec{u}\cdot(\vec{v}+\vec{w}) = \vec{u}\cdot\vec{v} + \vec{u}\cdot\vec{w}$$

6. Use Mathematica to prove that $\vec{u}$ and $\vec{v}$ are orthogonal to $\vec{u}\times\vec{v}$.

7. Use Mathematica to prove the identity

$$|\vec{u}\times\vec{v}|^2 = |\vec{u}|^2|\vec{v}|^2 - (\vec{u}\cdot\vec{v})^2.$$

Look at the terms involved, and appreciate how hard it would be to prove it by hand. This algebraic result is important, because it leads immediately to a proof of the fundamental geometric rule concerning the length of a cross product, namely,

$$|\vec{u}\times\vec{v}| = |\vec{u}||\vec{v}|\sin(\theta),$$

where θ $(0 \le \theta \le 2\pi)$ is the angle between $\vec{u}$ and $\vec{v}$.

8. For reasons of orthogonality that were discussed on page 178, it follows that $(\vec{u}\times\vec{v})\times\vec{w}$ is a linear combination of $\vec{u}$ and $\vec{v}$. Find scalars a, and b, such that

$$(\vec{u}\times\vec{v})\times\vec{w} = a\vec{u} + b\vec{v}.$$

Identify the scalars a and b that are found as **certain dot products**, and state a very interesting **identity** involving this **triple cross product**.

9 Use Mathematica to prove Rule

$$(\vec{u}\times\vec{v})\times\vec{w} = \vec{u}\times(\vec{v}\times\vec{w}) + (\vec{u}\times\vec{w})\times\vec{v}.$$

Use this rule to create a nontrivial example of vectors $\vec{u}$, $\vec{v}$, and $\vec{w}$ for which $(\vec{u}\times\vec{v})\times\vec{w} = \vec{u}\times(\vec{v}\times\vec{w})$. (By nontrivial, we mean that the final products should be nonzero.)

10. Use Mathematica to prove the following rule, known as the *Distributive Law of Cross Product Over Addition.*

$$\vec{u}\times(\vec{v}+\vec{w}) = \vec{u}\times\vec{v} + \vec{u}\times\vec{w}$$

11. Consider the parallelepiped formed by the three vectors $\vec{u} = 2\vec{i}-7\vec{j}+5\vec{k}$, $\vec{v} = 4\vec{i}+2\vec{j}+5\vec{k}$, $\vec{w} = 6\vec{i}+9\vec{j}-3\vec{k}$ based at a common vertex P. Determine the angle (in degree measure) between the main diagonal from P (to the opposite corner of the parallelepiped), and the diagonal from P, along each of the three faces adjacent to P.

12. Let $\vec{u}$, $\vec{v}$, and $\vec{w}$, be any three noncolinear vectors in three dimensional space, Let R be the parallelepiped formed by $\vec{u}$, $\vec{v}$, $\vec{w}$ based at a common vertex P. Let $\vec{f1}$, $\vec{f2}$, and $\vec{f3}$ be the diagonal vectors from P along the three faces of R adjacent to P, and finally, let Rf be the parallelepiped formed by $\vec{f1}$, $\vec{f2}$, and $\vec{f3}$. Use Mathematica to show that the volume of Rf is exactly twice the volume of R. (This can also be show by hand—in just a few lines—using the algebraic properties of the vector operations.)

13. Let $\vec{u}$, $\vec{v}$ $\vec{w}$ be the vectors defined by

$$\vec{u} = \frac{3}{\sqrt{14}}\vec{i} - \frac{1}{\sqrt{14}}\vec{j} + \frac{2}{\sqrt{14}}\vec{k},$$

$$\vec{v} = \frac{1}{\sqrt{3}}\vec{i} + \frac{1}{\sqrt{3}}\vec{j} - \frac{1}{\sqrt{3}}\vec{k},$$

$$\vec{w} = \frac{-1}{\sqrt{42}}\vec{i} + \frac{5}{\sqrt{42}}\vec{j} + \frac{4}{\sqrt{42}}\vec{k}.$$

Show that $\vec{u}$, $\vec{v}$ $\vec{w}$ are mutually orthogonal unit vectors. (Such a collection is called an orthonormal set). Then use Mathematica to show that any vector $\vec{p}$ in three dimensional space can be expressed in the form

$$\vec{p} = a\vec{u} + b\vec{v} + c\vec{w}$$

where a, b, and c are real numbers. In the process, find formulas for a, b, and c in terms of the components of $\vec{p}$.

The numbers a, b, and c can also be found by hand with very little computation, by an algebraic manipulation using the rules associated with the various vector operations. As a fitting conclusion to this problem, explain how a, b, and c can be found in this way. (The answer is obvious, once it is seen, but it can be puzzling at first glance. If you already have the formula generated by Mathematica, it might provide some insight.)

14. Find the equation of the plane through the points $A(5,-9,1)$, $B(16,-2,31)$, $C(-8,53,-17)$.

15. Find the equation of the plane which contains the point $A(7,15,-8)$ and the line defined parametrically by $x = 4+2t, y = -6+5t, z = 7-3t$.

16. Find the equation of the plane which contains the point $A(-4,9,6)$, is parallel to, and 2 units away from, the line L through the points $B(7,5,-3)$ and $C(-6,4,13)$.

17. Find the parametric equation of the line of intersection between the plane passing through $A(2,1,-5)$, $B(-4,11,8)$, $C(12,3,-2)$ and the plane passing through $P(14,23,18)$, $Q(-9,-1,-32)$, $R(-1,19,-6)$.

18. How would you define the angle between a plane and a nonparallel line? Define this term. Find the angle between the plane whose equation is $5x - 8y + 2z = 17$ and the line with equation $x = 7-t, y = 2+9t, z = -6+5t$.

19. Create a Mathematica command which computes the distance between two lines. Use an appropriate name. Create the command, so that if a,u, b, and v are Mathematica names (expressed as vectors) for the line through A with direction $\vec{u}$ and the line through B with direction $\vec{v}$, then LineDist(a,u,b,v) is the distance between the two lines. This is another command you may want to save in your own personal library. Use it to compute the distance between the line through $A(8,1,-4)$, $B(-2,5,3)$ and the line through $P(14,5,-6)$, $Q(-3.-17,2)$.

20. Captain Ralph has spotted a small but deadly object at position

$$P(232.627, 632.879, 3275.68)$$

traveling at great speed in a straight line with direction vector

$$\vec{v} = 1.2764\vec{i} - 0.26271\vec{j} + 1.7014\vec{k}$$

towards a spherical space station of radius 250 meters centered at

$$Q([258.257, 627.413, 3309.599]).$$

Will it hit the space station, and if not, how close will it come? The coordinate system is in kilometers.

21. A damaged space ship from United Earth Federation (UEF) is hiding from the dreaded Lizzard Warriors and attempting to make radio contact with its headquarters without being observed by the enemy. A UEF observer at position $P(73.1, -19.8, -42.6)$ has determined that a brief radio signal was broadcast from somewhere in the direction of the vector $\vec{u} = 3.2\vec{i} + 1.2\vec{j} - 5.1\vec{k}$ based at P. At the same time, another UEF observer at position $Q(-1.4, 81.9, -13.7)$ received a radio signal broadcast from somewhere in the direction of the vector $\vec{v} = 9.4\vec{i} - 3.0\vec{j} - 10.5\vec{k}$ based at Q. The coordinate system is expressed in kilometers. Determine the position of the damaged ship. There are observable errors in these measurements. Account for them.

Captain Ralph will launch a fast rescue mission only if the location data is accurate enough to justify the risk. The two direction lines used to locate the damaged ship should meet, but as long as they are less than 1 kilometer apart, the data will be considered reliable enough to begin the operation. Captain Ralph, always ready to spring into action, awaits your call. Should he rescue, or should he not?

22. Plot the following surfaces. **Try to roughly predict the basic shape before you ask Mathematica for a plot.** After all, how else would you know if Mathematica is behaving? Choose a range of values which is suitable.

How do you plot an equation like $x = f(y, z)$, for example? Actually, it is easy, if you do not mind that the x-axis is vertical—Mathematica does not mind. If you want the z-axis to be vertical (we usually do), then a plot option can be used to orient the plot as desired. In the following, plot the appropriate surfaces both ways.

Recall Mathematica's behavior concerning **fractional powers of negative numbers.** Some of the following plots will require the use of the If[ ] command to first define the real root.

a) $z = \sqrt[3]{x^2 + y^2}$ b) $z = \sqrt[3]{x^2 - y^2}$ c) $z = \sqrt[3]{(x - y)^2}$
d) $z = \sqrt[3]{x - y}$ e) $z = \cos(xy)$ f) $z = \ln(x^2 + y^2)$
g) $z = \frac{x^2+y^2}{x^2-y^2}$ h) $y = e^{x-z}$ i) $x = 4y^2 + 16z^2$

23. Plot the region bounded by the surfaces $z = 5 + \frac{100\cos(|x|+|y|)}{10+x^2+y^2}$ and $z = |x| + |y| - 10$

24. a) Plot the region bounded by the surfaces $f(x, y) = 200 - 9x^2 - 4y^2 - 18x - 8y$ and $g(x, y) = 4x^2 + 9y^2 + 8x - 18y - 200$.

b) Try to enhance this plot to show it in the best possible way. If possible, draw the region in true scale.

Project: Just for Fun

Find a three dimensional shape (and a corresponding mathematical formula) that you find to be visually interesting and appealing. Make a Mathematica plot and enhance it "to the max" (see plot[options] in Mathematica's Help File). If your computer has enough RAM use, for example, more points to generate a plot. Arrange for a color print of your "art." Frequently, color printers are on the campus network, in which case your file can be sent to a printer from your computer over the network. Send your print to the Metropolitan Museum of ..., no, on second thought, never mind. Hang it on your door and enjoy.

Project: Captain Ralph Shoots at a Mirror

A planar circular mirror with a radius of 10 meters is built, and great effort is taken to make the mirror absolutely flat. The mirror is then mounted to the ends of three, parallel, variable length rods which project from the outer wall of a space station. A heavy, fixed, mounting devise from the wall of the space station is attached to the center of the mirror's underside, which secures the center of the mirror in a fixed position, and the mirror is then allowed to pivot on its center, in a way which is determined by the lengths of the three adjustment rods. Since the center of the mirror is fixed, the lengths of two of the adjustable rods determine a unique setting of the mirror, and the length of the third rod is then adjusted to provide a secure position for the mirror.

The *mirror's target vector* is defined to be the unit vector, normal to the planar surface of the mirror, which, if based at the mirror's center, would point away from the space station.

The mirror must be mounted so that it's target vector is exactly the same as the unit vector with direction

$$\vec{n} = \sqrt{3}\vec{\imath} + \sqrt{7}\vec{\jmath} + 42\sqrt{2}\vec{k}$$

(relative to a certain three dimensional coordinate system discussed below), and the lengths of the rods are to be adjusted to product this direction. The adjustment must be made with great accuracy.

Since Captain Ralph happens to be in the right region, on his way back to the space station for lunch, he has been asked to help out. He is told to maneuver his space ship to a position which is roughly in line with the mirror's target vector and about 30 kilometers away. Here he is to shoot a beam of light from a laser gun in his ship at the mirror, adjust his position and shoot again, and continue doing this until the beam is reflected by the mirror and returned to exactly its original position at the gun. A record will then be made of the gun's coordinate position at this time, and this information will be used to adjust the mirror's position.

Being the best pilot in the fleet, Captain Ralph does the job in short order. While maneuvering his ship, he takes a test shot here and there, and soon sees the tell tale reflection of the laser beam appearing back in the gun's sight. As instructed, he marks his position (actually, the gun's position) as

$$P(873.57838, 1334.4355, 29957.581).$$

Here is some information concerning the rods and the coordinate system. All units are measured in *meters*. The coordinate system is arranged so that the fixed center of the mirror is at (0,0,0). The z-axis is parallel to the adjustment rods with a positive direction pointing away from the space station. The positive x and y axes are shown the the figure, along with cross sections of the three rods which are labeled a, b, and c. The adjustment rods are 9.5 *meters* away from the z-axis and are equally distributed around the z-axis as suggested in the picture.

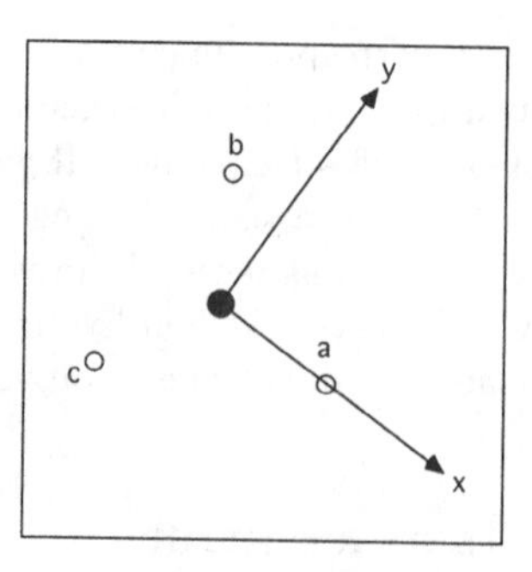

1. Exactly where the laser light beam strikes the mirror is unknown, and so the center of the mirror is used as the point where the light beam strikes. What should the adjustment be at each adjustment rod in order to produce a mirror position whose target vector has direction $\vec{n}$? How many centimeters should each rod be lengthened or shortened?

Remember that the coordinate system is in meters, and the adjustment specifications are to be made in centimeters. Good luck. This has to be accurate!

2. The laser light beam could, of course, strike any point on the mirror, not necessarily its center, and this could cause an error in positioning the mirror. Determine the size of the error in the following way. If the final setting of the mirror results in a target vector $\vec{n_0}$, then determine, in degree measure, the largest possible angle between $\vec{n_0}$ and the vector $\vec{n}$.

Chapter 10

Vector Valued Functions

Vector valued functions of a real variable t play an important roll in mathematics and its applications. Now that we have broken out of a one dimensional world, we can look forward to a variety of really interesting problems. If done by hand, calculations can quickly get out of hand, but, fortunately, we can look forward to a big help from Mathematica.

Most of Mathematica's commands are said to be "listable." We haven't talked much about this, but we have seen the consequences in every chapter. If a command Comd[ ] is listable and it operates on something like A to produce Comd[A], then it operates on a list $\{A_1,\ A_2, \ldots, A_n\}$ to form {Comd[A_1],ComdA_2], ... ,ComdA_n]}. In short: the result on a list is a list of the results. This is a powerful feature of Mathematica. In the last chapter, for example, the ordinary addition symbol (+) automatically turned into vector addition because addition, like most commands, is "listable." In this chapter, we will continue to enjoy the listable advantage of Mathematica as we study the consequences of differentiation and integraton on vectors.

10.1 Curves in Space

Parametrically defined curves $x = f(t),\ y = g(t)$ and $x = f(t),\ y = g(t),\ z = h(t)$ in the plane and in 3-dimensional space are also defined equivalently by the vector valued functions $\vec{r} = f(t)\vec{i} + g(t)\vec{j}$ and $\vec{r} = f(t)\vec{i} + g(t)\vec{j} + h(t)\vec{k}$. As vector valued expressions in Mathematica, a typical input statement would be entered as follows:

$r = \{f(t), g(t), h(t)\};$

Naturally, this can be enter as a function as well.

$r[t\_] := \{f(t), g(t), h(t)\};$

This may be a good time to refresh our memory about some of the **consequences of the delayed assignment operator (:=)**. Recall that if B is a previously defined Mathematica expression in the variable t, then the assignment

$r[t\_] := B;$

does not create the intended function. This matter was discussed on page 18.

Mathematica's **differentiation and integration commands are listable**, so the commands D[ ], Integrate[ ] for differentiating and integrating expressions automatically extend to vector valued expressions. We should expect the output to be in the form of vector valued expressions. The command Derivative[1][r] takes a function as input and returns the derivative of r as a function. This command has a very nice equivalent form, which allows us to mimic pencil and paper notation in an input cell. We demonstrate all this using an arbitrary vector valued expression.

```
In[1]:= r[t_]:={f[t],g[t],h[t]}
```

The symbol r above represents an arbitrary vector valued function. The symbol $r[t]$ represents an arbitrary vector valued expression (the value of r at t). We use the D[ ] command on an expression and the primed (′) notation on a function.

```
In[2]:= rp=D[r[t],t]
Out[2]= {f'[t], g'[t], h'[t]}
```

```
In[3]:= rp=r'
Out[3]= {f'[#1], g'[#1], h'[#1]}&
```

```
In[4]:= {rp[t],rp[s]}
Out[4]= {{f'[t], g'[t], h'[t]}, {f'[s], g'[s], h'[s]}}
```

The somewhat baffling and abstract output for r' can be avoided by ending the input cell with a semicolon. In addition, we don't have to pass through the name rp to use this derivative.

```
In[5]:= r'[t]
Out[5]= {f'[t], g'[t], h'[t]}
```

```
In[6]:= Integrate[r[t],t]
```

Out[6]= $\{\int f[t]dt, \int g[t]dt, \int h[t]dt\}$

Naturally, the integration command does not return the constant of integration. **It is of fundamental importance to realize that the constant of integration is an arbitrary vector valued constant**. Here is a potpourri of sample calculations.

```
In[7]:= r={t^2,t^3,t^4};R[t_]:={Cos[t],Log[t],E^t};
```

```
In[8]:= {rp=D[r,t],R'[t]}
```

Out[8]= $\{\{2\ t, 3\ t^2, 4\ t^3\}, \{-\sin[t], \frac{1}{t}, e^t\}\}$

```
In[9]:= {rp/.t->2,R'[Pi/2]}
```

Out[9]= $\{\{4, 12, 32\}, \{-1, \frac{2}{\pi}, e^{\pi/2}\}\}$

```
In[10]:= Integrate[r,t]
```

Out[10]= $\{\frac{t^3}{3}, \frac{t^4}{4}, \frac{t^5}{5}\}$

```
In[11]:= Integrate[R[t],t]
Out[11]= {sin[t], -t + t log[t], e^t}
```

If $\vec{r} = \vec{r}(t)$ defines a curve Γ, then the vector $\vec{r}\,'(t)$ (based at the point $\vec{r}(t)$) is tangent to the curve at that point. We can use the Arrow[ ] command to demonstrate this. The input statement Arrow[$\{a, b\}, \{c, d\}$] (a graphic primitive) draws a vector from (a, b) to (c, d). In our case, $\{a, b\} = \vec{r}(t)$, but what are the coordinates of (c, d)? Imagine the vector $\vec{r}(t)$ based at the origin. Place (geometrically) the base of $\vec{r}\,'(t)$ at the tip of $\vec{r}(t)$, and the tip of $\vec{r}\,'(t)$ is at the point (c, d). Now we see that $\{c, d, \} = \vec{r}(t) + \vec{r}\,'(t)$.

Example 10.1 *Plot the curve Γ defined parametrically by $x = 2\sin(t)$, $y = \cos(\pi t)(0 \leq t \leq 2\pi)$ Draw the position vectors and (forward) unit tangent vectors at the points on the curve corresponding to $t = 0.25$ and $t = 1.95$*

```
In[12]:= r[t_]:={2Sin[t],Cos[Pi*t]}
```

```
In[13]:= {base1=r[0.25],base2=r[1.95]}
Out[13]= {{0.494808, 0.707107},
          {1.85792, 0.987688}}
```

```
In[14]:= u1=r'[0.25]/Sqrt[r'[0.25].r'[0.25]]
Out[14]= {0.657363, -0.753574}
```

```
In[15]:= u2=r'[1.95]/Sqrt[r'[1.95].r'[1.95]]
Out[15]= {-0.83315, 0.553047}
```

```
In[16]:= {end1=base1+u1,end2=base2+u2}
Out[16]= {{1.15217, -0.046467}, {1.02477, 1.54073}}
```

```
In[17]:= <<Graphics`Arrow`
```

```
In[18]:= p1=Graphics[Arrow[{0,0},base1]];
         tan1=Graphics[Arrow[base1,end1]];
         p2=Graphics[Arrow[{0,0},base2]];
         tan2=Graphics[Arrow[base2,end2]];
         g=ParametricPlot[r[t],{t,0,2Pi},DisplayFunction->Identity];

ParametricPlot::ppcom :
  Function r[t] cannot be compiled; plotting
     will proceed with the uncompiled function.
```

```
In[19]:= Show[{g,p1,tan1,p2,tan2},DisplayFunction->$DisplayFunction,
         PlotRange->{{0,3},{-1.5,1.5}},AspectRatio->1];
```

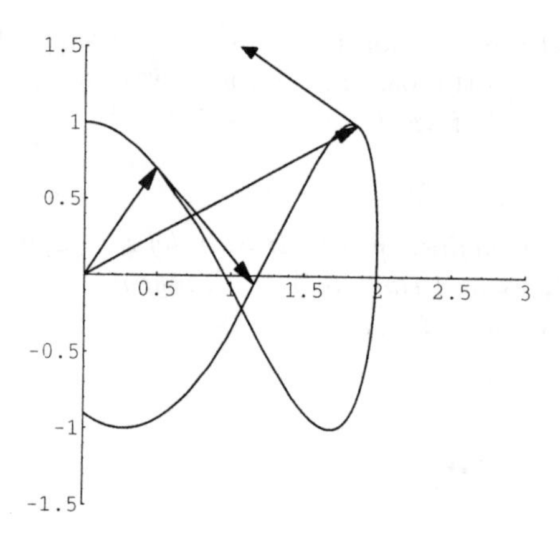

Example 10.2 *Plot the curve* Γ *defined by the vector valued position function*

$$r(\vec{t}) = 5\sin(2t)\vec{i} + (\sin(t) - t + 4\cos(3t))\,\vec{j} + (2t - 3\cos(t))\,\vec{k}, \;\; -2\pi \le t \le 2\pi.$$

Find the equation of the plane which passes through the point $r(\vec{3})$ *on* Γ *and which is parallel to both* $\vec{r}'(3)$ *and* $\vec{r}''(3)$

If $\vec{r}(t)$ describes the position of an object at time t, then, $\vec{r}'(t)$ and $\vec{r}''(t)$ turn out to be the velocity and acceleration vectors of the object at time t. We will study velocity and acceleration vectors soon enough, but for now, notice, just for the sake of interest, that the plane under consideration seems to describe, in a sense, the plane where all of the "action" is taking place near time $t = 3$ and near the point $r(\vec{3})$.

To plot the space curve Γ, we use a familiar command in a new way. The command **ParametricPlot3D[ ] can be used to plot both curves and surfaces in 3-dimensional space which are defined parametrically**. To plot a parametrically defined surface, two plot ranges, corresponding to two parametric variables are entered into the command. When only one parametric plot range is inserted in the command, Mathematica knows that the expression defines a curve rather than a surface.

```
In[20]:= r[t_]:={5Sin[2t],Sin[t]-t+4Cos[3t],2t-3Cos[t]}
```

```
In[21]:= ParametricPlot3D[r[t],{t,-2Pi,2Pi}]
```

```
ParametricPlot3D::"ppcom":
  Functionr[t] cannot be compiled; plotting
     will proceed with the uncompiled function.
```

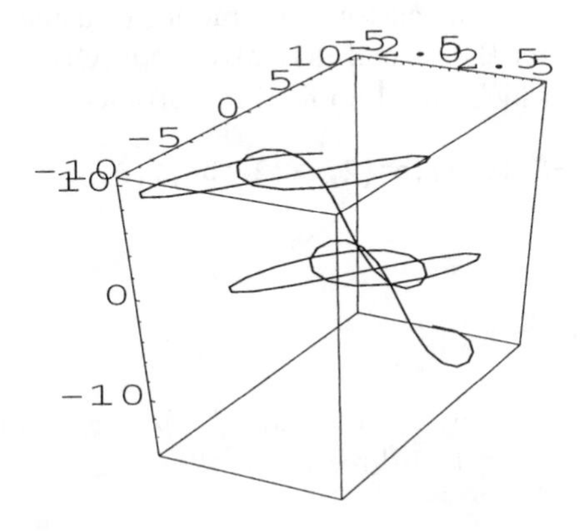

```
In[22]:= {u=N[r'[3]],v=N[r''[3]]}
Out[22]= {{9.6017, -6.93541,
           2.42336}, {5.58831,
           32.6596, -2.96998}}
```

```
In[23]:= n=u× v
Out[23]= {-58.5479, 42.0593,
          352.345}
```

```
In[24]:= X={x,y,z};p=N[r[3]]
Out[24]= {-1.39708, -6.5034,
          8.96998}
```

```
In[25]:= plane=Simplify[n.(X-p)==0]
Out[25]= -306.637 - 26.1478 x+
           8.41187 y + 36.211 z ==
          0
```

Example 10.3 *Find a vector valued function $\vec{r}(t)$, such that*

$$\vec{r}''(t) = \cos(3t)\vec{i} + (4+t^2)\vec{j} - e^{4t}\vec{k},$$

$$\vec{r}'(0) = 2\vec{i} - \vec{j} + 2\vec{k},\ \vec{r}(0) = \vec{i} + 6\vec{j} - \vec{k}.$$

Each time we antidifferentiate, we pick up another constant of integration. No big surprise there, but remember that in the current vector setting, the **constant of integration is a vector valued constant**. Recall that Mathematica leaves it up to us to attach the constant of integration. These constants, denoted by $c1$ and $c2$ below, are just unassigned names, but **they must be set up as vectors, or Mathematica would regard them as real numbers**. Evaluating these constants can be a tedious experience if Mathematica is not used effectively. It may be worth mentioning that this issue was discussed separately in Example 9.2.

In the next work session we follow our customary practice of using names like rp (p for prime) for the derivative and rpp for the second derivative of $\vec{r}$. Other unassociated names are used for intermediate objects which are of no final importance.

```
In[26]:= rpp={Cos[3t],4+t^2,-E^(4t)};a={2,-1,2};b={1,6,-1};
```

```
In[27]:= gp=Integrate[rpp,t]
```
Out[27]= $\left\{\frac{1}{3}\sin[3\,t],\ 4\,t+\frac{t^3}{3},\ -\frac{e^{4\,t}}{4}\right\}$

Without entering an input statement, use the "idea," $\vec{rp} = \vec{gp} + \vec{c1}$ for some constant vector $\vec{c1}$, and $\vec{rp} = \vec{a}$, when $t = 0$, to get $\vec{c1}$ directly as follows:

```
In[28]:= c1=a-(gp/.t->0)
```
Out[28]= $\left\{2,\ -1,\ \frac{9}{4}\right\}$

```
In[29]:= rp=gp+c1
```
Out[29]= $\left\{2+\frac{1}{3}\sin[3\,t],\ -1+4\,t+\frac{t^3}{3},\ \frac{9}{4}-\frac{e^{4\,t}}{4}\right\}$

```
In[30]:= g=Integrate[rp,t]
```
Out[30]= $\left\{\frac{1}{9}(18\,t-\cos[3\,t]),\ -t+2\,t^2+\frac{t^4}{12},\ -\frac{e^{4\,t}}{16}+\frac{9\,t}{4}\right\}$

Again, use the idea that $\vec{r} = \vec{g} + \vec{c2}$ for some constant vector $\vec{c2}$, and $\vec{r} = \vec{b}$, when $t = 0$, to get the following.

```
In[31]:= c2=b-(g/.t->0)
```
Out[31]= $\left\{\frac{10}{9},\ 6,\ -\frac{15}{16}\right\}$

```
In[32]:= f=g+c2
```
Out[32]= $\left\{\frac{10}{9}+\frac{1}{9}(18\,t-\cos[3\,t]),\ 6-t+2\,t^2+\frac{t^4}{12},\ -\frac{15}{16}-\frac{e^{4\,t}}{16}+\frac{9\,t}{4}\right\}$

```
In[33]:= {f/.t->0,D[f,t]/.t->0}
Out[33]= {{1, 6, -1}, {2, -1, 2}}
```

```
In[34]:= D[g,{t,2}]
```
Out[34]= $\left\{\cos[3\,t],\ 4+t^2,\ -e^{4\,t}\right\}$

10.2 Motion in Space

If $\vec{r}(t)$ is the position of an object at time t, then $\vec{v}(t) = \vec{r}\,'(t)$ is the object's velocity, and $\vec{a}(t) = \vec{v}\,'(t)$ is its acceleration at time t. The speed of the object at time t is $s(t) = |\vec{v}(t)|$. Mathematica can be an excellent tool to use in the solution of motion problems. We begin with a fairly straightforward example and follow that with a more interesting one.

Example 10.4 *A golf ball is hit with an initial speed of 140 feet per second and with an angle of inclination of 34°. The ball travels down a field which is level for the first 374 feet and which then rises at an angle of inclination of 22°. How far up the hill does the ball land, and what is its speed at impact?*

This can be realized as a two dimensional problem with a vertical y-axis. If $\vec{g}$ is the gravitational vector, and $\vec{v_0}$ is the initial velocity vector, then the velocity of the ball at time t is $\vec{v}(t) = t\vec{g} + \vec{v_0}$, and the position of the ball at time t is $\vec{r}(t) = t^2\vec{g}/2 + t\vec{v_0}$. At the moment of impact,

$$\vec{r}(t) = 374\vec{i} + s(\cos(22°)\vec{i} + \sin(22°)\vec{j},$$

where s is the distance up the hill, and t is the time at impact.

Recall that $Degree = Pi/180$ is a Mathematica defined constant. If θ is an angle in degree measure, then $\theta Degree$ (actually a multiplication) is the radian measure of the angle.

```
In[1]:= g={0,-32};v0={140Cos[34Degree],140Sin[34Degree]};

In[2]:= v=t*g+v0
Out[2]= {140 cos[34 °], -32 t + 140 sin[34 °]}

In[3]:= r=(t^2/2)*g+t*v0
Out[3]= {140 t cos[34 °], -16 t^2 + 140 t sin[34 °]}

In[4]:= p={374,0}+{s*Cos[20Degree],s*Sin[20Degree]}
Out[4]= {374 + s cos[20 °], s sin[20 °]}

In[5]:= st=FindRoot[r==p,{s,50},{t,10}]
Out[5]= {s → 127.311, t → 4.25306}

In[6]:= {HillDistance=s/.st,ImpactTime=t/.st}
Out[6]= {127.311, 4.25306}

In[7]:= ImpactVel=v/.t->ImpactTime
Out[7]= {140 cos[34 °], -57.811}

In[8]:= ImpactSpeed=Sqrt[ImpactVel.ImpactVel]
Out[8]= 129.666
```

Example 10.5 *An enemy space ship is spotted traveling along a wide sweeping arc. Captain Ralph hits the attack button on his console and his computer determines the enemy ship's current path, relative to a certain coordinate system expressed in kilometers. The curve appears on his computer screen defined parametrically as*

$$x = 3.1t + 85\cos(.01t) + 400, \; y = -12.8t, \; z = 13\cos(.03t) - 4.1t,$$

where t is time in seconds since he hit the attack button. The computer also determines the current path of his own fighter ship and displays the curve on his computer screen parametrically as

$$x = 340 - 2.1t, \; y = 197\sqrt{t+1} - .015t^2 - 2300, \; z = 0.032t^2 + 3.6t - 608.$$

"What a great computer," Ralph says to himself, as he prepares to launch a missile and attack the enemy ship.

First he sets the missile launch time to $t = 40$ seconds, and then he tries to determine the missile launch direction at $t = 40$. The missile will launch with a speed of 20.8 kilometers per second (relative to Ralph's ship). The missile has no power of its own, and so it will travel in a straight line with a constant velocity vector until it bumps (hopefully) into the enemy ship and explodes. What direction vector should Captain Ralph give the missile so that it will knock out the enemy ship? At what time will the missile reach the enemy ship, and how far will it be from Captain Ralph at that moment? Ralph, unfortunately, is a "shoot from the hips kind of guy," and not a very good student of calculus. He really needs our help.

We let $\vec{e}(t)$, $\vec{r}(t)$, and $\vec{m}(t)$, denote the vector valued position functions for the enemy ship, Ralph's ship, and the missile respectively. These letters will be prefixed with v to denote respective velocities. If Ralph points the missile in the direction of some unit vector $\vec{u}$, then the velocity vector of the missile will be

$$v\vec{m}(t) = 20.8\vec{u} + v\vec{r}(40).$$

The reasoning here is that the missile will acquire not only the launch velocity relative to Ralph, but the velocity of Ralph's ship at the firing time as well. We need to determine the vector $\vec{u}$ (three unknowns), and the time t (a fourth unknown) when $\vec{m}(t) = \vec{e}(t)$.

```
In[9]:= Clear["Global`*"]

In[10]:= e={3.1t+85Cos[.01t]+400,-12.8t,13Cos[.03t]-4.1t};

In[11]:= r={340-2.1t,197Sqrt[t+1]-.015t^2-2300,
           0.032t^2+3.6t-608};

In[12]:= u={u1,u2,u3};

In[13]:= vr=D[r,t]
```

Out[13]= $\left\{-2.1, -0.03\ t + \frac{197}{2\sqrt{1+t}}, 3.6 + 0.064\ t\right\}$

```
In[14]:= vm=20.8u+(vr/.t->40)
Out[14]= {-2.1 + 20.8 u1, 14.1831 + 20.8 u2, 6.16 + 20.8 u3}

In[15]:= m=(t-40)*vm+(r/.t->40)
Out[15]= {256. + (-40 + t) (-2.1 + 20.8 u1), -1062.58 + (-40 + t) (14.1831 + 20.8 u2),
          -412.8 + (-40 + t) (6.16 + 20.8 u3)}
```

We set the missile's position equal to the enemy ship's position, to get three equations in four unknowns (t and the three unknowns in the direction vector u). A fourth equation comes from setting the length of u (a unit vector) equal to one.

```
In[16]:= eq1=m[[1]]==e[[1]];eq2=m[[2]]==e[[2]];
         eq3=m[[3]]==e[[3]];eq4=u1^2+u2^2+u3^2==1;

In[17]:= S=FindRoot[{eq1,eq2,eq3,eq4},{t,60},{u1,0.5},{u2,0.5},{u3,0.5}]
```

```
Out[17]= {t → 61.75, u1 → 0.995586, u2 → -0.0802347, u3 → 0.0486983}
```

```
In[18]:= {t,u1,u2,u3}=({t,u1,u2,u3}/.S)
Out[18]= {61.75, 0.995586, -0.0802347, 0.0486983}
```

We check to see if e, the position of the enemy ship and m, the position of the missile, are the same at explosion time.

```
In[19]:= {m,e}
Out[19]= {{660.728, -790.4, -256.789}, {660.728, -790.4, -256.789}}
```

```
In[20]:= Aim={u1,u2,u3}
Out[20]= {0.995586, -0.0802347, 0.0486983}
```

```
In[21]:= ExplosionTime=t
Out[21]= 61.75
```

```
In[22]:= ExplosionDistance=Sqrt[(e-r).(e-r)]
Out[22]= 450.499
```

10.3 Geometry of Curves

The ideas of slope and concavity that we used to describe the geometry of the graph of a real valued function of a real variable are ideas that can be used to describe the geometry of a parametrically defined curve in two or three dimensional space in much the same way.

The geometry is described by the so-called "curvature" of a curve and by three vectors: the unit tangent vector $\vec{T}$, the principle unit normal vector $\vec{N}$ and the binormal vector $\vec{N}$. For the sake of brevity, they are introduced without explanation or interpretation. Visit our web site for a more thorough explanation of these ideas, or consult your main calculus text.

Let Γ be a smooth curve with arc length L, defined by the vector valued function

$$\vec{r} = \vec{r}(t), \ a \leq t \leq b. \tag{10.1}$$

The first vector to be introduced hardly needs an introduction. **The unit tangent vector** is defined to be

$$\vec{T} = \vec{T}(t) = \frac{1}{|\vec{r}\,'(t)|}\vec{r}\,'(t) = \frac{\vec{r}\,'(t)}{|\vec{r}\,'(t)|}. \tag{10.2}$$

By definition of a smooth curve, $\vec{r}\,'(t)$ exists and is never the zero vector, so the unit tangent vector is well defined.

The **principle unit normal vector** (or just unit normal vector) to the curve Γ defined by 10.1 is

$$\vec{N} = \vec{N}(t) = \frac{1}{|\vec{T}\,'(t)|}\vec{T}\,'(t) = \frac{\vec{T}\,'(t)}{|\vec{T}\,'(t)|}.$$

The last of the three vectors, the **binormal vector** is defined by

$$\vec{B} = \vec{T} \times \vec{N}.$$

For each t, the vectors $\vec{T}$, $\vec{B}$, and $\vec{N}$ are mutually orthogonal unit vectors (a so-called orthonormal set). Why? Actually, the answer is immediate—no computations are required to verify this result. **The collection $\{\vec{T}, \vec{B}, \vec{N}\}$, of vectors based at the point $\vec{r}(t)$ on the curve Γ is called the frame (or TNB frame) based at $\vec{r}(t)$.** This is a frame of mutually orthogonal unit vectors that moves along the curve with the point.

In order to define the curvature of the curve Γ, it is useful to think of Γ as being parametrized in terms of the arc length variable s $(0 \leq s \leq L)$. We can think of the position vector $\vec{r}$ as a function of t or as a function of s. A formula for $\vec{r}$ as a function of s exists as a concept, but it is usually impossible to find. It is, however, useful to think of this as a concept, even though we are forced to use the given formula $\vec{r} = \vec{r}(t)$ in order to perform calculations. With this in mind, we define the **curvature** of the curve Γ as

$$\kappa = \left|\frac{dT}{ds}\right| = \left|\frac{dT}{dt}\frac{dt}{ds}\right| = \frac{1}{|\vec{r}\,'(t)|}\left|\frac{dT}{dt}\right|.$$

The curvature is a measure of how quickly the unit tangent vector is turning. We do not want this to depend on the speed of a point moving along a curve, and so curvature is defined in terms of the rate of change of $\vec{T}$ with respect to s rather than t. Notice that $\frac{ds}{dt} = |\vec{r}\,'(t)|$ is just the speed of the point $\vec{r}(t)$ moving along the curve. The formula

$$\vec{N} = \frac{1}{\kappa}\frac{dT}{ds} = \frac{1}{\kappa}\frac{dT}{dt}\frac{dt}{ds} = \frac{1}{\kappa}\frac{dT}{dt}\frac{1}{|\vec{r}\,'(t)|}$$

is sometimes useful.

The radius of curvature is $\rho = 1/\kappa$. If Γ is the curve defined by 10.1, and if $\vec{T}$, $\vec{N}$, $\vec{B}$, κ, and ρ are all evaluated at the point $\vec{r}_0 = \vec{r}(t)$ on Γ, then the circle of curvature is the circle with radius ρ, centered at the point $\vec{r}_0 + \rho\vec{N}$, which lies in the plane through $\vec{r}_0$ with normal vector $\vec{B}$ (the plane formed by $\vec{T}$ and $\vec{N}$.

Incidentally, if formula 10.1 defines the motion of an object along a curve Γ, then the most useful way to present the acceleration vector, $\vec{a}$, acting on the object at time t is to express it in the form

$$\vec{a} = a_T\vec{T} + a_N\vec{N}.$$

The tangential component of acceleration, a_T, changes the speed of the object, and the normal component of acceleration, a_N, changes the direction of the object. Finally, the object's acceleration is related to the force acting on it by *Newton's Second Law* $\vec{F} = m\vec{a}$. These issues will be raised in the exercise set.

Example 10.6 *Let Γ be the curve defined by the vector valued position function*

$$\vec{r} = \sin(t)\vec{i} + (\cos(t) - \sin(t))\vec{j} + \cos(3t)\vec{k}, \ 0 \leq t \leq 2\pi.$$

Compute the vectors $\vec{T}$, $\vec{N}$, $\vec{B}$ at the point on Γ corresponding to $t = \pi/4$. Plot Γ along with the frame $\{\vec{T}, \vec{N}, \vec{B}\}$ based at $\vec{r}(\pi/4)$. Finally, determine the curvature of Γ at $t = \pi/4$.

Some attention has to be focused on the evaluation procedure. It is clear from the above formulas that we must compute $\vec{T}$ as an expression (or a function) in t, so that we can differentiate $\vec{T}$ to get $\vec{N}$. This is where computations frequently get out of hand. Once we differentiate $\vec{T}$, there is no longer a need to maintain t, as a variable, and so we immediately evaluate everything at $t = \pi/4$ to avoid complications.

As a reminder, we compute the length of a vector by taking advantage of the vector identity

$$|\vec{v}|^2 = \vec{v}\cdot\vec{v}$$

and using $\sqrt{\vec{v}\cdot\vec{v}}$ for the length of $\vec{v}$.

As a convenience, we can **create our own length command**. The word *length* is already used by Mathematica for something else, so a name like **size** might be a more appropriate length command. We may have use for it in the future as well. If you are building a personal library of special commands, this might be a good one to include.

The details of Mathematica's solution to this problem are bf available at our web site, but they are omitted from this manual to save space.

10.4 Exercise Set

Some New and Some Old Mathematica features and Commands

Animate[ ]	Array[ ]	Arrow[ ]
Cross[ ] §	D[ ]	Derivative[*n*][ ]
Dot[ ] §	FindRoot[ ]	FullSimplify[ ]
Get[ ] or <<	Graphics[ ]	<<Graphics`Arrow`
Hue[ ]	If[ ]	ImplicitPlot[ ]
Integrate[ ]	<<Graphics`ImplicitPlot`	MatrixForm[ ]
ParametricPlot3D[ ]	Part[ ] or [[ ]]	Plot3D[ ]
Show[ ]	Table[ ]	

Use D[f, x] to differentiate a vector valued expression f.

Use Derivative[f] or f' to differentiate a vector valued function f.

§ Use the infix forms for scalar, dot and cross products: $a\vec{v}$, $\vec{v}\cdot\vec{w}$, and $\vec{v}\times\vec{w}$ can be computed using input statements $a * v$, $v.w$, and $v \times w$ (enter \[Cross] for $\times$).

1. An object moves along a curve with position vector

$$\vec{r} = (3 - \sin(4t))\,\vec{i} + 2t\cos(t)\vec{j} + 5\sin(3t)\vec{k},\ \ 0 \le t \le 10.$$

Plot the curve. It may not look very good unless a larger number of points are used to form the plot (the default value is only 50 points). To plot more points, say 200, for example, use **numpoints=200** as a plot option. Use the mouse to move the plot around, so that it can be viewed from several directions. Find, in decimal form, at time $t = 7$, the velocity, acceleration, and speed of the object, the angle between the velocity and acceleration vectors, and the projection of the acceleration vector onto the velocity vector. Using just this decimal information at time $t = 7$, decide whether the object is speeding up or slowing down at this time.

2. Let γ be a circle in the xy-plane, of radius r centered at the point $A(0, R)$ where $0 < r < R$. A torus is formed by rotating γ about the x-axis. Two circular axes are naturally associated with the torus. One would be γ itself, and the other would be the circle β formed by rotating the point $A(0, R)$ about the x-axis.

If m and n are relatively prime integers (m, n have no common prime factors other than 1), then the following curves wrap m times around the torus in one "circular" way and n times around the torus in the other "circular" way. Set $r = 1$, $R = 4$, and experiment with this family of curves, by using a variety of n and m values. Which of the integers n or m controls the number of times the curves wrap in the direction of γ, and which controls the number of times the curves wrap in the direction of β. Use the plot option **numpoints=200** or some other suitably large number to produce curves which look nice.

$$\begin{aligned} x &= r\sin(2\pi mt) \\ y &= (R + r\cos(2\pi mt))\cos(2\pi nt) \\ z &= (R + r\cos(2\pi mt))\sin(2\pi nt) \end{aligned}$$

3. In a rectangular coordinate system with units in feet, a projectile is shot from a gun at position $P(2,-3,6)$ into a constant force field with acceleration vector $\vec{a} = 4.2\vec{i} - 1.8\vec{j} + 29.7\vec{k}$ in ${}^{ft}/_{sec^2}$. No other forces are acting on the projectile. The muzzle velocity of the gun (the initial speed of the projectile) is 2,000${}^{ft}/_{sec}$, and the gun is pointed in the direction of the vector $\vec{g} = 7.2\vec{i} + 12.4\vec{j} + 8.82\vec{k}$ Find the velocity and position of the particle at time t in seconds. How close does the projectile come to a target located at $Q(120, 213, 89)$? The muzzle velocity of the gun is a constant 2,000${}^{ft}/_{sec}$, but its direction can be altered. Find a direction (expressed as a unit vector) so that the projectile will hit the target.

4. A force (in pounds) of

$$\vec{F} = 8\sqrt{t}\,\vec{i} - 3t\vec{j} - \frac{7}{t^2+4}\vec{k}, \; 0 \le t \le 60$$

is exerted on an object of mass $m = 12 slugs$, where t is time in seconds. The position and velocity of the object at time $t = 0$ are

$$\vec{v}_0 = 2\vec{i} + 7\vec{j} + 13\vec{k}, \; \vec{r}_0 = 8\vec{i} - 33\vec{j} - 9\vec{k}.$$

Find the velocity and position of the object for time t in the interval $[0, 60]$. (Force is related to acceleration by *Newton's Law*, $\vec{F} = m\vec{a}$. When the force $\vec{F}$ is expressed in pounds, and mass m is expressed in slugs, then acceleration $\vec{a}$ is expressed in ${}^{ft}/_{sec^2}$.)

5. Let $\vec{f}(t)$ and $\vec{g}(t)$ be vector valued functions. State a *Product Rule* for the derivative of $\vec{f}(t) \cdot \vec{g}(t)$ and use Mathematica to verify the rule.

6. Let $\vec{f}(t)$ and $\vec{g}(t)$ be vector valued functions. State a *Product Rule* for the derivative of $\vec{f}(t) \times \vec{g}(t)$, and use Mathematica to verify the rule.

7. Let $\vec{f}(t)$, be vector valued function, and let $c(t)$ be a scalar (real valued) function. State a *Product Rule* for the derivative of $c(t)\vec{f}(t)$, and use Mathematica to verify the rule.

8. A projectile is shot from the top of a 100 ft tall building, at an angle of elevation of 27° (above horizontal) and at an initial speed of 243${}^{ft}/_{sec}$. How high does the projectile go? What is its speed and angle at impact? How much time does it take to land, and how far down range is it at impact? Assume a level field surrounding the building.

9. A baseball is hit 3 ft above ground, down the third base line, at an angle of elevation of 26°, and at an initial speed of 140${}^{ft}/_{sec}$. Will it clear a 30 ft. high fence which is 340 ft from home plate down the third base line?

10. Let Γ be the spiral

$$\vec{r} = \vec{r}(t) = a\cos(2\pi nt)\vec{i} + a\sin(2\pi nt)\vec{j} + 2\pi nbt\vec{k}, \; 0 \le t \le 1,$$

where a and b are positive constants and n is a positive integer. Show that Γ is a curve of constant curvature κ and determine the value of κ. Find the center of the circle of curvature corresponding to the point $\vec{r}(t)$ on Γ. Before asking Mathematica to perform these computations, try to guess their expected values.

11. Determine the frame vectors $\vec{T}$, $\vec{N}$, and $\vec{B}$ at the point $t = 2/3$ on the curve Γ defined by

$$x = \sin(4\pi t), \; y = (4 + \cos(4\pi t))\cos(10\pi t), \; z = (4 + \cos(4\pi t))\sin(10\pi t)$$

The curve Γ is one of the curves discussed in Problem 2. Determine the curvature, the radius of curvature, and the center of the circle of curvature at the point $t = 2/3$. If t is time, the parametric representation for Γ describes the motion of an object along the curve. Find the tangential, a_T, and normal, a_N components of acceleration at $t = 2/3$.

Project: Slamming Sam's Home Run Attempt

A baseball field is shown in the following figure. The angle at home plate (labeled H in the picture) is 90° and the three outfield fences are all 30 ft. tall. All of the other information concerning the field is shown in the picture. Slamming Sam hits the ball 2.8 ft above ground, at an angle of elevation of 25°, and at an initial speed of 152 $ft/_{sec}$. From a "bird's eye" view, far above the playing field, the ball travels in a direction exactly midway between second and third base, as shown in the picture. Does Slamming Sam have another home run? If not, describe, in very exact language, how close it is to being a home run.

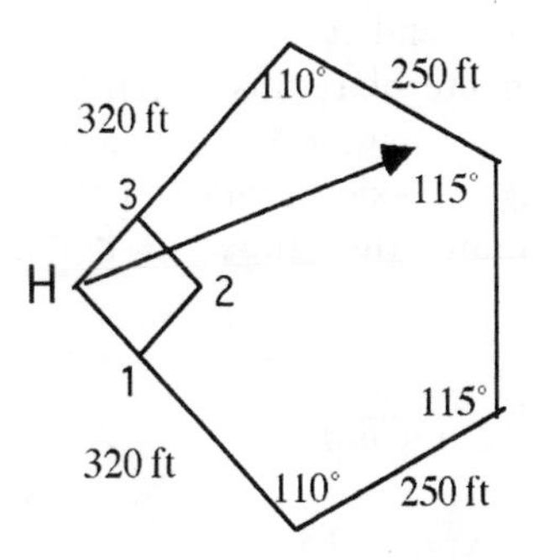

Project: Captain Ralph's Computer Failure (Version 1)

Captain Ralph is on a routine patrol a few thousand miles away from his home space station where the origin of a three dimensional coordinate system is located. (Coordinates are expressed in miles.) Looking for action, he grins as he spots an enemy ship. His display monitor shows that his coordinate position is $P(405, -3201, 95)$. He hits the attack button on his computer, and his ship begins to travel, starting at $t = 0$, along the attack curve

$$\vec{r}(t) = (405 - t^2)\vec{i} + (5t - 3201)\vec{j} + (95 + t^3)\vec{k}$$

(t is in seconds), firing off a projectile at the enemy ship $t = 2$ seconds later. The projectile travels along a straight line having the same direction as Captain Ralph's velocity vector at time $t = 2$, with a constant acceleration vector of 1 $mile/sec^2$ (naturally, in the same direction as Captain Ralph's velocity vector at time $t = 2$). The projectile has an initial speed which is 14 miles per second faster than Captain Ralph's speed at the firing time.

Unfortunately, Ralph's computer makes a mistake, and it programs the projectile to explode prematurely 4 seconds later. When the projectile explodes, it will annihilate everything within a 100 mile radius, and, in the mean time, Captain Ralph has no choice but to continue along the same attack curve for another 5 seconds, before he regains control of his ship. Will he make it? Will he be 100 miles away from the projectile when it explodes? Come on Captain Ralph! Weeeeeeeeeee ... neeeeeeeeeeed ... youuuuuuuuuu!

Project: Captain Ralph's Computer Failure (Version 2)

This version is the same as the first, except for the way that the bomb explodes. As every fighter pilot for the United Earth Federation knows, bombs do not always explode in spherical patterns. Acceleration warps and flattens the explosive region.

By a circular ellipsoid we mean the surface generated by revolving an ellipse about its major or minor axis. The axis of the circular ellipsoid is the line segment on the axis of rotation terminating at points on the surface. The center of the ellipse is the midpoint of the axis. The diameter of the ellipsoid is the diameter of the circle formed by intersecting the ellipsoid with a plane passing through the center which is perpendicular to the axis. It is a straightforward exercise to show that if d is the diameter, and l is the length of the axis of a circular ellipsoid, then the volume of the enclosed region is

$$V = \frac{\pi l d^2}{6}.$$

As you can see, when $l = d = 2r$ (r =radius), this turns into the familiar formula for the volume of a sphere of radius r.

When Captain Ralph fires a projectile and it travels at a constant velocity, its region of destruction is the region inside a sphere of radius 100 miles centered at the point of explosion. However acceleration warps the region of destruction into a circular ellipsoid centered at the explosion point, having an axis parallel to the acceleration vector. The volume of the region of destruction remains the same, but it flattens so that l decreases, and d increases, according to the rule

$$l = \frac{d}{1 + 2|\vec{a}|},$$

where $\vec{a}$ denotes the acceleration vector expressed in ${}^{miles}/_{sec^2}$.

Outside of this change in the explosive pattern, this version of the problem is the same as the first. Determine whether or not Captain Ralph survives this computer failure.

You have all the tools needed to solve this problem, but this is definitely a nontrivial project. A planar ellipse is defined in terms of the sum of the distances between a point on the ellipse and its two focal points. This definition can be used very effectively to decide whether a point P is inside or outside an ellipse. Now try to turn this problem into a planar problem.

Chapter 11

Multivariate Differential Calculus

11.1 Functions of Several Variables

In a pencil and paper mathematical environment, expressions and functions as often treated as if they were equivalent, even though there is, philosophically, a difference between the two ideas. We know from our experiences with functions of one variable, that Mathematica, not only appreciates the difference between a function and an expression, but it will not allow us to get careless and treat these ideas as being equivalent.

The same is true of functions and expressions in two or more variables. We will work with both of these Mathematica data types, but their differences must be appreciated. If f is a Mathematica function of x and y, then $f[x, y]$—a Mathematica expression—is the value of f, at x, y. If g is a Mathematica expression in x and y, then g itself is the value of (what we're thinking of as) the function at x, y, and $g[x, y]$ makes no sense to Mathematica.

Here is a sample of calculations. As a reminder of the significance of local variables and the delayed assignment operator (:=), we assign a value to one of the variables, just to observe the consequences.

```
In[1]:= f[x_,y_]:=x^2*Sin[y]
```

```
In[2]:= {f[3,2Pi/3],Log[f[y,x*y]]}
```

Out[2]= $\left\{\frac{9\sqrt{3}}{2}, \log\left[y^2 \sin[x\ y]\right]\right\}$

```
In[3]:= w=8;
```

```
In[4]:= g[u_,v_,w_,x_]:=u+v^2+w^3+x^4
```

```
In[5]:= {g[u,v,w,x],g[u,v,0,x]}
```

Out[5]= $\left\{512 + u + v^2 + x^4, u + v^2 + x^4\right\}$

```
In[6]:= {g[4,3,2,1],g[1,2,3,f[u,v+x]]}
```

Out[6]= $\{22, 32 + u^8 \sin[v + x]^4\}$

```
In[7]:= p[x_,y_]:=Which[x^2<y,25-x^2-y^2,y<0,y^2,True,0]
```

```
In[8]:= {p[3,10],p[5,-2],p[6,2]}
Out[8]= {-84, 4, 0}
```

The **new command Which[ ]** needs an explanation. We have used the similar If[ ] command several times before. The Which[ ] command is used when there are more conditions. These two commands are set up in basically the following way:

If[$test_1$, $value_1$, *value otherwise*]

Which[$test_1$, $value_1$, $test_2$, $value_2$, ..., $test_n$, $value_n$]

Testing proceeds from left to right, and $value_j$ is selected as soon as the first $test_j$ evaluates to *True*. The arguments $test_j$, $value_j$ can only be entered in pairs, and so this set up **does not allow us to enter *value otherwise* as a last argument, unless it has a test associated with it**. There is, however, a way around this problem. Since the condition *true* always evaluates to "true," we enter the last *test/value* pair in the form:

Which[$test_1$, $value_1$, $test_2$, $value_2$, ..., *True*, *value otherwise*]

Expressions in two variables were plotted in Chapter 9. Remember to pull down the **input Menu** and use the **3D ViewPoint Selector...** to insert the *Viewpoint* option inside the Plot3D[ ] command.

```
In[9]:= Plot3D[p[x,y],{x,-10,10},{y,-10,10},ViewPoint->{2.261, 2.262, 1.104}];
```

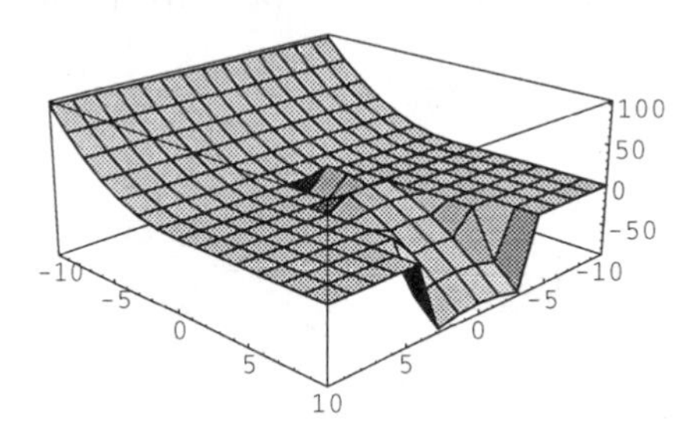

We cannot graph a function of three or more variables, at least not visually. The graph of a function of n variables can still be defined as a certain subset of $(n+1)$-dimensional space, and such a graph can be studied mathematically, but obviously, we can only see it with our eyes when $n = 1, 2$. Nevertheless, there are a few "tricks," most notably, level sets and sections, that can be used to bring our sense of vision to bear on some of these graphs. These techniques are even useful to use on the 3-dimensional graph of function of two variables.

The level curves, $f(x, y) = k$ ($k = k_1, k_2, \ldots, k_n$) for a function of x and y, can be plotted using the command **ContourPlot[ ]**. This allows us to view a graph living in 3-dimensional space, by looking at certain parts of the graph in 2-dimensional space. An individual level curve $f(x, y) = C$ can also be plotted using the familiar **ImplicitPlot[ ]** command, which resides in the Graphics`ImplicitPlot` package.

The graph of a function of the form $f(x,y,z)$ lives in 4-dimensional space, but the level surfaces of f are surfaces of the form $f(x,y,z) = k$ ($k = k_1, k2, \ldots, k_n$), which live in 3-dimensional space. We have already plotted surfaces of this form using Mathematica's **ContourPlot3D[ ]** command . It would be difficult to see more than one such level surface at a time, so this command should be used sparingly. Recall that the default value for the range, $k = k_1, k_2, \ldots, k_n$, of k values used in this command is just $k = 0$.

Another effective way of reducing dimensions, by looking at parts of a graph, is to plot **sections**. The graph of $z = f(x,y)$ lives in three dimensional space, but the intersection of this graph with a vertical plane is a curve which can be plotted as a two dimensional graph. Such plots are frequently called **sections of a graph**. The most convenient sections to plot are those of the form $z = f(x,C)$ or $z = f(C,y)$, where C is an arbitrary constant. These sections are the intersections of $z = f(x,y)$ with vertical planes parallel to the x or y axes.

Sections of $g(x,y,z)$ are particularly useful, since the graph of a function of three variables lives in four dimensional space. The three dimensional sections of the form $w = g(x,y,C)$, $w = g(x,C,z)$, $w = g(C,y,z)$ can be be viewed using the Plot3D[ ] command, and the two dimensional sections of the form $w = g(x,C_1,C_2)$, $w = g(C_1,y,C_2)$, $w = g(C_1,C_2,z)$ can be viewed using the Plot[ ] command.

An animation of the sections of $z = f(x,y)$ can be created using Mathematica's Animate[ ] command. For example,

$$\text{Animate[Plot}[f[x,c],\{x,x_{min},x_{max}\}],\{c = c_{min},c_{max}\}]$$

will draw the intersection of $z = f(x,y)$ with the plane $y = C$, for 24 values (the default number) of C over the interval $[c_{min}, c_{max}]$. **To see the animation, double click on the first picture.** The three dimensional sections of $w = g(x,y,z)$ can be viewed by entering an input statement similar to the following:

$$\text{Animate[Plot3D}[g[x,y,c],\{x,x_{min},x_{max}\},\{y,y_{min},y_{max}\}],\{c = c_{min},c_{max}\}]$$

These are memory intensive activities. When the animation command is used, it is **essential to maintain a fixed set of plotting ranges** in all direction. Otherwise the comparisons between frames may well be pointless. Appropriate plot options, such as *PlotRange*, *BoxRatios*, *AspectRatio*, among others should be used. Look at the output created by the following work session. The animation needs to be experienced first hand, of course, so its output is not included.

```
In[10]:= f[x_,y_]:=(25-x^2-y^2)*Sin[x^2+y^2]
```

```
In[11]:= ContourPlot[f[x,y],{x,-2,2},{y,-2,2}];
```

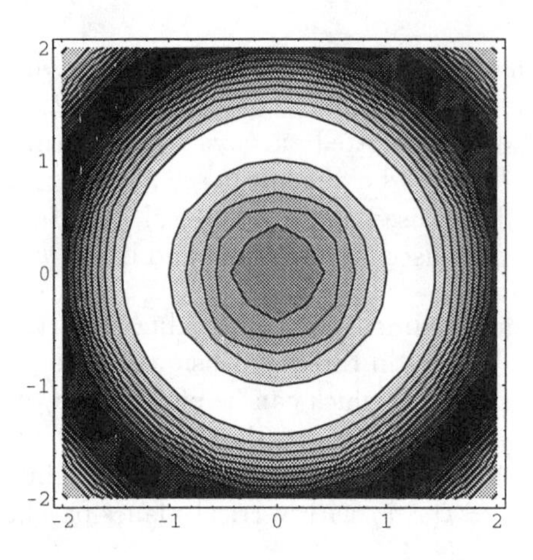


```
In[12]:= Animate[Plot[f[x,c],{x,-5,5},PlotRange->{-25,25}], {c,0,5}];
```

```
In[13]:= <<Graphics'ImplicitPlot'
```

```
In[14]:= ImplicitPlot[f[x,y]==1,{x,-5,5},{y,-5,5}]
```

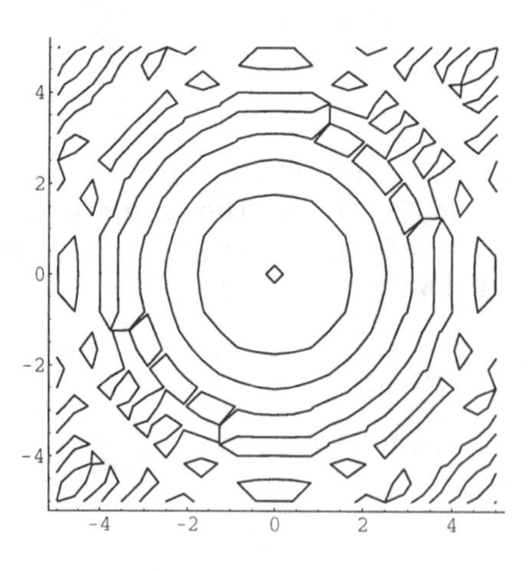

This plot is clearly wrong! The function f becomes increasingly chaotic as the point (x, y) moves further away from the origin. Mathematica seems to be able to handle this function reasonably well for x and y between -2 and 2, but not for x and y in the larger range. This **should have been expected**, if not for x, and y between -5 and 5, then at least in some slightly larger range. Why should this behavior by Mathematica be expected? **Think about this question. It is important to understand the limitations of computer software**.

```
In[15]:= g[x_,y_,z_]:=(10-z)*x^2+z*y^2;
```

```
In[16]:= Animate[Plot3D[g[x,y,c],{x,-5,5},{y,-5,5},
         PlotRange->{-200,400}],{c,-5,5}]
```

11.2 Limits with Two or More Variables

Limits of functions are frequently easy to compute by observation. We are all familiar with the circumstances which allow for the immediate evaluation of a limit of a function of one variable, and these same circumstances allow for the immediate evaluation of a limit of a function of several variables. It is only when a function has one of the so-called indeterminate forms at a point that its limit becomes more difficult to compute. In the case of a function of one variable, *L'Hopital's Rule* can often be used to compute a limit. **This rule, unfortunately, does not extend to functions of more than one variable.** Establishing the existence of a such a limit is fundamentally more difficult for a function of two or more variables than it is for a function of just one variable. To show that the limit of $f(x, y)$ as $(x, y) \to (a, b)$ exists, we must show that $f(x, y)$ approaches the same number L, regardless of how $(x, y) \to (a, b)$. In the one variable case, there were only two approach paths. In the case of two or more variables, there are infinitely many approach paths.

Mathematica has a familiar command for computing the limit of a function of one variable, but it has no command for computing the limit of a function of more than one variable. Don't blame Mathematica for this shortcoming. This is a difficult matter to deal with.

If a limit is not **indeterminate**, then we don't need a limit command to evaluate the limit. If it is indeterminate, and if the limit exists, it can be very difficult to prove that it exists. There are interesting special manipulations that work on special problems, but there is no useful strategy that works on a large class of problems.

On the other hand, if a limit is indeterminate, and the limit does not exist, it is usually possible to come up with unequivocal evidence to support this conclusion. This is demonstrated in the next work session, where we consider some limits which do not exist. Notice how exotic a function of just two variables can be, near such an indeterminate point.

```
In[17]:= f[x_,y_]:=(x-2)*(y-7)/(y-x-5)
```

Notice that the expression $f(x, y)$ has indeterminate form $(0/0)$ as $(x, y) \to (2, 7)$. There are no zero producing terms to factor out of the numerator and denominator, so how do we determine (without *L'Hopital's Rule*, whether the limit exists or not, and what its value is if it exists? To investigate this matter, we let $(x, y) \to (2, 7)$ along the line having slope m, where m is an arbitrary real number. The equation for this line is $y = m(x - 2) + 7$.

```
In[18]:= f1=f[x,m*(x-2)+7]
```

Out[18]= $\dfrac{m\ (-2 + x)^2}{2 + m\ (-2 + x) - x}$

```
In[19]:= f2=Simplify[f1]
```

Out[19]= $\dfrac{m\ (-2 + x)}{-1 + m}$

Because of the term $(x - 2)$ in the numerator, it follows that the limit of $f(x, y)$ is 0 along every linear path leading to (2,7), except along the line corresponding to $m = 1$, a line which is not in the domain of f (its denominator is 0 on this line). This would suggest that the limit exists and equals 0 (at least if we restrict (x, y) to the domain of the expression). It is important to realize, however, that the above work **does not prove that the limit exists**. In fact, it turns out, as we show next, that **the limit does not exist**.

In our next calculation, we let $(x, y) \to (2, 7)$ along a path which pushes (x, y) closer to the boundary line $y = x + 5$ where f is undefined. We let $(x, y) \to (2, 7)$ along a quadratic passing through (2,7), which has slope 1 at $x = 2$ (the same slope as $y = x + 5$). It is easy to create such quadratics. Just add a term of the form $a(x-2)^2$, where a is an arbitrary constant, to the line $y = x + 5$.

```
In[20]:= f3=f[x,x+5+a*(x-2)^2]
```

Out[20]= $\frac{-2 + a\ (-2 + x)^2 + x}{a\ (-2 + x)}$

```
In[21]:= f4=Simplify[f3]
```

Out[21]= $-2 + \frac{1}{a} + x$

Now we have a clear proof that the limit does not exist. By using different values for a, we see that the limit varies as we let $(x, y) \to (2, 7)$ along different quadratic paths.

Example 11.1 *Show that the limit does not exist at the origin for the function f defined by*

$$f(x, y) = \frac{sin(x^2 y)}{sin(x^3 + y^3)}$$

When a limit does not exist, it is usually a straightforward process to confirm this. All we have to do is show that $f(x, y)$ tends to different values as $(x, y) \to (0, 0)$ along two different paths. In this example, We approach the origin along the line $y = mx$, where m is any fixed real number.

```
In[22]:= f[x_,y_]:=Sin[x^2*y]/Sin[x^3+y^3]
```

```
In[23]:= f1=f[x,m*x]
```

Out[23]= $\csc\left[x^3 + m^3\ x^3\right]\ \sin\left[m\ x^3\right]$

```
In[24]:= Limit[f1,x->0]
```

Out[24]= Limit$\left[\csc\left[x^3 + m^3\ x^3\right]\ \sin\left[m\ x^3\right],\ x \to 0\right]$

We discussed the existence of an **enhanced limit command** back in Chapter 9, but we have not yet actually used it. Let's load the appropriate package and try again.

```
In[25]:= <<Calculus`Limit`
```

```
In[26]:= Limit[f1,x->0]
```

Out[26]= $\frac{m}{1 + m^3}$

This limit value is clearly different for different values of m, and so the full limit does not exist.

On the other hand, suppose we show that the limit of a function $f(x, y)$ as $(x, y) \to (a, b)$ is the same number L along a variety of paths. If the full limit exists, its value is obviously L, but have we shown that the full limit exists? **Clearly, the answer is no**! We saw dramatic evidence in a previous work session, that such a limit may fail to exist, regardless of how many paths were used in the analysis. Showing that the value of a limit is path independent can be very difficult for a function of more than one variable, if it has an indeterminate form at the point.

On that less than happy note, we move on to other matters. Actually, these limit complications are seldom encountered. It is enough to be aware of the circumstances which create this unresolved limit problem. In addition, limits frequently have directions attached to them, so that path independence is not even an issue.

11.3 Partial Derivatives

We already have a Mathematica command for computing a partial derivative of an expression. If f is an expression, the familiar Mathematica input statement D$[f, x]$, that we have been using, differentiates the expression f with respect to x and treats all unassigned letters in f as constants. In other words, it computes the partial derivative of f with respect to x. The output, recall, is a Mathematica expression (rather than a Mathematica function).

If f is a Mathematica function of one variable, then Derivative[1]$[f]$ (or f') is the derivative of f in the form of a Mathematica function. We have all experienced how convenient it is to have the derivative of a function returned as a function rather than as an expression.

The same command can be used to produce partial derivatives as functions. If f is a function in the variables $x_1, x_2, \ldots, x_n$, then Derivative$[j_1, j_2, \ldots, j_n][f]$ represents—in the form of a Mathematica function—the result of differentiating f, j_1 times with respect to x_1, j_2 times with respect to x_2, and so on.

Mathematica makes no mention of the order in which the differentiation is performed, and so it must assume that the outcome does not depend on order. There are, by the way, examples of functions of the form $f(x, y)$ where the mixed partials $f_{xy}(x, y)$ and $f_{yx}(x, y)$ are not the same, but if $f(x, y)$ has continuous second order partial derivatives, than the so-called mixed partial derivatives are always the same. This result, which is usually proved in a more advanced course, is at least mentioned in most calculus books.

This brings up an important general issue that has not been mentioned often. **Mathematica, generally, tends to assume that the objects it deals with are "nice" enough to justify what it wants to do**. This matter was discussed before, in connection with the use of the PowerExpand[ ] command. This matter over mixed partial derivatives is a good example of what can happen, when we work with objects that are not nice enough. Rather that create mathematical uncertainty, however, we hasten to add that most objects are "nice enough."

To begin our work, we take an elementary function and look at a sample of derivatives computed in these two way.

```
In[1]:= f[x_,y_,z_]:=x^3*y^4*z^5
```

```
In[2]:= {D[f[x,y,z],y],D[f[x,y,z],x,y,y],D[f[x,y,z],{z,4}]}
Out[2]= {4 x^3 y^3 z^5, 36 x^2 y^2 z^5, 120 x^3 y^4 z}
```

```
In[3]:= g1=Derivative[0,1,0][f]
Out[3]= 4 #1^3 #2^3 #3^5&
```

There is no need to see this abstract output, so we avoid looking at it in the next input statement.

```
In[4]:= g2=Derivative[1,2,0][f];g3=Derivative[0,0,0,4][f];
```

```
In[5]:= {g1[x,y,z],g2[x,y,z],g3[x,y,z]}
Out[5]= {4 x^3 y^3 z^5, 36 x^2 y^2 z^5, 120 x^3 y^4 z}
```

```
In[6]:= g3[3,-2,1]
Out[6]= 51840
```

```
In[7]:= Derivative[0,0,0][f][x,y,z]
Out[7]= x^3 y^4 z^5
```

Notice above that the 0-*th* derivative of f is just f itself. This is a definition that is usually made, for the sake of convenience, during the study of Taylor Series.

Example 11.2 *Compute the first and second order partial derivatives of the function f defined by*

$$f(x,y) = \frac{x^2 y}{4+y^2}.$$

Verify the first order partials by using the definition of a partial derivative as the limit of a difference quotient.

```
In[8]:= f[x_,y_]:=x^2*y/(4+y^2)
```

```
In[9]:= D[f[x,y],x]
```

Out[9]= $\frac{2\ x\ y}{4+y^2}$

```
In[10]:= D[f[x,y],y]
```

Out[10]= $-\frac{2\ x^2\ y^2}{(4+y^2)^2}+\frac{x^2}{4+y^2}$

```
In[11]:= D[f[x,y],y,x]
```

Out[11]= $-\frac{4\ x\ y^2}{(4+y^2)^2}+\frac{2\ x}{4+y^2}$

```
In[12]:= D[f[x,y],x,y]
```

Out[12]= $-\frac{4\ x\ y^2}{(4+y^2)^2}+\frac{2\ x}{4+y^2}$

```
In[13]:= D[f[x,y],x,x]
```

Out[13]= $\frac{2\ y}{4+y^2}$

```
In[14]:= D[f[x,y],y,y]
```

Out[14]= $\frac{8\ x^2\ y^3}{(4+y^2)^3}-\frac{6\ x^2\ y}{(4+y^2)^2}$

```
In[15]:= qx=(f[x+h,y]-f[x,y])/h
```

$$Out[15]= \frac{-\frac{x^2 y}{4+y^2} + \frac{(h+x)^2 y}{4+y^2}}{h}$$

The first order partial derivatives are verified, by using the definition of a partial derivative, as the limit of a certain difference quotient.

```
In[16]:= qx1=Simplify[qx]
```

$$Out[16]= \frac{(h + 2 x) y}{4 + y^2}$$

The limit of $qx1$ as $h \to 0$ is now obvious.

```
In[17]:= Limit[qx1,h->0]
```

$$Out[17]= \frac{2 x y}{4 + y^2}$$

```
In[18]:= qy=(f[x,y+k]-f[x,y])/k
```

$$Out[18]= \frac{-\frac{x^2 y}{4+y^2} + \frac{x^2 (k+y)}{4+(k+y)^2}}{k}$$

```
In[19]:= qy1=Simplify[qy]
```

$$Out[19]= \frac{x^2 \left(-\frac{y}{4+y^2} + \frac{k+y}{4+(k+y)^2}\right)}{k}$$

```
In[20]:= qy2=Together[qy1]
```

$$Out[20]= -\frac{x^2 (-4 + k y + y^2)}{(4 + y^2) (4 + k^2 + 2 k y + y^2)}$$

The limit of $qy2$ as $k \to 0$ is now obvious.

```
In[21]:= Limit[qy2,k->0]
```

$$Out[21]= -\frac{x^2 (-4 + y^2)}{(4 + y^2)^2}$$

Is this the same answer as we obtained by using the differentiation command? In order to check, we write our first answer in another form.

```
In[22]:= Together[D[f[x,y],y]]
```

$$Out[22]= \frac{4 x^2 - x^2 y^2}{(4 + y^2)^2}$$

The next example is offered just to demonstrate the convenience of using the command Derivative[ ] to return a partial derivative as a function. The required calculations are straightforward enough, but if the derivatives are computed individually, as expressions, the details could become somewhat tedious. We have made it a point to do this next example with a minimal amount of input activity.

Example 11.3 *Let f be the function defined by $f(x,y) = \cos(2x - 3y)$. Determine the polynomial function p defined by*

$$p(x,y) = f(0,0) + f_x(0,0)x + f_y(0,0)y + \frac{1}{2}f_{xy}(0,0)xy$$
$$+\frac{1}{2}f_{yx}(0,0)xy + \frac{1}{2}f_{xx}(0,0)x^2 + \frac{1}{2}f_{yy}(0,0)y^2.$$

Evaluate $f(x,y)$, $p(x,y)$ at $(x,y) = (0.2, 0.2)$.

As we pointed out earlier, Mathematica will assume that the two mixed second order partial derivatives $f_{xy}(x,y)$ and $f_{yx}(x,y)$ are the same, so they don't have to be computed separately. The example is stated in this way, simply to create "good form." Notice that all of the second order partials contain a multiplier of 1/2.

By using Mathematica's Derivative[ ] and Do[ ] commands, we can compute the polynomial directly in one step.

```
In[23]:= f[x_,y_]:=Cos[2x-3y]
```

```
In[24]:= Do[p[j,k]=Derivative[j,k][f][0,0],{j,0,2},{k,0,2}]
```

This short program has created all of the coefficients for our polynomial (except for the divisions by 2).

```
In[25]:= p=p[0,0]+p[0,1]*x+p[1,0]*y+p[1,1]*x*y+p[2,0]*x^2/2+p[0,2]*y^2/2
```

Out[25]= $1 - 2\ x^2 + 6\ x\ y - \frac{9\ y^2}{2}$

The function f and the polynomial p, both agree at $(x,y) = (0,0)$. All of the first and second order partials of f agree with those of p at $(x,y) = (0,0)$ as well. This is evident by observation, but it is easy to verify using Mathematica. Not surprisingly, with all this agreement, the polynomial $p(x,y)$ is a good approximation of $f(x,y)$ for values of (x,y) close to $(x,y) = (0,0)$. Look carefully, and you can see the similarity here with the notion of a Taylor polynomial of a function of one variable. We finish our example by comparing the values of $f(x,y)$ and its approximating polynomial p at a point close to $(0,0)$.

```
In[26]:= {f[0.2,0.2],p/.{x->0.2,y->0.2}}
Out[26]= {0.980067, 0.98}
```

Let *eqn* (or *expr* = 0) denote an equation in two or more variables. It is natural to think of solving the equation for one of the letters, w let's say, in terms of the other letters. If the solution exists, we say that the equation *eqn* defines w **implicitly** as a function of the other variables. If we actually find the solution, then we have w **explicitly** as a function of the other variables, but most of the time the equation *eqn* is **too difficult to actually find the solution**. Graphical considerations suggest, however, that solutions almost always exist, even though we may not be able to find them. We will look at some of this graphical evidence in a moment.

Implicit differentiation is a technique that can be used to compute a derivative (or partial derivative) of such a function without finding a formula for the function. This technique was used early in this manual, when we assumed that an equation *eqn* in x and y, defined y implicitly as a function of x, and we computed the derivative, $y'(x)$, using this method. **It may help to review Example 2.1**. The situation is much the same when *eqn* is an equation in more than two variables. The only real difference is that the derivatives we compute become partial derivatives.

Example 11.4 *Assume that the equation $x^2y + e^{xz}y^2 + 4xz - 2yz^2 = 8$ defines z implicitly as a differentiable function $z = f(x, y)$ of x and y in some neighborhood of the point $(x_0, y_0) = (5, -2)$. Use implicit differentiation and Mathematica's D[ ] command to compute the partials of z with respect to x and y at $(5, -2)$. The equation may define, implicitly, more than one such function $z = f(x, y)$ in a neighborhood of the point $(5, -2)$. In this case, compute the partials for each such function.*

Intuitively, each function $z = f(x, y)$ represents one of the solutions obtained by solving the equation for z in terms of x and y. There may be several such functions defined by the equation. The assumption that they exist and are differentiable is really quite reasonable. The graph of the given equation is a surface, which may overlap the (x, y)-plane several times. If we graph the equation over a small enough rectangle R in the (x, y)-plane centered at the point (5,-2), we will see the parts of the whole surface which lie over the small rectangle R. It is reasonable to expect that each one of these parts (if there is more than one) represents the graph of a differentiable function of the form $z = f(x, y)$. These are the graphs of the functions which are defined implicitly in a neighborhood of $(x_0, y_0) = (5, -2)$ by the given equation.

We begin our Mathematica work session by forming this graph. To do this, pick (by guess work) a small x-range and y-range centered about the point (5,-2). All that is left to do is pick a range of z-values for the plot. This requires a little thought, for we don't want to exclude parts of the surface that may represent the kinds of functions we are looking for. How many solutions are there, and where are they likely to occur?

It helps to look at the equation in z, which is obtained by letting $x = 5$ and $y = -2$. This is denoted by *eq0* in the Mathematica work session below. Clearly the term $4z^2$ dominates this equation for negative values of z, and the term e^{5z} quickly dominates the equation for positive values of z. The left hand side of *eq0* is clearly larger (much larger) than 8 for z outside of the interval $-10 < z < 1$, and so any solution must be in this interval. We add some wiggle room to this range of z values and plot the surface accordingly. Surely, with this range of z-values, we are not excluding any part of the surface that may generate another solution to this problem.

```
In[27]:= eq=x^2*y+Exp[x*z]*y^2+4x*z-2*y*z^2==8
```

Out[27]= $x^2\ y + e^{x\ z}\ y^2 + 4\ x\ z - 2\ y\ z^2 == 8$

```
In[28]:= eq0=eq/.{x->5,y->-2}
```

Out[28]= $-50 + 4\ e^{5\ z} + 20\ z + 4\ z^2 == 8$

```
In[29]:= <<Graphics`ContourPlot3D`
```

```
In[30]:= ContourPlot3D[eq[[1]]-8,{x,3,7},{y,-4,0},{z,-20,10},
         BoxRatios->{1,1,1},ViewPoint->{2.296,-2.338, 0.845},];
```

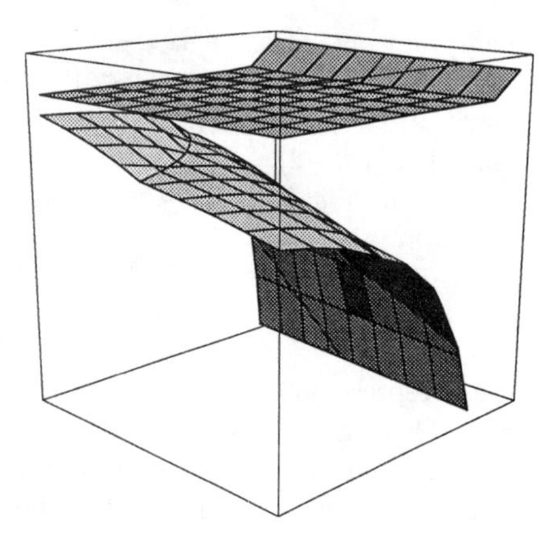

Remember to pull down the **Input Menu** and use the **3D ViewPoint Selector** to insert the *ViewPoint* option.

Notice that **two distinct surfaces** have appeared. Each one is the graph of a function $z = f(x, y)$ defined implicitly by the given equation.

Finding the two values of z corresponding to $x = 5$ and $y = -2$ is next on our agenda. We need to insert a start-up value for z (an initial guess), in order to solve $eq0$ for z using the FindRoot[ ] command. Can you look at the above graph and supply Mathematica with a good start-up z-value for either one of the two solution? Perhaps, but here is a more reliable method. Write the equation $eq0$ in the form $expr(z) = 0$. The left hand side of this new equation is an expression in z which can be graphed. The solutions we are looking for are the z-intercepts of this graph.

```
In[31]:= h=eq0[[1]]-eq0[[2]]
Out[31]= -58 + 4 e^(5 z) + 20 z + 4 z^2
```

```
In[32]:= Plot[h,{z,-10,1}];
```

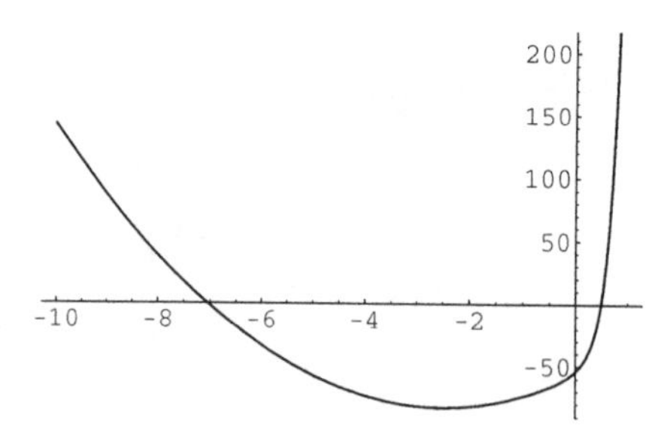

From this plot, we see that there is a solution between 0 and 1 and a solution between -8 and -6.

```
In[33]:= s0=FindRoot[h==0,{z,-7}]
Out[33]= {z → -7.05522}
```

```
In[34]:= s1=FindRoot[h==0,{z,1}]
Out[34]= {z → 0.493439}
```

```
In[35]:= z0=(z/.s0);z1=(z/.s1);
```

Before we differentiate, we must tell Mathematica that z depends on x and y. Unassigned letters are always independent of each other,. Mathematica will simply return 0 for D$[z, x]$, and D$[z, y]$ unless it is told that z depends on x and y.

```
In[36]:= eq1=eq/.z->z[x,y]
Out[36]= x^2 y + e^(x z[x,y]) y^2 + 4 x z[x, y] - 2 y z[x, y]^2 == 8
```

Notice that we used a substitution rather than an assignment in the last input statement. The assignment $z = z(x, y)$ may seem like a reasonable strategy , but **it will not work.** Assigning z to an expression involving z is a "circular" assignment and definitely prohibited as long as z is initially unassigned.

```
In[37]:= eq2=D[eq1,x]

Out[37]= 2 x y+4 z[x, y]+4 x z^(1,0)[x, y]-4 y z[x, y] z^(1,0)[x, y]+e^(x z[x,y]) y^2 (z[x, y]+
x z^(1,0)[x, y]) == 0


In[38]:= eq3=D[eq1,y]

Out[38]= x^2 + 2 e^(x z[x,y]) y - 2 z[x, y]^2 + 4 x z^(0,1)[x, y] + e^(x z[x,y]) x y^2 z^(0,1)[x, y] -
4 y z[x, y] z^(0,1)[x, y] == 0
```

Now that we have the partials, we simplify these equations with more manageable names. We use names like ZX for the derivative of z with respect to x.

```
In[39]:= eq4=eq2/.{D[z[x,y],x]->ZX,z[x,y]->z}
Out[39]= 2 x y + 4 z + 4 x ZX-4 y z ZX+e^(x z) y^2 (z + x ZX) == 0


In[40]:= eq5=eq3/.{D[z[x,y],y]->ZY,z[x,y]->z}
Out[40]= x^2 + 2 e^(x z) y - 2 z^2 + 4 x ZY+e^(x z) x y^2 ZY-4 y z ZY == 0


In[41]:= S4=Solve[eq4,ZX]
```

$$\text{Out[41]= } \left\{\left\{ ZX \to -\frac{2\ x\ y + 4\ z + e^{x\ z}\ y^2\ z}{4\ x + e^{x\ z}\ x\ y^2 - 4\ y\ z} \right\}\right\}$$

```
In[42]:= S5=Solve[eq5,ZY]
```

$$\text{Out[42]= } \left\{\left\{ ZY \to -\frac{x^2 + 2\ e^{x\ z}\ y - 2\ z^2}{4\ x + e^{x\ z}\ x\ y^2 - 4\ y\ z} \right\}\right\}$$

```
In[43]:= ZX=(ZX/.S4[[1]]);ZY=(ZY/.S5[[1]]);
```

All that is left to do is to evaluate these partial derivatives at the points $(5, -2, z_0)$, and $(5, -2, z_1)$. Hopefully, our names, like $PX0$, for the partial with respect to x using the z-coordinate of z_0, are transparent enough.

```
In[44]:= {z0,PX0=(ZX/.{x->5,y->-2,z->z0}),PY0=(ZY/.{x->5,y->-2,z->z0})}
Out[44]= {-7.05522, -1.32323, -2.04579}
```

```
In[45]:= {z1,PX1=(ZX/.{x->5,y->-2,z->z1}),PY1=(ZY/.{x->5,y->-2,z->z1})}
Out[45]= {0.493439, -0.020186, 0.0871825}
```

The *Chain Rule* for functions of several variables bears a striking resemblance to its one variable form, but it is somewhat more involved. While it can be written in matrix form as one rule, it is usually presented in calculus as a list of rules. Mathematica understands how to use the *Chain Rule* even in its most abstract setting.

Example 11.5 *Let $f(u,v,w)$ be a differentiable function of u, v, and w, and let $u = u(x,y)$,$v = v(x,y)$, and $w = w(x,y)$, be differentiable functions of x, and y, so that*

$$h(x,y) = f((u(x,y), v(x,y), w(x,y))$$

is a differentiable function of x and y. Compute the partials of $h(x,y)$ with respect to x and y.

```
In[46]:= Clear["Global`*"]
```

```
In[47]:= h=f[u[x,y],v[x,y],w[x,y]];
```

```
In[48]:= D[h,x]
Out[48]= w^(1,0)[x, y] f^(0,0,1)[u[x, y], v[x, y], w[x, y]]+

         v^(1,0)[x, y] f^(0,1,0)[u[x, y], v[x, y], w[x, y]]+

         u^(1,0)[x, y] f^(1,0,0)[u[x, y], v[x, y], w[x, y]]
```

```
In[49]:= D[h,y]
Out[49]= w^(0,1)[x, y] f^(0,0,1)[u[x, y], v[x, y], w[x, y]]+

         v^(0,1)[x, y] f^(0,1,0)[u[x, y], v[x, y], w[x, y]]+

         u^(0,1)[x, y] f^(1,0,0)[u[x, y], v[x, y], w[x, y]]
```

11.4 Directional Derivatives and the Gradient

The Mathematica command for computing the gradient, ∇f, of a real valued expression f of several variables is Grad[ ]. This and several other commands reside in the **Calculus`VectorAnalysis` package**, which must be loaded before the commands are used. Not surprisingly, we must tell Mathematica what variables to use in the differentiation process, but in this command there is an added twist. When we compute $\nabla f(x,y,z)$ by hand, we have a common understanding that x is the first variable, y is the second, and z is the third. This ordering of the variables is a critical issue in the definition of the gradient vector. **Since Mathematica has no order preference for the variables used in an expression, we must, not only specify the variables, but specify their ordering as well**

This is done in an interesting way. Not to change the story line, but we should begin by saying that there are many different kinds of coordinate systems used in mathematics. Most of our work is performed in the **Cartesian Coordinate System**, but you have probably been introduced to the **Cylindrical Coordinate System** and the **Spherical Coordinate System**. These and several other less common coordinate systems are defined in the **Calculus`VectorAnalysis` package**. The formulas for the gradient vector change drastically depending on the coordinate system used. **When we use the Grad[ ] command, we must enter the coordinate system used, as well as the letters used and their ordering**. If we plan to use the Cartesian Coordinate System and the variables x, y, z in that order, we use the Cartesian[ ] command and enter Cartesian[x, y, z].

An alternative strategy is to **set the coordinate system once and for all (in this Mathematica work session), by using the SetCoordinates[ ] command**. Once this is done, simple input statements like Grad[f] can be used without entering a second argument. As a final note, we add that the the gradient command in Cartesian Coordinates is so simple to create, that one could easily avoid Mathematica's version by creating one's own version. Here is a sample of calculations. Using an arbitrary function allows us to take a close look at what is being differentiated.

```
In[1]:= <<Calculus'VectorAnalysis'
```

```
In[2]:= Grad[f[x,y,z],Cartesian[x,y,z]]
Out[2]= {f^(1,0,0)[x, y, z], f^(0,1,0)[x, y, z], f^(0,0,1)[x, y, z]}
```

```
In[3]:= SetCoordinates[Cartesian[b,p,v]]
Out[3]= Cartesian[b, p, v]
```

```
In[4]:= Grad[f[b,p,v]]
Out[4]= {f^(1,0,0)[b, p, v], f^(0,1,0)[b, p, v], f^(0,0,1)[b, p, v]}
```

```
In[5]:= f=x^2*Cos[y*z^3];
```

```
In[6]:= Grad[f]
Out[6]= {0, 0, 0}
```

This unintended output for ∇f is the consequence of differentiating f with respect to b, p, and v in that order.

```
In[7]:= SetCoordinates[Cartesian[x,y,z]];
```

```
In[8]:= Grad[f]
Out[8]= {2 x Cos[y z^3], -x^2 z^3 Sin[y z^3], -3 x^2 y z^2 Sin[y z^3]}
```

Mathematica does not have a command for computing the directional derivative of a real valued function. It hardly seems necessary to create one, since it is determined so easily from the gradient vector. In three dimensions, the directional derivative of a real valued function f of x, y, and z at a point $P_0 = (x_0, y_0, z_0)$ in the direction of the unit vector $\vec{u}$ is

$$\mathcal{D}_{\vec{u}}(f)(P_0) = \nabla f(P_0) \cdot \vec{u}.$$

A similar formula holds for a function of two variables, or for that matter, for a function of any number of variables.

Example 11.6 *An engineer is laying out the direction for a road which must be built through a mountainous area. A flat map of the region is being studied, and a two dimensional grid* *has been placed over the map with units in miles, with the positive x-axis pointing east, and the positive y-axis pointing north. The area being studied extends roughly 5 miles in each direction from the center of the map. The elevation h, in feet above sea level, at the point (x, y) on the map is*

$$h = 200000\frac{\cos(x + 0.2 * y + 2)\cos(0.3 * y - 0.1 * x - 1)}{50 + x^2 + y^2} + 3200.$$

The grade of the road must less than 4% at each point. Starting at a scenic point located at $P_0(-2.1, -0.4)$, determine the range of directions which will produce an upward grade of between 2% and 4%. Express the answer as an angle from the positive x-axis (in degree measure).

The elevation function, h, is expressed in feet, and x, and y are in miles. Since these are appropriate units for elevation, we will not express h in terms of miles. At the appropriate point, however, an adjustment to miles will have to be made.

```
In[9]:= h[x_,y_]:=200000(Cos[x+0.2y+2]*Cos[0.3y-0.1x-1])/
        (50+x^2+y^2)+3200
```

```
In[10]:= Plot3D[h[x,y],{x,-5,5},{y,-5,5}];
```

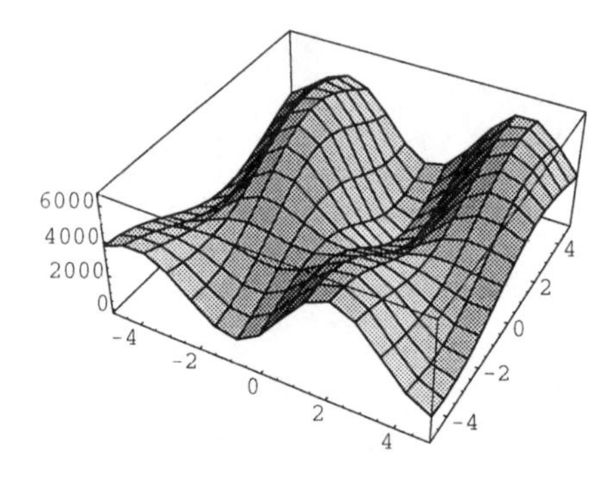

This plot is not a necessary part of the problem, but it is interesting to look at the mountains that the road is running through. Remember that **we have not yet turned elevation from feet to miles**, and even if we did, **it would have no effect on the plot, because Mathematica would adjust the picture for "best fit"**. **The mountains are not nearly as dramatic as they appear to look on the plot**, but, as you can see from the evaluations below, there are still some striking changes in elevation. In a moment, we will suggest a way to look more accurate at the hills, but before we do, the above plot, with its "magnifying glass" on elevation, is probably the best way to see the twists and turns. Turn the plot around to get a good look at the hills. Adjust your view point to get a "birds eye view" looking straight down on the land. From this point of view, you can see approximately where the high and low points are on the land. Better yet, create a contour plot. This will show the level curves (constant elevation curves), which will provide better insight into exactly where the hills and valleys are. The plot is shown below. Evaluate h at some of these points, and you will see some striking changes in elevation.

```
In[11]:= ContourPlot[h[x,y],{x,-5,5},{y,-5,5}];
```

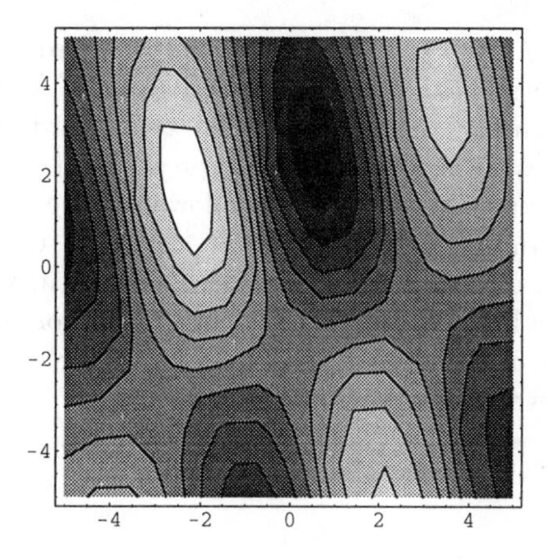


```
In[12]:= N[{h[-2.4,1.6],h[.4,2.2],h[-2.1,-.4]}]
Out[12]= {6485.26, -24.5418, 5413.05}
```

Now, let's get an **accurate picture of the land area involved**. We must **plot $h(x, y)/5280$**. In addition, we **must tell Mathematica to provide a true scale**, rather than a "best fit" picture. The options used to create this picture are shown below. Surprisingly, the plot looks somewhat flat for land with such a change in elevation. It may help to remember that the map covers 100 square miles of land.

```
In[13]:= Plot3D[h[x,y]/5280,{x,-5,5},{y,-5,5},BoxRatios->{10,10,1},
        ViewPoint->{2.051, -2.555, 0.845}];
```

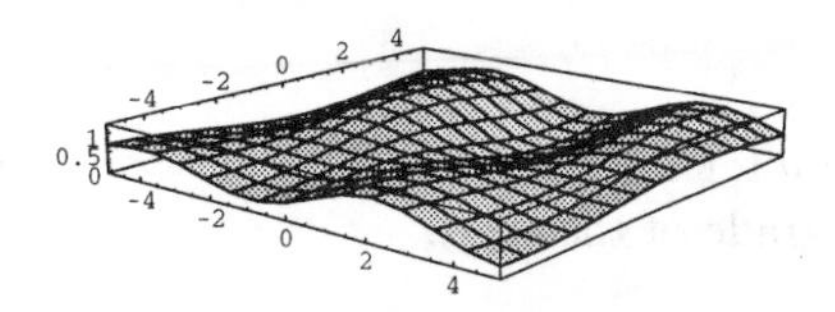

By looking directly down on a contour map, it becomes clear that much of the elevation change occurs near a line through (-4,0) and (4,4), or, in other words, along the line $x = -4 - 2y$. An interesting final look at this land, can be obtained by plotting the section $h(x, y)/5280$, with $x = -4 - 2y$. Mathematica must be told to maintain the same scale on the elevation axis. The plot is omitted, but its input statement should look like the following.

```
In[14]:= Plot[h[-4-2y,y]/5280,{y,-5,5},PlotRange->{-5,5}];
```

The grade of the road is its vertical rise over its horizontal run, or in other words, just the slope of a certain tangent line to the surface—the tangent line determined by the direction of the road. To specify the direction of the road, simply project this tangent line, with its forward direction, down onto the x, y-plane, and let $\vec{L}$ denote a direction vector for this line with its forward direction. Then the unit vector

$$\vec{u} = \frac{1}{|\vec{L}|}\vec{L} = \cos(\theta)\vec{\imath} + \sin(\theta)\vec{\jmath},$$

where θ is the angle from $\vec{i}$ to $\vec{L}$ (or $\vec{u}$), will be called the (forward) direction of the road. The grade of the road is, essentially, the directional derivative of h at P_0 in the direction of $\vec{u} = \cos(\theta)\vec{i} + \sin(\theta)\vec{j}$. Naturally, this number must be divided by 5280, which turns elevation, h, from feet into miles. It is convenient to express $\vec{u}$ in terms of θ.

A visual idea of what to expect for a range of allowable directions can be found by looking at our contour plot of the region. At the point $P_0(-2.1, -0.4)$, the tangent line to the contour through P_0 has the direction of a 0% grade. Consequently, a slight turn up the hill from the tangent line is the kind of direction we can expect from our calculations.

The command Grad[ ] can only be used for functions of three variables. This is no problem. We just create our own gradient command.

```
In[15]:= G[f_]:={D[f,x],D[f,y]}
```

```
In[16]:= g=G[h[x,y]]
```

Out[16]= $\left\{ -\frac{400000\ x\ \cos[1+0.1\ x-0.3\ y]\ \cos[2+x+0.2\ y]}{(50+x^2+y^2)^2} - \frac{20000.\ \cos[2+x+0.2\ y]\ \sin[1+0.1\ x-0.3\ y]}{50+x^2+y^2} - \frac{200000\ \cos[1+0.1\ x-0.3\ y]\ \sin[2+x+0.2\ y]}{50+x^2+y^2}, -\frac{400000\ y\ \cos[1+0.1\ x-0.3\ y]\ \cos[2+x+0.2\ y]}{(50+x^2+y^2)^2} + \frac{60000.\ \cos[2+x+0.2\ y]\ \sin[1+0.1\ x-0.3\ y]}{50+x^2+y^2} - \frac{40000.\ \cos[1+0.1\ x-0.3\ y]\ \sin[2+x+0.2\ y]}{50+x^2+y^2} \right\}$

```
In[17]:= g2=(g/.{x->-2.1,y->-.4})
Out[17]= {288.356, 967.024}
```

```
In[18]:= u={Cos[θ ],Sin[θ ]}
Out[18]= {cos[θ], sin[θ]}
```

The forward direction vector $\vec{u}$ for the road, which we described above, is entered in the next input cell, followed by the grade of the road.

```
In[19]:= grade=g2.u/5280
```

Out[19]= $\frac{288.356\ \cos[\theta] + 967.024\ \sin[\theta]}{5280}$

```
In[20]:= Plot[grade,{θ ,-Pi,Pi}];
```

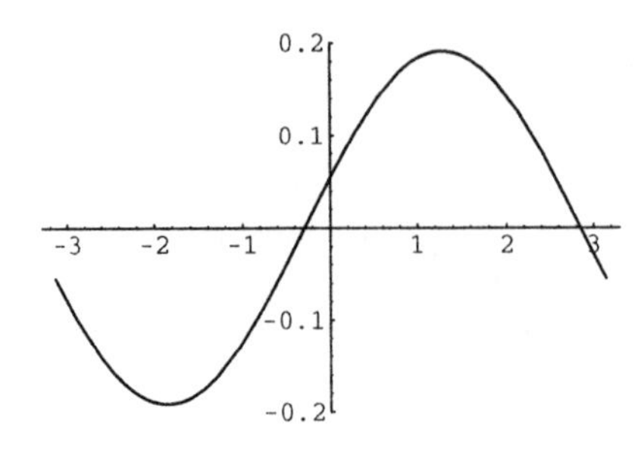

By looking at this plot, we can see approximately what angles θ will return a grade which is between 2% and 4% percent.

```
In[21]:= s1=FindRoot[grade==.02,{θ ,0}]
Out[21]= {θ → -0.184955}
```

```
In[22]:= s2=FindRoot[grade==.04,{θ ,0}]
Out[22]= {θ → -0.07894}
```

```
In[23]:= s3=FindRoot[grade==.04,{θ ,2.8}]
Out[23]= {θ → 2.64094}
```

```
In[24]:= s4=FindRoot[grade==.02,{θ ,2.8}]
Out[24]= {θ → 2.74696}
```

```
In[25]:= a=(θ /.s1);b=(θ /.s2);c=(θ /.s3);d=(θ /.s4);
```

To build a road with a grade of between 2% and 4%, we must choose a direction angle (in degrees) between the two entries in the following list or in the second list which follows.

```
In[26]:= DirectionAngle1={a*180/Pi,b*180/Pi}
Out[26]= {-10.5972, -4.52293}
```

```
In[27]:= DirectionAngle2={c*180/Pi,d*180/Pi}
Out[27]= {151.315, 157.389}
```

If f is a real valued function of several variables, then the gradient vector, ∇f evaluated at a point P_0, and based at the point P_0, points in the direction in which f increases at its greatest rate.

A vector valued function which associates a vector (of the same dimension) at each point in a vector space is called a **vector field**, and if the vector field is of the form ∇f, then the vector field is called a **gradient field**. There are many vector fields that are not gradient fields, but they are usually studied in a more advanced course in vector calculus. Gradient fields, on the other hand, are quite straight forward, and are an important component of our current study. Plotting a large sample of vectors in a vector field, ∇f, can supply a wealth of information about the underlying function f itself. **Mathematica's commands for plotting two and three dimensional gradient fields are PlotGradientField[ ] and PlotGradientField3D[ ]**, respectively. . These commands reside in the **Graphics`PlotField` and Graphics`PlotField3D` packages**.

Example 11.7 *plot the gradient field for* $f(x, y) = -4 + 6x + 4y - 2xy$

```
In[28]:= f=-4+6x+4y-2x*y
Out[28]= -4 + 6 x + 4 y - 2 x y
```

```
In[29]:= <<Graphics'PlotField'
```

```
In[30]:= PlotGradientField[f,{x,-10,10},{y,-10,10},Axes->True];
```

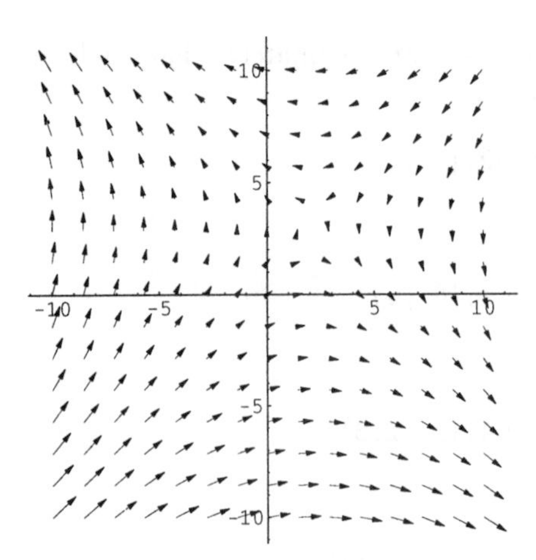

Gradient fields are used frequently in applications. Heat energy, for example, flows in the direction of the greatest rate of decrease in temperature. If $T(x, y, z)$ is the temperature at the point (x, y, z) in a region, then the gradient field $-\nabla T$ models the flow of heat energy in the region.

Since ∇f always points in the direction of the greatest rate of increase in the values of $f(x, y)$, it follows that **∇f is always orthogonal to the level sets of f**. Thus if S is a surface described by an equation of the form $f(x, y, z) = 0$, then the vector $\nabla f(x_0, y_0, z_0)$ based at (x_0, y_0, z_0) is orthogonal to the surface S. A similar statement can be made about a curve defined by an equation of the form $f(x, y) = 0$. The graph of a function $f(x, y)$ of two variables can be viewed as the graph of the equation $F(x, y, z) = z - f(x, y) = 0$. This level set of $F(x, y, z)$ has ∇F as a normal vector. Keep in mind that gradient vectors are not normal to graphs of functions, but they are instead normal to the level sets of the function.

Example 11.8 *Find the equation of the tangent plane to the graph of*

$$f(x, y) = e^{3x} \cos(2y)$$

at the point on the graph corresponding to $(x, y) = (0.7, 2.8)$. *Graph* $f(x, y)$ *along with its tangent plane.*

```
In[31]:= f=.; f[x_,y_]:=Exp[3x]*Cos[2y]
```

```
In[32]:= Q={.7,2.8,f[.7,2.8]}
Out[32]= {0.7, 2.8, 6.3334}
```

```
In[33]:= F=z-f[x,y]
Out[33]= z - e^(3 x) Cos[2 y]
```

This turns the graph of f into the level set $F = 0$. In this example, we use the SetCoordinates[ ] command, to make it easier to compute gradient vectors. We have to load a special package to use these commands.

```
In[34]:= <<Calculus`VectorAnalysis`
```

```
In[35]:= SetCoordinates[Cartesian[x,y,z]];
```

With a coordinate system established, we don't have to specify the variables, nor their order in the next command.

```
In[36]:= g=Grad[F]
```

Out[36]= $\{-3\ e^{3\ x}\ \cos[2\ y],\ 2\ e^{3\ x}\ \sin[2\ y],\ 1\}$

```
In[37]:= n=(g/.{x->Q[[1]],y->Q[[2]]})
Out[37]= {-19.0002, -10.3101, 1}
```

If P is an arbitrary point in the plane, then the vector from Q (a fixed point in the plane) to P is orthogonal to the normal vector n.

```
In[38]:= P={x,y,z}
Out[38]= {x, y, z}
```

```
In[39]:= eq=n.(P-Q)==0
Out[39]= -6.3334 - 19.0002 (-0.7 + x) - 10.3101 (-2.8 + y) + z == 0
```

```
In[40]:= tanplane=Simplify[eq]
Out[40]= 35.8349 - 19.0002 x - 10.3101 y + z == 0
```

```
In[41]:= s=Solve[tanplane,z]
Out[41]= {{z -> -35.8349 + 19.0002 x + 10.3101 y}}
```

```
In[42]:= h=z/.s[[1]]
Out[42]= -35.8349 + 19.0002 x + 10.3101 y
```

```
In[43]:= p1=Plot3D[f[x,y],{x,0,1},{y,0,4},DisplayFunction->Identity];
         p2=Plot3D[h,{x,0,1},{y,0,4},DisplayFunction->Identity];
```

```
In[44]:= Show[{p1,p2},DisplayFunction->$DisplayFunction,
         ViewPoint->{3.000, 0.881, 1.294},AxesLabel->{"x","y","z"}];
```

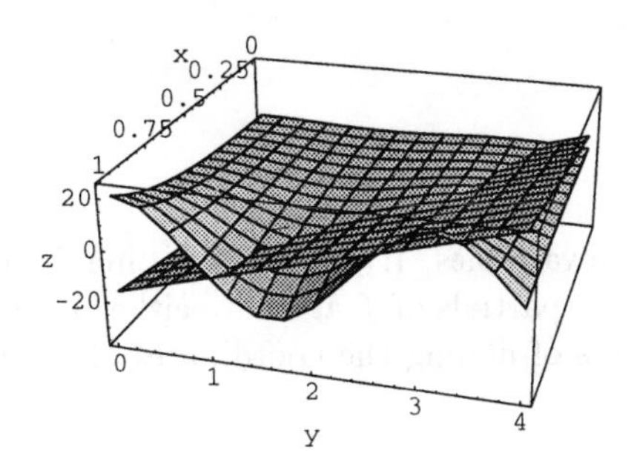

Example 11.9 *Find the equation of the plane tangent to the surface defined by the equation*

$$3x^2 - y^2 + 2z^2 = 8x + 5y - 7z + 15$$

at the point $P(-3, 10, 6)$ *on the surface.*

```
In[45]:= eq=3x^2-y^2+2z^2==8x+5y-7z+15
Out[45]= 3 x^2 - y^2 + 2 z^2 == 15 + 8 x + 5 y - 7 z
```

```
In[46]:= f=eq[[1]]-eq[[2]]
Out[46]= -15 - 8 x + 3 x^2 - 5 y - y^2 + 7 z + 2 z^2
```

```
In[47]:= Q={-3,10,6}
Out[47]= {-3, 10, 6}
```

```
In[48]:= f/.{x->Q[[1]],y->Q[[2]],z->Q[[3]]}
Out[48]= 0
```

This verifies that the point is on the surface.

```
In[49]:= g=Grad[f]
Out[49]= {-8 + 6 x, -5 - 2 y, 7 + 4 z}
```

```
In[50]:= n=(g/.{x->Q[[1]],y->Q[[2]],z->Q[[3]]})
Out[50]= {-26, -25, 31}
```

If (x, y, z) is an arbitrary point in the plane, then the vector $\vec{v}$ from $(-3, 10, 6)$ (a fixed point in the plane) to (x, y, z) is orthogonal to the normal vector $\vec{n}$.

```
In[51]:= P={x,y,z};
```

```
In[52]:= eq=n.(P-Q)==0
Out[52]= -26 (3 + x) - 25 (-10 + y) + 31 (-6 + z) == 0
```

```
In[53]:= plane=Simplify[eq]
Out[53]= -14 - 26 x - 25 y + 31 z == 0
```

11.5 Optimization

Let f be a function of two or more variables. If f has a local maximum or local minimum at a point P in its domain, then the partials of f at P are either zero or fail to exist. This provides us with an excellent means of finding the coordinates of the local maximums and minimums.

Example 11.10 *Find the critical points of* $f(x, y) = x^2 - 4xy^2 + 2x + 5y - 8y^3$, *and in each case, decide whether it is a local maximum point, a local minimum point or a saddle point.*

We simply compute the partial derivatives and set them equal to zero. This gives us a system of two nonlinear equations in the unknowns x and y. Mathematica is happy to solve such a system, but we have to insert start-up values into the FindRoot[ ] command, and then it will only give us one solution. Are there others? Where are they? By just looking at a system of equations, however, it is usually hard to tell how many solutions exist and hard to approximate their location. Plotting tools can frequently be used to supply these answers. In the present example, it is easier to simply solve one equation for x in terms of y and substitute that into the other equation. This process of elimination, as long as it can be done, is probably the most reliable way to gather all of the solutions.

If (a,b) is a critical point of $f(x,y)$, then we can decide whether (a,b) is a local maximum point, local minimum point, or saddle point by looking at the sign of $f_{xx}(a,b)$ and the sign of $M(a,b)$, where

$$M(x,y) = \frac{\partial^2 f}{\partial x^2}(a,b)\frac{\partial^2 f}{\partial y^2}(a,b) - \left(\frac{\partial^2 f}{\partial x \partial y}(a,b)\right)^2 .$$

A graph of f can also be used to make these decisions, but sometimes, graphical evidence can be quite subtle. The so-called*Second Derivative Test* might be more reliable.

```
In[54]:= f[x_,y_]:=x^2-4x*y^2+2x+5y-8y^3

In[55]:= {xeq=D[f[x,y],x]==0,yeq=D[f[x,y],y]==0}
Out[55]= {2 + 2 x - 4 y^2 == 0, 5 - 8 x y - 24 y^2 == 0}

In[56]:= s1=Solve[xeq,x]
Out[56]= {{x -> -1 + 2 y^2}}

In[57]:= a=x/.s1[[1]]
Out[57]= -1 + 2 y^2

In[58]:= eq=yeq/.s1[[1]]
Out[58]= 5 - 24 y^2 - 8 y (-1 + 2 y^2) == 0

In[59]:= s2=NSolve[eq,y]
Out[59]= {{y -> -1.6866}, {y -> -0.347143}, {y -> 0.533741}}

In[60]:= {pt1={a,y}/.s2[[1]],pt2={a,y}/.s2[[2]],pt3={a,y}/.s2[[3]]}
Out[60]= {{4.68923, -1.6866}, {-0.758984, -0.347143}, {-0.430242, 0.533741}}

In[61]:= fxx[x_,y_]=D[f[x,y],x,x]
Out[61]= 2

In[62]:= M[x_,y_]=D[f[x,y],x,x]*D[f[x,y],y,y]-D[f[x,y],x,y]^2
Out[62]= 2 (-8 x - 48 y) - 64 y^2
```

Notice that the **function $M(x,y)$ was defined without the delayed assignment operator (:=)**. We talked about problems associated with delayed assignment on page 18. If we tried to define a function, let's say g, by an delayed assignment such as $g[x\_]$:=

$D[x\hat{}2, x]$, then a value such as $g[3]$ would, because of delayed assignment, evaluate to $D[3\hat{}2, 3]$. Similar problems would be encountered, if we used (:=) in the assignment for $M(x, y)$.

It is **quite acceptable to define a function of x and y with an assignment, rather than a delayed assignment.** We merely have to recognize that the **x and y on the right hand side of the assignment will immediately take on values if they were previously assigned values.**

```
In[63]:= {M[pt1[[1]],pt1[[2]]],M[pt2[[1]],pt2[[2]]],M[pt3[[1]],pt3[[2]]]}
Out[63]= {-95.1694, 37.7569, -62.5875}
```

```
In[64]:= saddle1={pt1[[1]],pt1[[2]],f[pt1[[1]],pt1[[2]]]}
Out[64]= {4.68923, -1.6866, 7.95992}
```

```
In[65]:= relmin1={pt2[[1]],pt2[[2]],f[pt2[[1]],pt2[[2]]]}
Out[65]= {-0.758984, -0.347143, -1.9771}
```

```
In[66]:= saddle2={pt3[[1]],pt3[[2]],f[pt3[[1]],pt3[[2]]]}
Out[66]= {-0.430242, 0.533741, 1.26718}
```

A Plot3D[ ] image of $f(x, y)$ is not very revealing. The behavior of $f(x, y)$ near these special points is quite subtle. More striking graphical evidence verifying these results can be achieved by graphing the level curves of $f(x, y)$. We chose a range which just barely includes all three critical points.

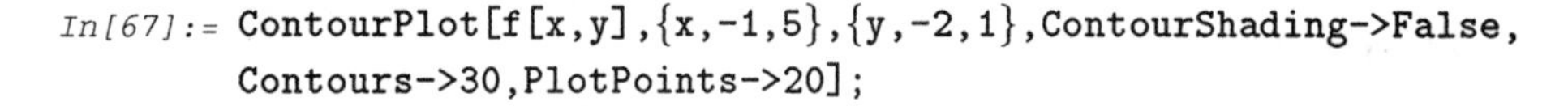

```
In[67]:= ContourPlot[f[x,y],{x,-1,5},{y,-2,1},ContourShading->False,
        Contours->30,PlotPoints->20];
```

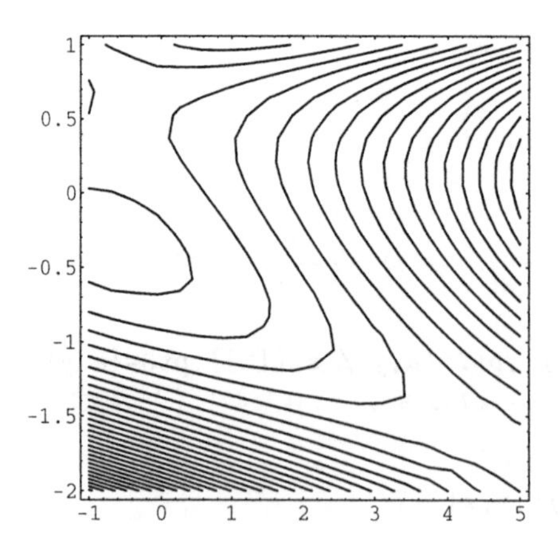

Notice the shape of the level curves near the local minimum and near the saddle points. **Notice how the level curves "split" near the saddle points.**

In closing, notice that the local minimum point is not an absolute minimum, because $f(0, y)$ clearly takes on large negative values when y is large and positive.

Example 11.11 *Find the critical points of*

$$f(x,y) = \frac{\ln(4 + x^2 + 3y^2)}{4 + 3x^2 + y^2},$$

and in each case determine whether it is local maximum point, a local minimum point, or a saddle point. Find the absolute maximum and absolute minimum if they exist.

The function f is clearly positive valued and goes to 0 as x and/or y go to ∞. For this reason, the absolute minimum does not exist. By the same token, the absolute maximum must exist. If D is a large closed disk, then f, being continuous on a closed and bounded set, must have an absolute maximum on D, and since f is small on the bounding circle, the absolute maximum must be at an interior point of D, a local maximum. This maximum is clearly the absolute maximum over the entire x, y-plane.

Following the standard practice, we set the partials equal to zero, and try to solve the resulting system of equations.

```
In[68]:= f[x_,y_]:=Log[4+x^2+3y^2]/(4+3x^2+y^2)
```

```
In[69]:= fx=D[f[x,y],x]
```

$$\text{Out[69]=} \quad \frac{2\,x}{(4 + 3\,x^2 + y^2)\,(4 + x^2 + 3\,y^2)} - \frac{6\,x\,\log[4 + x^2 + 3\,y^2]}{(4 + 3\,x^2 + y^2)^2}$$

```
In[70]:= fy=D[f[x,y],y]
```

$$\text{Out[70]=} \quad \frac{6\,y}{(4 + 3\,x^2 + y^2)\,(4 + x^2 + 3\,y^2)} - \frac{2\,y\,\log[4 + x^2 + 3\,y^2]}{(4 + 3\,x^2 + y^2)^2}$$

As we shall see, the system of two equations obtained by setting $fx = 0$ and $fy = 0$ is not nearly as complicated as it looks, but Mathematica, nevertheless, needs our help. Notice, to begin with, that there is an obvious solution: $x = 0$, $y = 0$. By plotting the level curves of $f(x, y)$, we can estimate where the critical points are.

```
In[71]:= ContourPlot[f[x,y],{x,-2,2},{y,-2,2},
         ContourShading->False,PlotPoints->20];
```

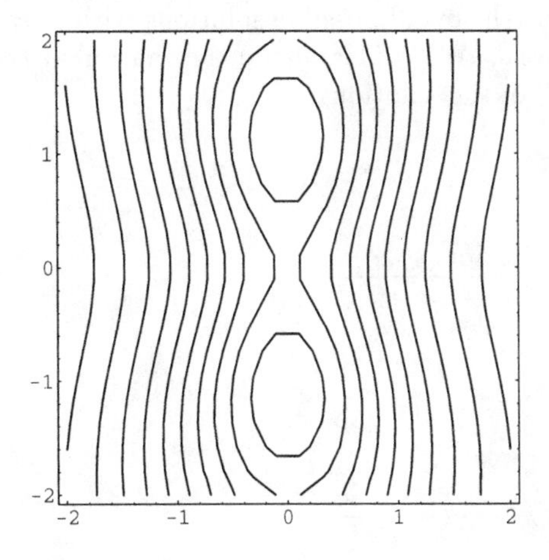

```
In[72]:= s1=FindRoot[{fx==0,fy==0},{x,-0.5,0.5},{y,0.5,1.5}]
Out[72]= {x → 2.33375 10^-7, y → 1.10927}
```

```
In[73]:= a=0;b=y/.s1[[2]];
```

The three points (0,0), (a, b), and $(a, -b)$ are critical points. To decide whether they correspond to local maximums, local minimums, or saddle points, we form next, the function M used in the *Second Derivative Test.* Notice that it is introduced as a function. Just as in the last example, $M(x, y)$ was defined **without the delayed assignment operator**. See the previous example for an explanation.

```
In[74]:= M[x_,y_]=D[f[x,y],x,x]*D[f[x,y],y,y]-D[f[x,y],x,y]^2;
```

The output is so complicated, that it was omitted.

```
In[75]:= {N[M[0,0]],M[a,b]}
Out[75]= {-0.0796486, 0.0569361}
```

As the contour plot suggested, the point (0,0) is a saddle point.

```
In[76]:= Derivative[2,0][f][a,b]
Out[76]= -0.397711
```

It follows from the values of $M(a, b)$ and $f_{xx}(a, b)$ that the point (a, b), and hence the point $(a, -b)$, are local maximum points. It looks like these two points also correspond to absolute maximums as well, but have we done enough to draw such a conclusion?

The complexity of the equations $fx = 0$, $fy = 0$ really does call for further investigation. Fortunately, the equations are not as complicated as they appear to be. If there is a solution with both x and y nonzero, then the common x and y terms in these equations can be canceled. The resulting two equations have many similar terms, and it is easy to show by hand that the system has no solution. It follows that the only solutions have an x or y coordinate which is zero. To search for all possible solutions with $x = 0$ or with $y = 0$, we plot the following two expressions. We could pursue the matter further, but this is solid evidence that we have found all of the solutions.

```
In[77]:= fx0=fx/.y->0
```

Out[77]= $$\frac{2\,x}{(4+x^2)\,(4+3\,x^2)} - \frac{6\,x\,\log[4+x^2]}{(4+3\,x^2)^2}$$

```
In[78]:= fy0=fy/.x->0
```

Out[78]= $$\frac{6\,y}{(4+y^2)\,(4+3\,y^2)} - \frac{2\,y\,\log[4+3\,y^2]}{(4+y^2)^2}$$

```
In[79]:= Plot[fx0,{x,-10,10}];
```

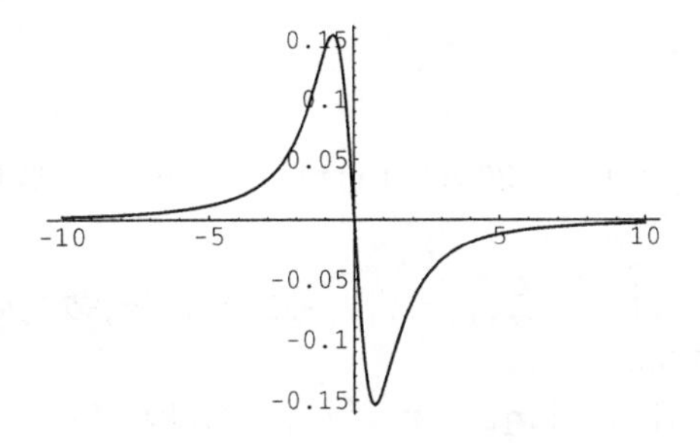


```
In[80]:= Plot[fy0,{y,-10,10}];
```

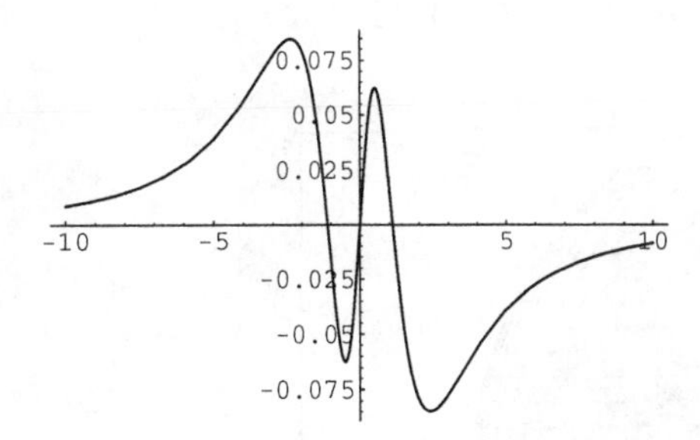

It is easy to imagine a system of equations of the form $fx = 0$, $fy = 0$, whose complete solution would not be available in such an elementary fashion. Each system must be considered on its own merits. With Mathematica's FindRoot[ ] command we need to get close to a desired solution. As an alternate strategy, we could have, for example, plotted the two equations using the ImplicitPlot[ ] command and then located the intersection points between the two curves by eye sight. Then, we would have points, which would be close enough to use in Mathematica's FindRoot[ ] command.

To find the local or absolute maximum or minimum points of a function of more than two variables is intrinsically more difficult. There is a version of the second derivative test similar to the so-called M-Test that we used above, but it is difficult to use when more than two variables are involved. Mathematica may have to struggle in order to solve a system of three or more equations in three or more unknowns. Showing that we have all of the solutions, or at least, the desired solution can be challenging. It is more difficult to use Mathematica's plotting tools to get useful information.

Example 11.12 *Find the minimum distance between the surface*

$$x = 4 - \frac{y^2}{10} - \frac{z^2}{30},$$

and the curve defined parametrically by

$$x = 16 - 4\sin(t),\ y = 9 + 2\cos(3t),\ z = 5 + 7\cos(2t).$$

Locate the point on the surface and the point on the curve where the minimum occurs.

The distance between a point on the surface, and a point on the curve can be expressed as a function of y, z, and t. **Rather than minimize this square root function, we minimize its square instead.** Surely $\sqrt{expr}$ will be a minimum at the same point that $expr$ is a minimum.

```
In[81]:= f[y_,z_]:=4-y^2/10-z^2/30
```

```
In[82]:= p[t_]:=16+4Sin[t];q[t_]:=9+2Cos[3t];r[t_]:=-5+7Cos[2t]
```

```
In[83]:= p1=ParametricPlot3D[{f[y,z],y,z},{y,-20,20},{z,-20,40},
         DisplayFunction->Identity];
         p2=ParametricPlot3D[{p[t],q[t],r[t]},{t,0,2Pi},
         DisplayFunction->Identity];
```

```
In[84]:= Show[{p1,p2},DisplayFunction->$DisplayFunction,
         ViewPoint->{3.172, -1.174, 0.109}];
```

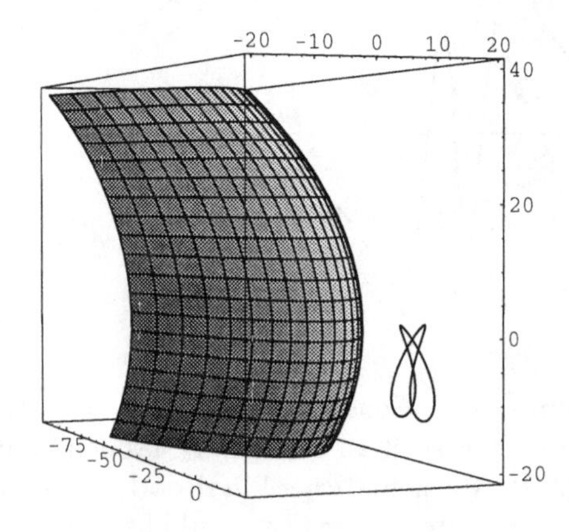

We define the square of the distance function next.

```
In[85]:= F=(f[y,z]-p[t])^2+(y-q[t])^2+(z-r[t])^2
```

Out[85]= $(5+z-7\ \cos[2\ t])^2+(-9+y-2\ \cos[3\ t])^2+\left(-12-\frac{y^2}{10}-\frac{z^2}{30}-4\ \sin[t]\right)^2$

```
In[86]:= Fy=D[F,y]
```

Out[86]= $2\ (-9+y-2\ \cos[3\ t])-\frac{2}{5}\ y\ \left(-12-\frac{y^2}{10}-\frac{z^2}{30}-4\ \sin[t]\right)$

```
In[87]:= Fz=D[F,z]
```

Out[87]= $2\ (5+z-7\ \cos[2\ t])-\frac{2}{15}\ z\ \left(-12-\frac{y^2}{10}-\frac{z^2}{30}-4\ \sin[t]\right)$

```
In[88]:= Ft=D[F,t]
```

Out[88]= $-8\ \cos[t]\ \left(-12-\frac{y^2}{10}-\frac{z^2}{30}-4\ \sin[t]\right)+$
$28\ (5+z-7\ \cos[2\ t])\ \sin[2\ t]+12\ (-9+y-2\ \cos[3\ t])\ \sin[3\ t]$

To find the critical points, we need some decent start-up values for y, z, and t in the next system of equations. By looking at the graph, the values of $y = 0$, and $z = 0$ appear to be at least reasonable start-up values. A value for t, however, alludes us. For lack of a better strategy, we make a wild guess.

```
In[89]:= s=FindRoot[{Fy==0,Fz==0,Ft==0},{y,0},{z,0},{t,0}]
Out[89]= {y → 2.22561, z → -5.37221, t → 1.73976}
```

```
In[90]:= {y0,z0,t0}=({y,z,t}/.s)
Out[90]= {2.22561, -5.37221, 1.73976}
```

```
In[91]:= P={f[y0,z0],y0,z0}
Out[91]= {2.54264, 2.22561, -5.37221}
```

```
In[92]:= Q={p[t0],q[t0],r[t0]}
Out[92]= {19.943, 9.9709, -11.6041}
```

The points P and Q could correspond to any local maximum or minimum of the function F. We can see by looking at the above graph that there are several such points, and there is no reason to suspect that we have found the desired ones.

How we proceed from here varies from one problem to the next. In our case, we determine the distance between the point P on the surface and an arbitrary point on the curve. This defines a function of t which can be plotted. The value t that we computed above will correspond to a local maximum or minimum point on this graph. If it is the desired point, it will correspond to a absolute minimum point.

```
In[93]:= Plot[Sqrt[F/.{y->y0,z->z0}],{t,0,2Pi}];
```

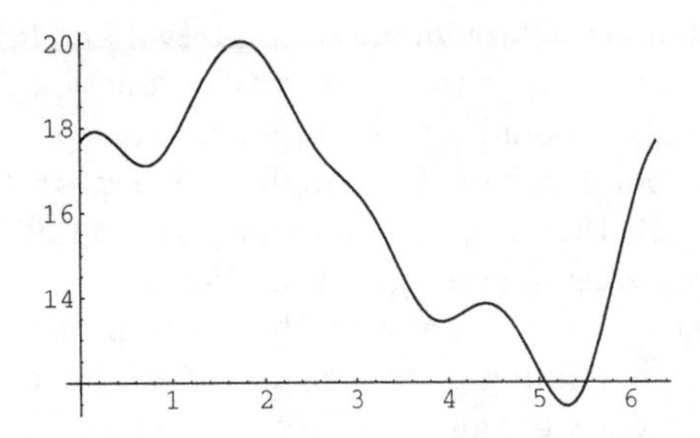

As expected, this plot has several local maximums and minimums. We can see now that our first answer was wrong.

```
In[94]:= s1=FindRoot[{Fy==0,Fz==0,Ft==0},{y,0},{z,0},{t,5.2}]
Out[94]= {y → 2.39749, z → -4.0174, t → 5.3762}
```

```
In[95]:= {y0,z0,t0}=({y,z,t}/.s1)
Out[95]= {2.39749, -4.0174, 5.3762}
```

```
In[96]:= P={f[y0,z0],y0,z0}
Out[96]= {2.88722, 2.39749, -4.0174}
```

```
In[97]:= Q={p[t0],q[t0],r[t0]}
Out[97]= {12.8494,7.17433,-6.68554}
```

```
In[98]:= a=Sqrt[F/.{y->y0,z->z0,t->t0}]
Out[98]= 11.3658
```

It looks like we may have found an answer to the problem. If so, then the minimum distance is the value $a = 11.3658$, between the point P on the surface and Q on the curve.

Have we, however, established that this answer corresponds to the absolute minimum and not just another local minimum? Certainly not! If the point P on the surface is the correct point, then we have indeed established that our answer is correct. However, what if the curve takes another dip closer yet to the surface at some point on the surface far removed from P? This is an important, but very difficult issue to resolve, and there is no swift, clean mathematical way to proceed. We cannot prove, in a pure mathematical sense, that our answer is correct, but on a practical level, we can offer very convincing evidence.

A thorough search for other answers will be conducted. It will help if we restrict our search to as small a window as possible, and so our approach begins with another look at the plot showing the curve and surface together. The distance function, F, (actually distance squared) depends on the variables y, z, and t, and we wish to establish small ranges for each of these variables and to restrict our search to these intervals. A range for the variable t is easy to establish. We simply use the interval $0 \leq t \leq 2\pi$. By turning the plot around and looking at it carefully from different directions, it is possible to establish the ranges $0 \leq y \leq 6$ and $-10 \leq z \leq 2$.

Now that we have a fairly narrow band of possible solutions, we blanket the entire region with a large number of points, evaluate F at each of the points, and check to see if there are any values of F smaller than $a^2 = 11.3658^2$. If we are careful how we do this, it will not take very much computer time.

We partition the interval $0 \leq y \leq 6$, the interval $-10 \leq z \leq 2$, and the interval $0 \leq t \leq 2\pi$ into points which are equally distributed throughout each interval and 0.1 units apart. This creates roughly $60 \times 120 \times 60 = 432,000$ partition points. The function F is then evaluated at each of the corresponding 432,000 points. It is important to speed things up, so we suppress output unless we find a point, where F is smaller than $a^2 = 11.3658^2$. To speed things up slightly more, we use 129.181 instead of its equivalent 11.3658^2. Of course, we could speed things up much more by using fewer partition points. The problem seems to be more sensitive to variations in t, so we could maintain the above partition on the t-interval and use coarser partitions on the remaining intervals. Using all 432,000 partition points, the following program took 371 seconds to run on a Power Macintosh 7600/132.

```
In[99]:= Do[If[N[F]<129.181,Print[{y,z,t}],Null],
          {y,0,6,.1},{z,-10,2,.1},{t,0,2Pi,.1}]
```

When this program is run, there is no output. This should mean that the answer we obtained above is correct. Can the program be trusted? It could be run again with $a^2 = 11.36582848^2$ replaced by a slightly larger number. The program should (and does) begin to return points y, z, and t close to the values where we expect the minimum to occur.

In a pure mathematical sense, the minimum could still occur at some point in between these values, but in a practical sense, surely we have enough evidence to declare our answer to be correct.

In the last example, we were able to easily express the function we wished to minimize—distance—as a function of several independent variables. In many optimization problems it is either difficult or impossible to express the function being optimized in terms of independent variables. In these cases, the method of Lagrange can be used to optimize the function.

Suppose we wish to optimize a real valued function $f(x, y, z)$ where x, y, and z are not independent variables, but must satisfy the constraint $g(x, y, z) = 0$. We introduce a new variable λ, called a Lagrange multiplier and define the function

$$F(x, y, z, \lambda) = f(x, y, z) + \lambda g(x, y, z),$$

where x, y, z, and λ are regarded as independent variables. According to the method of Lagrange, the optimum value of the original function f, if it exists, will occur at a critical point of F, that is to say, at a point (x, y, z, λ), where the four partials of F are zero (or fail to exist). Setting the partial of F with respect to λ equal to zero simply returns the constraint $g(x, y, z) = 0$.

More generally, to optimize $f(x_1, x_2, \ldots, x_n)$, subject to the constraints

$$g_1(x_1, x_2, \ldots, x_n) = 0,\ g_2(x_1, x_2, \ldots, x_n) = 0, \ldots, g_p(x_1, x_2, \ldots, x_n) = 0,$$

we introduce the Lagrange multipliers $\lambda_1, \lambda_2, \ldots, \lambda_p$ and define the function

$$F(x_1, x_2, \ldots, x_n, \lambda_1, \lambda_2, \ldots, \lambda_p) = f(x_1, x_2, \ldots, x_n) + \sum_{j=1}^{p} \lambda_j g_j(x_1, x_2, \ldots, x_n),$$

where the variables $x_1, x_2, \ldots, x_n, \lambda_1, \lambda_2, \ldots, \lambda_p$ are regarded as $n+p$ independent variables. Then the optimum value of f, if it exists, will occur at a critical point of F.

Example 11.13 *Find the minimum distance between the curves*

$$8x^2 + 3y^2 + 80x - 42y + 251 = 0,\ 108x^2 + 68xy + 57y^2 + 340x + 570y + 1225 = 0,$$

and the points on each curve where the minimum occurs.

We begin by plotting the two curves. Each is an ellipse, and the plot can be used to get an approximate locations for the two points (one on each ellipse), which are closest to each other. This kind of information is frequently needed in order to help Mathematica solve more complicated problems.

Let $P(x, y)$ be an arbitrary point on the first curve, and let $Q(u, v)$ be an arbitrary point on the second. The points P and Q are obviously different points, and so variables other than x and y must be used for one of the points. We minimize the distance between P and Q, using the method of Lagrange. Just as in the previous problem, minor complications can be avoided by minimizing, instead, the square of the distance between P and Q.

```
In[100]:= curve1=8x^2+80x+3y^2-42y+251==0;
          curve2=108x^2+68x*y+57y^2+340x+570y+1225==0;

In[101]:= <<Graphics`ImplicitPlot`

In[102]:= p1=ImplicitPlot[curve1,{x,-15,10},{y,-10,15},
          DisplayFunction->Identity];
          p2=ImplicitPlot[curve2,{x,-15,10},{y,-10,15},
          DisplayFunction->Identity];

In[103]:= Show[{p1,p2},DisplayFunction->$DisplayFunction];
```

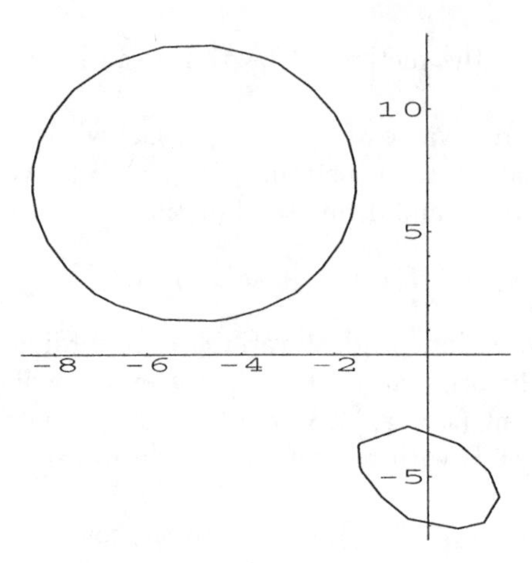

Looking at this plot, we can see approximately where the points generating a minimum distance occur. This will help, when we try to solve certain equations. Notice that the **plot was drawn to scale**[1]. Mathematica doesn't always do that. If it were not drawn to scale, the **warped picture could severely change our estimates of the points** where a minimum distance occurs.

```
In[104]:= g1=curve1[[1]];g2=(curve2[[1]]/.{x->u,y->v});
```

```
In[105]:= f=(x-u)^2+(y-v)^2
Out[105]= (-u + x)^2 + (-v + y)^2
```

The expression f is the square of the distance between P and Q.

The variables λ_1 and λ_2 can be used in a Mathematica work session by using the unassigned names \[Lambda]1 and \[Lambda]2 as input variables. They will turn into Greek letters when we type the closing bracket. We have used them before, but in this example, we take an easier approach, and simply let r and s denote the Lagrange multipliers.

```
In[106]:= F=f+r*g1+s*g2
Out[106]= s (1225 + 340 u + 108 u^2 + 570 v + 68 u v + 57 v^2) + (-u + x)^2 +
          (-v + y)^2 + r (251 + 80 x + 8 x^2 - 42 y + 3 y^2)
```

```
In[107]:= eq1=D[F,x]==0
Out[107]= 2 (-u + x) + r (80 + 16 x) == 0
```

```
In[108]:= eq2=D[F,y]==0
Out[108]= 2 (-v + y) + r (-42 + 6 y) == 0
```

```
In[109]:= eq3=D[F,u]==0
Out[109]= s (340 + 216 u + 68 v) - 2 (-u + x) == 0
```

```
In[110]:= eq4=D[F,v]==0
```

[1]Mathematica draws the plot to scale, but it is not reproduced that way in this manual, because it would take up too much vertical space.

```
Out[110]= s (570 + 68 u + 114 v) - 2 (-v + y) == 0

In[111]:= eq5=g1==0;eq6=g2==0;
```

To solve this system of 6 equations in 6 unknowns, we need some start-up values. They are not hard to get for x, y, u, and v. As we mentioned above, all we have to do is look at the above picture. Start-up values for r and s, on the other hand, are not really available. Should we just use some random numbers and see what happens? This is a rather unscientific approach, but let's try. The Lagrange technique **requires** r, $s \neq 0$, so we must choose nonzero numbers.

```
In[112]:= FindRoot[{eq1,eq2,eq3,eq4,eq5,eq6},
            {x,-3.67},{y,2},{u,-1},{v,-3},{r,1},{s,1}]

Out[112]= {x → -3.82194, y → 1.68031, u → -1.04392,
           , v → -3.02391, r → 0.294767, s → 0.0609799}
```

Is this our solution? Considering how we entered start-up values for r and s, it is hard to trust the accuracy of this result. Fortunately, **a more scientific approach is at hand!** Since we do not have start-up values for the Lagrange multipliers r and s, we choose to eliminate these variables. Incidentally, these variables **can always be eliminated**, because Lagrange multipliers always appear as first power terms, and so the equations involved can always be solved for them. We solve the first equation for r and substitute its value into the second equation. A similar approach is used on the third and forth equations to eliminate s.

```
In[113]:= R=Solve[eq1,r]
```

$$\text{Out[113]= } \left\{\left\{r \to -\frac{-u + x}{8\ (5 + x)}\right\}\right\}$$

```
In[114]:= S=Solve[eq3,s]
```

$$\text{Out[114]= } \left\{\left\{s \to -\frac{u - x}{2\ (85 + 54\ u + 17\ v)}\right\}\right\}$$

```
In[115]:= eq2=(eq2/.R[[1]])
```

$$\text{Out[115]= } 2\ (-v + y) - \frac{(-u + x)\ (-42 + 6\ y)}{8\ (5 + x)} == 0$$

```
In[116]:= eq4=(eq4/.S[[1]])
```

$$\text{Out[116]= } -\frac{(570 + 68\ u + 114\ v)\ (u - x)}{2\ (85 + 54\ u + 17\ v)} - 2\ (-v + y) == 0$$

```
In[117]:= Sol=FindRoot[{eq2,eq4,eq5,eq6},{x,-3.67},{y,2},{u,-1},{v,-3}]

Out[117]= {x → -3.82194, y → 1.68031, u → -1.04392, v → -3.02391}
```

The solution is the same, but it is comforting to know that it is grounded on more scientific principles.

```
In[118]:= {x,y,u,v}=({x,y,u,v}/.Sol)
Out[118]= {-3.82194, 1.68031, -1.04392, -3.02391}
```

The two points P and Q, and the distance between the curves are as follows:

```
In[119]:= {P={x,y},Q={u,v}}
Out[119]= {{-3.82194, 1.68031}, {-1.04392, -3.02391}}
```

```
In[120]:= CurveSeparation=Sqrt[f]
Out[120]= 5.46325
```

```
In[121]:= {r=(r/.R[[1]]),s=(s/.S[[1]])}
Out[121]= {0.294767, 0.0609799}
```

Before we walk away from this problem, we should check to see if all six equations are satisfied for these values of x, y, u, v, r, and s.

```
In[122]:= {eq1,eq2,eq3,eq4,eq5,eq6}
Out[122]= {True, False, False, True, False, True}
```

Three of the equations are not satisfied. **Is this a small decimal approximation problem that we can ignore, or does it represent a genuine lack of equality?** It would seem reasonable to evaluate $eq2[[1]]$, the left hand side of the second equation, to see if it evaluates to a small, negligible number, but **this will not work. It's too late to do this!** With values in $eq2$ for all the variables, **the value of $eq2$ is now "False." It's no longer an equation**, so $eq[[1]]$ makes no sense. We are forced to return x, y, u, and v, to their unassigned status, before we can evaluate the left hand sides of these equations.

```
In[123]:= Clear[x,y,u,v]
```

```
In[124]:= {eq2[[1]],eq3[[1]],eq5[[1]]}/.Sol
Out[124]= {-1.42109 10^-14, 8.88178 10^-16, 5.32907 10^-15}
```

If f is a continuous real valued function in a closed and bounded region D, then f has an absolute maximum and an absolute minimum in D. That is is say, there are points (x_1, y_1) and (x_2, y_2) in D such that

$$f(x_1, y_1) \leq f(x, y) \leq f(x_2, y_2)$$

for all (x, y) in D. The absolute maximum or minimum could occur at an interior point (x_0, y_0) of D, in which case, (x_0, y_0) would be a critical point of f. On the other hand, the absolute maximum or minimum could occur at a boundary point of D, in which case no special role is played by the partials of f.

To solve such a problem, we collect all the critical points of f interior to the region D. Points on the boundary of D are simply points that satisfy some constraint condition $g(x, y) = 0$. We use either substitution methods or the method of Lagrange to find all of the candidates for maximum or minimum on the boundary of D. **Since we know the absolute maximum and minimum exist, once we have a collection of all the candidates interior to D and on its boundary**, all we have to do to finish the problem is evaluate f at each of the candidates. The absolute maximum and minimum values of f will be the largest and smallest of these numbers.

Example 11.14 *Let D be the region in the x,y-plane which is bounded by the curves $y = 16 - x^2$ and $y = x^2 - 16$, and let f be defined by*

$$f(x,y) = \frac{8(x-3)^2 + 5(y+1)^2 + 1}{x^2 + 2y^2 + 1}.$$

Find the absolute maximum and absolute minimum of f in D and the points where they occur.

An explanation of our confusing array of names may make our solution more readable. As usual, the letter "p" is used in a name (p for **p**rime) to denote a derivative. The names ubf and lbf refer to the **u**pper and **l**ower **b**oundary values of f.

```
In[125]:= f[x_,y_]:=(8(x-3)^2+5(y+1)^2+1)/(x^2+2y^2+1)
```

```
In[125]:= Plot3D[f[x,y],{x,-4,4},{y,-16,16}];
```

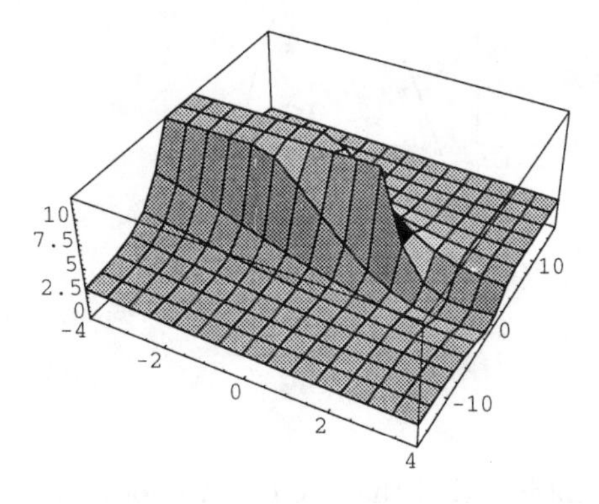


```
In[127]:= fpx=D[f[x,y],x]
```

$$\text{Out[127]=} \frac{16\ (-3+x)}{1+x^2+2\ y^2} - \frac{2\ x\ (1+8\ (-3+x)^2+5\ (1+y)^2)}{(1+x^2+2\ y^2)^2}$$

```
In[128]:= fpy=D[f[x,y],y]
```

$$\text{Out[128]=} \frac{10\ (1+y)}{1+x^2+2\ y^2} - \frac{4\ y\ (1+8\ (-3+x)^2+5\ (1+y)^2)}{(1+x^2+2\ y^2)^2}$$

```
In[129]:= eqx=Numerator[Together[fpx]]==0
```

$$\text{Out[129]=}\ 2\ \left(-24-70\ x+24\ x^2-10\ x\ y-48\ y^2+11\ x\ y^2\right) == 0$$

```
In[130]:= eqy=Numerator[Together[fpy]]==0
```

$$\text{Out[130]=}\ -2\ \left(-5-5\ x^2+151\ y-96\ x\ y+11\ x^2\ y+10\ y^2\right) == 0$$

The absolute maximum appears to occur near the origin, but what about the minimum? What should we use for start-up values. **An excellent way to find approximate solutions is to plot the equations *eqx* and *eqy*, and see where they intersect.** Each intersection point is a solution to the system. **This is worth remembering!**

```
In[131]:= <<Graphics`ImplicitPlot`

In[132]:= p1=ImplicitPlot[eqx,{x,-4,4},DisplayFunction->Identity];

In[133]:= p2=ImplicitPlot[eqy,{x,-4,4},DisplayFunction->Identity];

In[134]:= Show[{p1,p2},DisplayFunction->$DisplayFunction];
```

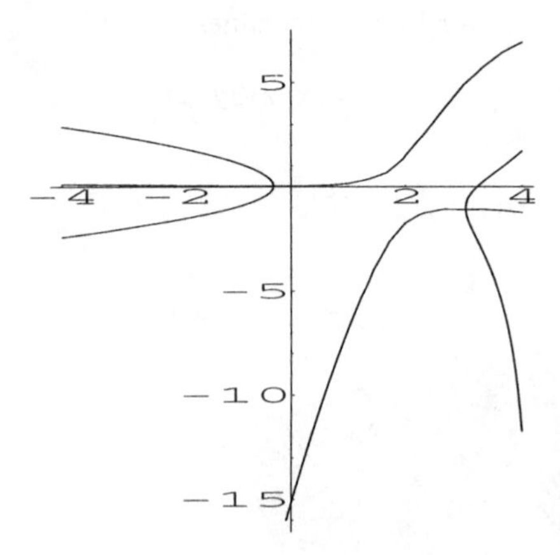

We can now see two points of intersection that we can use for start-up values.

```
In[135]:= s1=FindRoot[{eqx,eqy},{x,3},{y,-1}]
Out[135]= {x → 3.03116, y → -1.03401}

In[136]:= s2=FindRoot[{eqx,eqy},{x,-.5},{y,0}]
Out[136]= {x → -0.309376, y → 0.0300931}

In[137]:= {a1,b1}=({x,y}/.s1);{a2,b2}=({x,y}/.s2);
```

We have two critical points interior to the region. The absolute maximum or absolute minimum could also occur at a boundary point. We search for these boundary points next, by replacing y, in $f(x, y)$ by its values on the boundary curves.

```
In[138]:= ubf=f[x,16-x^2]
```

Out[138]= $$\frac{1+8\ (-3+x)^2+5\ (17-x^2)^2}{1+x^2+2\ (16-x^2)^2}$$

```
In[139]:= lbf=f[x,x^2-16]
```

Out[139]= $$\frac{1+8\ (-3+x)^2+5\ (-15+x^2)^2}{1+x^2+2\ (-16+x^2)^2}$$

```
In[140]:= ubfp=D[ubf,x]
```

$$Out[140]= \frac{16\ (-3+x)-20\ x\ (17-x^2)}{1+x^2+2\ (16-x^2)^2} - \frac{(2\ x-8\ x\ (16-x^2))\ \left(1+8\ (-3+x)^2+5\ (17-x^2)^2\right)}{(1+x^2+2\ (16-x^2)^2)^2}$$

```
In[141]:= lbfp=D[lbf,x]
```

$$Out[141]= \frac{16\ (-3+x)+20\ x\ (-15+x^2)}{1+x^2+2\ (-16+x^2)^2} - \frac{(2\ x+8\ x\ (-16+x^2))\ \left(1+8\ (-3+x)^2+5\ (-15+x^2)^2\right)}{(1+x^2+2\ (-16+x^2)^2)^2}$$

```
In[142]:= eq1=Numerator[Together[ubfp]]==0
```

$$Out[142]= 6\ \left(-4104+4176\ x-504\ x^2-314\ x^3+48\ x^4+3\ x^5\right) == 0$$

```
In[143]:= eq2=Numerator[Together[lbfp]]==0
```

$$Out[143]= -2\ \left(12312-2628\ x+1512\ x^2-338\ x^3-144\ x^4+31\ x^5\right) == 0$$

The equations *eq1* and *eq2* are polynomial equations, so we use a better equation solver. This command has not been used for some time, so it is worth pointing out, that the **NSolve[ ] command finds all of the decimal roots of a polynomial equation**.

```
In[144]:= s3=NSolve[eq1,x]
Out[144]= {{x → -20.5309}, {x → -3.97094}, {x → 1.34057},
           {x → 3.02921}, {x → 4.13206}}
```

```
In[145]:= s4=NSolve[eq2,x]
Out[145]= {{x → -4.00088}, {x → -0.0116391 - 2.31425 i},
           {x → -0.0116391 + 2.31425 i}, {x → 3.82983}, {x → 4.83949}}
```

```
In[146]:= g[x_]:=16-x^2
```

The upper and lower boundaries consist of points (x, y) with $y = \pm g(x)$ and $-4 \leq x \leq 4$. Only 4 of the 6 values of x in the above list of candidates are real and in the right interval.

```
In[147]:= {a3=(x/.s3[[2]]),b3=g[a3]}
Out[147]= {-3.97094, 0.231659}
```

```
In[148]:= {a4=(x/.s3[[3]]),b4=g[a4]}
Out[148]= {1.34057, 14.2029}
```

```
In[149]:= { {a5=(x/.s3[[4]]),b5=g[a5]}"
Out[149]= {3.02921, 6.82391}
```

```
In[150]:= { {a6=(x/.s4[[4]]),b6=g[a6]}"
Out[150]= {3.82983, 1.33243}
```

If we maximize and minimize a function over an interval, its end points are always candidates, regardless of any properties of derivatives. Consequently, we must consider the points corresponding to $x = \pm 4$ as two more candidates.

```
In[151]:= {a7,b7}={-4.0,0};{a8,b8}={4.0,0};
```

We now have a complete list of candidates. The absolute maximum and minimum exist, because $f(x, y)$ is continuous on the set D, so all we have to do is evaluate $f(x, y)$ at each of the candidates and pick out the largest and smallest values. The first two values correspond to interior points of D, the next three values correspond to points on the upper boundary of D, the next value corresponds to a point on the lower boundary of D, and the last two correspond to the corner points of D.

```
In[152]:= {f[a1,b1],f[a2,b2],f[a3,b3],f[a4,b4],
          f[a5,b5],f[a6,b6],f[a7,b7],f[a8,b8]}

Out[152]= {0.0822268, 85.5755, 23.5449, 2.9014,
              2.97243, 1.75406, 23.4118, 0.823529}
```

The absolute maximum value, $M = 85.5755$, occurs at the interior point $(a2, b2) = (-0.309376, 0.0300931)$, and the absolute minimum value of $m = 0.0822268$ occurs at $(a1, b1) = (3.03116, -1.03401)$.

11.6 Exercise Set

Some New and Some Old Mathematica features and Commands

Animate[ ]	<<Calculus`VectorAnalysis`	Cartesian[ ]
ContourPlot[ ]	ContourPlot3D	D[ ]
Derivative[ ][ ]	Do[ ]	Grad[ ]
Cross[ ] §	Dot[ ] §	FindRoot[ ]
Get[ ] or <<	<<Graphics`ImplicitPlot`	If[ ]
ImplicitPlot[ ]	<<Graphics`PlotField`	<<Graphics`PlotField3D`
NSolve[ ]	ParametricPlot3D[ ]	Part[or [[ ]]
Plot[ ]	Plot3D[ ]	PlotGradientField[ ]
PlotGradientField3D[ ]	SetCoordinates[ ]	Show[ ]
Which[ ]		

Plot options

AspectRatio− > 1	AxesLabel− > {"x","y","z"}
BoxRatios− > { }	DisplayFunction− >Identity
DisplayFunction− >$DisplayFunction	Mesh− >False
PlotPoints− > n (a positive integer)†	PlotRange− > { }
ViewPoint− > { } ‡	ViewVertical− > { }

§ Use the infix forms for scalar, dot and cross products: $a\vec{v}$, $\vec{v}\cdot\vec{w}$, and $\vec{v}\times\vec{w}$ can be computed using input statements $a * v$, $v.w$, and $v \times w$ (enter \[Cross] for $\times$).

† the default value for PlotPoints is $n = 25$. Don't make it too big.

‡ Use the **3D ViewPoint Selector...** in the **Input Menu** to insert the ViewPoint option. See page 184 for details.

Use the identity $|\vec{v}|^2 = \vec{v}\cdot\vec{v}$ to compute the length of a vector $\vec{v}$.

1. This problem may require more memory than is available, if all of its parts are done in one Mathematica work session. Keep an eye on the memory bar in the lower right hand corner of the work sheet. Consider the function

$$f(x,y) = (83 - 5x - 7y)e^{-0.57x^2 - 0.43y^2 + 0.48xy + 0.24x + 5.82y - 26}.$$

a) Plot f so that all of the significant features are the graph are shown.
b) Display a contour plot by using the same plot window used in part a), making appropriate adjustments and dragging the plot to a view point which looks down from above.
c) Display another contour plot, over a much smaller range, and use it to estimate (visually) the x and y coordinates of the absolute maximum and minimum points of f. (Use the contourplot( ) command in the plots package.)

Let (x_1, y_1) and (x_2, y_2) denote the approximate coordinates of the maximum and minimum points (respectively) found in this manner. These points will be referred to in subsequent parts of the problem.
d) Display two vertical sections of the graph which pass through the point (x_1, y_1). One of the vertical sections should be parallel to the x-axis, and the other should be parallel to the y-axis.
e) Display a vertical section which passes through the points (x_1, y_1) and (x_2, y_2).
f) Create an animation which plots sections of the graph of f, which are parallel to the x-axis, which start on one side of the point (x_0, y_0) and end on the other side.
g) This animation is more involved. Create an animation which plots vertical sections of the graph of f which are all perpendicular to the line L passing through the points (x_1, y_1) and (x_2, y_2). Arrange the sections so that they are all centered over the line L. The sections should start on one side of both points and end of the other side of both points.

2. Use Mathematica's differentiation commands to compute, directly, the partials of $f(x,y) = x^4 + 5x^2y^3 - 8y^3 + 7x^3$. Confirm that the answers are correct by using the definition of a partial derivative as a limit of a certain difference quotient. Simplify the difference quotients so that their limits are obvious.

3. Use Mathematica's differentiation commands to compute directly the partials of

$$f(x,y,z) = \frac{x^2 \cos(x+y)}{x^2 + y^3 + z^4}$$

at the point $(x_0, y_0, z_0) = (5, -8, 3)$. Confirm that the answers are correct by using the definition of a partial derivative as a limit of a certain difference quotient.

4. Use Mathematica's D[ ] command to compute the first and second order partial derivatives of $f(x,y) = \cos(x^2y)e^{3y-5x}$. Compute the partials again using Mathematica's Derivative[ ] command. Sometimes, an expression is desired as output, and sometimes a function is desired as output. Both commands are useful. Evaluate all of the partials, as decimals, at the point $(x,y) = (8, 14)$.

5. The equation $x^3y^2 + 5xz^2y^3 + 4yz^3 - 7x^2 = 17z + 9$ defines z implicitly as a function $z = f(x,y)$ of x, and y. Use the D[ ] command and implicit differentiation to determine (as decimal numbers) the partials of z with respect to x and y at the point $(x_0, y_0) = (5, 7)$. The function $z = f(x,y)$ may not be uniquely defined. Evaluate the partials for every such determination $z = f(x,y)$.

Before you can differentiate z, implicitly, with respect to x, or y, Mathematica must be told that z depends on x and y. These issues are discussed in Example 11.4.

6. A general (nice) equation in the variables x, y, and z is modeled by the equation $F(x,y,z) = 0$, where F is any real valued differentiable function of three variables. It

is natural to attempt to solve such an equation for z (for example) in terms of x and y. When such a solution exists, even if it can not be found, we say that $F(x,y,z)=0$ defines z implicitly as a function of x and y.

If $F(x_0,y_0,z_0)=0$ and if the partial $F_z(x_0,y_0,z_0)\neq 0$, then it turns out that there is a differentiable function $g(x,y)$, with $z_0=g(x_0,y_0)$, such that $z=g(x,y)$ solves the equation $F(x,y,z)=0$ in some neighborhood of (x_0,y_0). Use the *Chain Rule* (Mathematica knows it well) to find formulas for the partials of g with respect to x and y, at x_0, y_0. In the process, you will find a reason for the condition $F_z(x_0,y_0,z_0)\neq 0$. (Hint: Before you differentiate, replace z by $g(x,y)$. Mathematica must be told that z depends on x and y.)

7. Building on the last problem, a general system of two (nice) equations in three variables x, y, and z is modeled by the system of equations

$$F(x,y,z)=0,\ G(x,y,z)=0,$$

where F and G are differentiable functions of three variables. It is natural to think of this as a system of two equations in two unknowns, (think of one of the variables as a given number, rather than as an unknown), and to attempt to solve such a system for the unknowns, say y and z (for example) in terms of x. When such a solution exists, even if it can not be found, we say that the system $F(x,y,z)=0$, $G(x,y,z)=0$ defines y and z implicitly as functions of x. Assuming that such a solution $y=p(x)$, $z=q(x)$ exists, take advantage of Mathematica's understanding of the *Chain Rule* to find formulas for $\frac{dy}{dx}=p'(x)$, $\frac{dz}{dx}=q'(x)$ in terms of the partials of F and G. In the process, if

$$F(x_0,y_0,z_0)=0,\ G(x_0,y_0,z_0)=0,$$

find a condition on the partials of F and G which must be met in order for the solution $y=p(x)$, $z=q(x)$ with $y_0=p(x_0)$, $z_0=q(x_0)$ to exist. Mathematica must be told that y and z depend on x, so substitute $y=p(x)$, $z=q(x)$ into these equations before you begin to differentiate. After differentiating, it might help to simplify notation. The symbols *yp* and *zp* are good names for D[$p(x)$, x], and D[$q(x)$, x].

8. Find the directional derivative of the function

$$f(x,y,z)=\frac{e^{2zx}}{y^2x^3},$$

at the point $P(4,9,2)$ in the direction $\vec{w}=7\vec{i}-15\vec{j}+3\vec{k}$. Use this to approximate (as a decimal) the change in f as the point (x,y,z) moves a distance of 0.02 units from P in the direction of $\vec{w}$. Compare this to the actual change in f (expressed as a decimal).

9. Let $f(x,y,z)=3x^2-y^2+2z^2-8x-5y+7z-6$. Find the point(s) on the level surface $f(x,y,z)=0$, if they exist, where ∇f has the same direction and sense as the vector $\vec{n}=2\vec{i}-\vec{j}+3\vec{k}$.

10. Suppose that the effect of three economic variables p, r and t on the value, V, of the U.S. dollar (The trading rate of the dollar in yen) is being studied, where the variables p, r, and t are the U.S. prime (annual) interest rate, the U.S monthly index of inflation, both as decimal rates, and the monthly U.S. trade deficit, in billions of dollars, respectively. Suppose that, in the short term, the effects of these three variables on V is thought to be

$$V(p,r,t)=101+3543p^2+841r-0.0174t^3.$$

Suppose that the prime interest rate is currently 7.4%, that the monthly index of inflation is 0.38%, and that the monthly trade deficit is 14 billion dollars, so that $p_0=0.074$, $r_0=$

0.0038, and $t_0 = 14$. By adjusting the values of p, r, and t, economists want to bring the value of the dollar down, but not at too fast of a rate, relative to the other variables, for fear of creating an unstable economy. Suppose that a desirable rate of decrease is 0.23 with respect to any combination of the variables. Use gradient vectors and directional derivatives to describe a range of allowable directions

$$\vec{u} = u_p\vec{i} + u_r\vec{j} + u_t\vec{k},$$

in which to move the variables p, r, and t. Here $\vec{u}$ denotes a unit vector in the p, r, t-coordinate system. To specify allowable values for the vector $\vec{u}$, write your answer as a system of two equations in the unknowns u_p, u_r, and u_t that the components of the unit vector $\vec{u}$ must satisfy in order to be an allowable direction.

The U.S. monthly trade deficit is difficult to control. Suppose we assume that it remains constant at $t = t_0 = 14$ billion dollars in the above adjustment. Determine the corresponding direction vectors $\vec{u}$ (there are two). For one of these two direction vectors, determine the parametric equation of the line through $P_0(p_0, r_0, t_0)$ with direction $\vec{u}$. Specify a few reasonable values of p, r, and t, along this line and compute the corresponding values of $V(p, r, t)$.

11. Suppose that a hill is modeled by the graph of the function

$$z = f(x, y) = (8.5 - 0.0763x^2 - 0.274y^2 + 1.4x - 0.98y)\cos(5x + 7y) + 500,$$

where the elevation $z = f(x, y)$ is in feet, and the coordinates of the point (x, y) are in miles. A projectile is fired from the point on the hill corresponding to $(x, y) = (13.423, -37.25438)$ with an initial speed of 97 ft/sec in a direction perpendicular to the hill. How long does it take for the projectile to fall back down and strike the hill? What are the coordinates of the contact point? Assume that the only force acting on the projectile after it is fired is gravitational acceleration $\vec{g} = -32\vec{k}$ in ft/sec^2.

12. Show that every tangent plane to the graph of $f(x, y) = \frac{x^2 - 3y^2}{5x + 4y}$ passes through the origin.

13. Let f and g be real valued differentiable functions of x, y, and z. Use Mathematica to prove that $\nabla(fg) = f\nabla g + g\nabla f$.

14. Find the critical points of $f(x, y) = x^4 - y^3 - 84x^2 + 22y^2 - 133y + 2431$ and the values of f at the critical points. Use the *Second Derivative Test* (if it applies) at each critical point to decide whether it corresponds to a local maximum, local minimum, or saddle point.

15. Find the critical points of $f(x, y) = (x^4 + y^2)e^{1 - x^2 - y^2}$, and the values of f at the critical points. Use the *Second Derivative Test* (if it applies) at each critical point to decide whether it corresponds to a local maximum, local minimum, or saddle point. If the *Second Derivative Test* does not apply, try to provide some evidence in support of some kind of conclusion.

16. Find the absolute maximum of the function f in problem 1, and the point where it occurs. Absolute maximums do not always exist. How do we know one exists in this case? Provide convincing evidence that it exists and that it must be the candidate of choice. Does the absolute minimum exist? Explain your answer.

17. Find the absolute maximum and absolute minimum of

$$f(x, y, z) = x^2 + 2x + y^2 + y + z^2 - 6z + 2,$$

on and inside the ellipsoid

$$\frac{x^2}{4} + \frac{y^2}{9} + \frac{z^2}{25} = 1$$

Be sure to collect all possible points where a maximum or minimum could occur. Absolute maximums and minimums do not always exist. How do we know they exist in this case? Provide convincing evidence that they exist, and that they must be the candidates of choice.

18. A propane storage tank in the shape of a cylinder with hemispheric ends must hold 10,000 cubic meters of gas. What dimensions will minimize the amount of material needed to build the tank.

Project: Implicit Differentiation

The system of equations

$$x = u\ln(yv), \quad y = v\ln(5 + xu^2),$$

defines $u = f(x, y)$ and $v = g(x, y)$ implicitly as functions of x and y. Use the D[ ] command and implicit differentiation to compute (as decimals) the values of the partials of u and v with respect to x and y at the point $(x, y) = (4, 8)$. Show that the values of u and v corresponding to $x = 4$, and $y = 8$ are uniquely determined. (This does not imply that the functions f and g are uniquely determined near the point $(x, y) = (4, 8)$, but it makes it seem likely.).

Intuitively, this problem should seem reasonable. Think of x and y as given (rather than as unknowns) and the system of two equations in the two unknowns u and v should be solvable for u and v. With a simpler set of equations, one might be able to actually solve for u and v in terms of x and y. In that case, the partials could be computed directly. In the present case, the equations are too complicated to actually solve for u and v in terms of x and y, and so the partials must be computed by implicit differentiation.

The remarks concerning implicit differentiation in problem 11.6, and the issues discussed in Example 11.4 are relevant here as well. Mathematica can be used in a very elegant way to solve this problem, but you must be careful in assigning names and in using the differentiation commands.

Project: Least Squares Regression Analysis

1. Suppose that experimental or statistical data is collected in the form of points (data points), (x_1, y_1), (x_2, y_2),..., (x_n, y_n). After plotting the points, suppose that they tend to cluster near some straight line in the x, y-plane. The so-called "least squares regression line" is the line $y = f(x) = ax + b$, which "best fits" the data points in the sense that

$$S = \sum_{j=1}^{n} (f(x_j) - y_j)^2 = \sum_{j=1}^{n} (ax_j + b - y_j)^2$$

is as small as possible. In other words, it is the line for which the sum of the squares of the error terms is as small as possible. Individual error terms are squared so that they are always positive. Consequently, they always accumulate in the sum. Simply summing the error terms (without squaring) could lead to a small sum (or zero) even if the individual error terms are quite large.

Determine formulas for a and b in terms of the data points. These are classic formulas, which you will probably find in your main text book. (Hint: Think of S as a function of a and b, and minimize S with respect to a and b.) Mathematica's bracket notation—$x[j]$, $y[j]$, $(j = 1, \ldots, n)$—can be used to label the data points, and Mathematica's sum( ) command can be used to form the sums.

2. Suppose that the data points (x_1, y_1), (x_2, y_2),..., (x_n, y_n), appear to almost fit a quadratic $y = ax^2 + bx + c$. Find formulas for a, b, and c in the same "least-squares" way that was used in the previous problem. Mathematica's sum( ) command should be used to form the sums. Test your results on a data set consisting of 10 points. To generate a meaningful data set in a convenient way, start with a specific function of the form $f(x) = Ax^2 + Bx + C + 0.2\sin(x)$ which is a quadratic together with a small alternating "chaotic" term. Use it to generate the y-coordinates of the 10 data points. If x and y are names of Mathematica sequences, which represent the data points, then the jth point of x or y can be conveniently accessed using Mathematica's bracket notation $x[j]$ or $y[j]$. Check to see if the "least squares" values of a, b, and c are close to the values of A, B, and C.

3. Suppose that the data points (x_1, y_1, z_1), (x_2, y_2, z_2),..., (x_n, y_n, z_n) appear to lie close to a plane $z = f(x, y) = ax + by + c$. The corresponding "mean-square" error term would take the form

$$S = \sum_{j=1}^{n}(f(x_j, y_j) - z_j)^2 = \sum_{j=1}^{n}(a(x_j + by_j + c - z_j)^2.$$

Find formulas for a, b, c in terms of the data points which would minimize S.

Test your results on a data set consisting of 10 points. Generate a data set in much the same way that a data set was generated in the preceding problem, by using a function of the form $f(x, y) = Ax + By + Cz + 0.2\sin(xyz)$. Enter it as a Mathematica sequence, so that computations can be performed with Mathematica's sum( ) command. Check to see if the "least squares" values of a, b, and c, are close to the values of A, B, and C.

Project: The Gradient in Cylindrical Coordinates

Let $f(r, \theta, z)$ be a real valued differentiable function of the variables r, θ, and z in a cylindrical coordinate system. Prove (using Mathematica) that the gradient vector

$$\nabla f = \frac{\partial f}{\partial x}\vec{i} + \frac{\partial f}{\partial y}\vec{j} + \frac{\partial f}{\partial z}\vec{k}$$

can be expressed in the cylindrical form

$$\nabla f = \frac{\partial f}{\partial r}\vec{e_r} + \frac{1}{r}\frac{\partial f}{\partial \theta}\vec{e_\theta} + \frac{\partial f}{\partial z}\vec{e_z},$$

where $\vec{e_r}$, $\vec{e_\theta}$, $\vec{e_z}$ are mutually orthogonal unit vectors in the r, θ, and z directions. More specifically, if $P(r, \theta, z)$ is a point in the cylindrical coordinate system, then $\vec{e_r}$ is the horizontal unit vector based at P projecting radially away from the z-axis, $\vec{e_\theta}$ is the horizontal unit vector based at P, tangent the the cylinder $x^2 + y^2 = r^2$ at P and pointing in the counterclockwise direction (towards increasing θ values), and $\vec{e_z}$ is the vertical unit vector $\vec{k}$ based at P.

Mathematica can compute, directly, gradient vectors in cylindrical coordinates. Find the command, and use it to create the above formula.

Project: Finding Local Extreme Values

It can be very difficult to find a local or absolute maximum or minimum of a differentiable function $f(x, y, z)$ of three variables, and the complexity of the problem grows enormously as the number of variables increases beyond three. It may well be a simple matter to get Mathematica to compute partial derivatives and set them equal to zero, but getting Mathematica to solve the system of equations is another matter.

Since the gradient vector always points in the direction in which f increases at its greatest rate, ∇f can be used to find the approximate locations of the candidates for local maximums and minimums.

To search for a local maximum, we start with a point (x_0, y_0, z_0), that serves as our initial guess. We can then use Mathematica's gradient and plotting tools to move from (x_0, y_0, z_0) in the direction of ∇f as far as possible towards a local maximum in this direction. (Mathematica's elementary Plot[ ] command should be used here to plot a function of one variable, regardless of how many variables appear in f.) This will determine a point (x_1, y_1, z_1) which is closer than (x_0, y_0, z_0) to a local maximum. The procedure can be repeated as often as necessary until a point sufficiently close to a local maximum is determined. A similar procedure can be used to find approximate locations of local minimum points.

In this problem, find (approximately) the coordinates of a point located in the first octant of three dimensional space, which is a local minimum of

$$f(x, y, z) = 48xy + 17yz + x^2z + \frac{83000}{xy^2z}.$$

Compute the partial derivatives and set them equal to zero. Can Mathematica solve this system of equations? Probably not!

Use a gradient search method to find an approximate location of the local minimum. Assign names to each successive approximation (x_j, y_j, z_j) determined by this method. Mathematica will treat these points as points in a sequence, if you use names like $X[j] := vector([x_j, y_j, z_j])$ (notice the brackets on $X[j]$). You can then use Mathematica's Table[ ] command to see how the points $X[1], X[2], \ldots,$ seem to be converging and how the values of f at these points seem to be decreasing and leveling off to some value.

Now return to the system of equations obtained by setting all three partial derivatives equal to zero. Use the approximations obtained with this gradient approach as start-up values in the FindRoot[ ] command. Mathematica may still balk, but if it produces a solution, check to see if it improves the accuracy of the solution.

If f is a real valued differentiable function of $n > 3$ variables, then ∇f is defined in the same expected way. It turns out that ∇f always points in the direction in which f increases at the greatest rate. This important feature, which holds regardless of how many variables are involved, can be used to create a gradient search method for local maximums and minimums, which is very similar to the method used in this problem.

Project: A Summer Job

1. It's summer, you have three months to make money before school starts, and you have just been offered \$7,000, to deliver 10,000 tropical birds from a warehouse on the water front to a dealer in a town 40 miles away. Rather than let someone else strike first at the opportunity, you readily accept the job—\$7000 seems like so much money for such a small job. But then, after you accept, your enthusiasm is tempered by rising doubts about costs. You have agreed to pay for all of the expenses involved.

Arrangements are made to rent a small flat bed truck that will cost \$20 for each run from the warehouse to the dealer and back. In order to transport the birds, you have agreed to build (and pay for) a rectangular cage meeting certain specifications. For the welfare of the birds, no more than 10 birds per cubic foot will be allowed in the cage. In order to fit safely on the truck bed, the cage cannot exceed 6 feet in width, 12 feet in length, or 3 feet in height. All that's left to do now is to arrange for the construction of the rectangular cage.

The material for the sides and and top will cost \$10 per square foot, the material for the bottom will cost \$14 per square foot. The two ends, which must contain special capture mechanisms, will cost \$24 per square foot. To maintain a constant temperature in the box, two special heating strips must be mounted on the bottom of the box running its entire length. The heating material costs \$8 per linear foot. Finally, perches for the birds, will be installed at a cost of \$4 per cubic foot. When the delivery service has been completed, the dealer has agreed to buy the cage for \$400, regardless of what it costs you to build.

A bigger cage means fewer trips, but a bigger cage is much more expensive to build. What should the dimensions of the cage be in order to minimize total costs? What is the cost of the cage? How many trips must be made? How much money is left for you?

The system of equations obtained by setting all of the partial derivatives equal to zero is complicated, but Mathematica is up to the task of supplying at least one solution. Are there other solutions to this system of equations? This issue, difficult to resolve, will be taken up in the next exercise. In this exercise, assume that there are no other critical points (interior to the region under consideration). This makes the problem easier. Don't forget, however, that the absolute minimum could occur at a boundary point rather than at a critical point.

2. It seems probable that the point we found in the first part of this problem is the minimum point, but if you are skeptical (as a good scientist should be), you may not feel comfortable assuming that there are no other critical points for this problem. Mathematica is ready and willing to help supply you with more evidence. Write a simple program, which evaluates the cost function at lots of points, searching (probably without success) for smaller values of the cost function. Write the program thoughtfully, so that you don't display output that you don't need to see. This will speed up the program enormously, when it is run (see Example 11.12). Evidence generated in this way will not prove that you already have the absolute minimum (if, in fact, you do), but, by spreading a large number of points around the region, it certainly can be used to supply very convincing evidence to support such a conclusion.

Project: Managing a Large Apartment Building

Suppose you are the manager of a large apartment building with 1450 units. Currently, the rent is $p_0 = 758$ dollars per month for each unit, and there are $x_0 = 1288$ units occupied. Fixed overhead costs are \$114,000 per month (mostly for salaries, property taxes, and mortgage payments), monthly utility costs average \$47 per occupied unit, and monthly (essential) repair costs average \$11 per occupied unit. In addition, you have a special monthly budget to pay for advertising and optional remodeling and repair projects. This budget amounts to \$50 for each occupied unit, and it is up to you to decide how much you wish to spend on each of these two items, but you are expected to spend the full amount each month. Currently, these costs are $a_0 = 23000$ dollars per month for advertising, and $r_0 = 41400$ per month per unit for optional remodeling and repair projects. It is also up to you to decide how much rent to charge for each unit.

Let p denote the monthly rental charge for each unit, let a denote the total monthly advertising cost, and let r denote the monthly cost for optional remodeling and repair. Then, these variables are currently

$$(p, a, r) = (p_0, a_0, r_0) = (758, 23000, 41400)$$

You are considering some changes in these variables. Naturally, the building owners are expecting your choices to generate as large a profit as possible under the circumstances.

The results of a market analysis are available. The number of units occupied, denoted by x, will naturally decrease (increase) from $x_0 = 1288$, if the monthly rent p increases (decreases) from p_0. By the same token, x will change as a changes and as r changes. It is

reasonable to assume that the relation between x, and the variables p, a, and r is linear, as long as p, a, and r do not move too far from p_0, a_0, and r_0, respectively. A market analysis suggests that these rates of change are (in units per dollar)

$$\frac{\partial x}{\partial p} = -1.6, \ \frac{\partial x}{\partial a} = 0.005, \ \frac{\partial x}{\partial r} = 0.011.$$

Furthermore, this linear relation, with these rates of change is expected to be valid for all p, a, and r in the intervals

$$|p - p_0| \leq 100, \ |a - a_0| \leq 10000, \ |r - r_0| \leq 10000.$$

The three variables, p, a, and r are not independent variables. It turns out that total monthly profit, P, can be expressed in terms of just two of the three variables. Find an expression for P in terms of p, and r. Determine, the values of p, and r, within these allowable intervals, which will generate the largest profit. For these values of p and r, give values for all of the economic terms involved in this problem.

A drastic change in the values of p, a, and r might be regarded as risky, in spite of the higher profits which would be generated by this model. A more conservative approach should also be considered. Suppose that you only wish to move the variables p and r slightly from their current values of p_0 and r_0, perhaps to test the waters before you take a more dramatic plunge. Determine a direction to move the variables p and r from p_0 and r_0, which will increase P at the greatest rate.

Chapter 12

Multiple Integration

Double, triple, and even higher order integrals play an important role in mathematics and in its applications. In this chapter, we will see that Mathematica's familiar integration commands, Integrate[ ], and NIntegrate[ ] can be used in a very straightforward way to compute all of these integrals.

12.1 Double Integrals

If $f(x, y)$ is a bounded real valued function of x and y which is defined on a rectangle $R = [a, b] \times [c, d]$, then the double integral of f over R is defined as the limit

$$\int\int_R f(x, y)\, dxdy = \lim_{||\mathcal{P}|| \to 0} S(\mathcal{P}, f),$$

where $S(\mathcal{P}, f)$ is the Riemann sum

$$S(\mathcal{P}, f) = \sum_{j=1}^{n} f(x_j, y_j)\Delta x_j \Delta y_j.$$

This definition is a bit abbreviated, and other details should be mentioned. In particular, $\mathcal{P} = \{R_1, R_2, \ldots, R_n\}$ is a partition of R into n subrectangles, $\Delta x_j, \Delta y_j$ denote the dimensions of R_j, and (x_j, y_j) represents an arbitrary point in R_j. The limit is taken as $||\mathcal{P}||$, the maximum of all of the dimensions, $\Delta x_j, \Delta y_j$ in $\mathcal{P}$, tends to zero, and the limit must be independent of the choices of the points (x_j, y_j) $(j = 1, \ldots, n)$. Taking the limit as $||\mathcal{P}|| \to 0$ is simply a convenient way of saying that all of the dimensions $\Delta x_j, \Delta y_j$ tend to 0. When Riemann sums are actually constructed, a double indexing scheme for the subrectangles and points is usually more convenient. The details can be found in any standard calculus text, along with a theorem stating that the double integral of a continuous function over a rectangle always exists.

The definition of a double integral could be used to create a numerical technique for approximating its value. Rather than move in this direction, however, we shall simply take advantage of Mathematica's own numerical techniques. Still, this definition is important to understand for other reasons. It gives intuitive meaning to the individual symbols in $\int\int_R f(x, y)\, dxdy$. Think of $dx\, dy$ as the area of a infinitesimally small rectangle of dimensions dx by dy located at the point (x, y) in R. Multiply this area by the value of f at that same point (x, y) to get $f(x, y)dxdy$. The symbol $\int\int_R f(x, y)\, dxdy$ represents, intuitively, the sum of all the terms $f(x, y)dxdy$ as (x, y) ranges over all of the points in R. This interpretation helps immensely in setting up integrals and also in setting up applications.

Mathematica is an excellent environment to help us gain some insight into this definition, and so we begin our study of double integrals with an example involving approximation by Riemann sums.

Example 12.1 *Let $f(x, y) = xe^{xy}$, and let R be the rectangle $R = [-1, 2] \times [1, 3]$. Approximate the value of $\int\int_R f(x, y)dxdy$ with a Riemann sum obtained by partitioning R into 20^2 subrectangles using a grid of 20 equally spaced horizontal lines and 20 equally spaced vertical lines and by letting (x_j, y_j) be the midpoint of the j-th subrectangle for each $j = 1, 2, \ldots, 400$.*

Setting up the partition points and the corresponding midpoints is easily accomplished with Mathematica's Table[ ] command. The coordinates of the midpoints of the subrectangles are easier to specify with a double indexing scheme.

```
In[1]:= f[x_,y_]:=x*Exp[x*y]
```

```
In[2]:= lx=(2-(-1))/20;ly=(3-1)/20;
```

```
In[3]:= X=Table[-1+j*lx,{j,0,20}]
```

Out[3]= $\left\{-1, -\frac{17}{20}, -\frac{7}{10}, -\frac{11}{20}, -\frac{2}{5}, -\frac{1}{4}, -\frac{1}{10}, \frac{1}{20}, \frac{1}{5}, \frac{7}{20}, \frac{1}{2}, \frac{13}{20}, \frac{4}{5}, \frac{19}{20}, \frac{11}{10}, \frac{5}{4}, \frac{7}{5}, \frac{31}{20}, \frac{17}{10}, \frac{37}{20}, 2\right\}$

```
In[4]:= Y=Table[1+j*ly,{j,0,20}]
```

Out[4]= $\left\{1, \frac{11}{10}, \frac{6}{5}, \frac{13}{10}, \frac{7}{5}, \frac{3}{2}, \frac{8}{5}, \frac{17}{10}, \frac{9}{5}, \frac{19}{10}, 2, \frac{21}{10}, \frac{11}{5}, \frac{23}{10}, \frac{12}{5}, \frac{5}{2}, \frac{13}{5}, \frac{27}{10}, \frac{14}{5}, \frac{29}{10}, 3\right\}$

We set up the x-coordinates and y-coordinates of the partition midpoints next.

```
In[5]:= mx=Table[(X[[j]]+X[[j+1]])/2,{j,1,20}]
```

Out[5]= $\left\{-\frac{37}{40}, -\frac{31}{40}, -\frac{5}{8}, -\frac{19}{40}, -\frac{13}{40}, -\frac{7}{40}, -\frac{1}{40}, \frac{1}{8}, \frac{11}{40}, \frac{17}{40}, \frac{23}{40}, \frac{29}{40}, \frac{7}{8}, \frac{41}{40}, \frac{47}{40}, \frac{53}{40}, \frac{59}{40}, \frac{13}{8}, \frac{71}{40}, \frac{77}{40}\right\}$

```
In[6]:= my=Table[(Y[[j]]+Y[[j+1]])/2,{j,1,20}]
```

Out[6]= $\left\{\frac{21}{20}, \frac{23}{20}, \frac{5}{4}, \frac{27}{20}, \frac{29}{20}, \frac{31}{20}, \frac{33}{20}, \frac{7}{4}, \frac{37}{20}, \frac{39}{20}, \frac{41}{20}, \frac{43}{20}, \frac{9}{4}, \frac{47}{20}, \frac{49}{20}, \frac{51}{20}, \frac{53}{20}, \frac{11}{4}, \frac{57}{20}, \frac{59}{20}\right\}$

```
In[7]:= S=N[Sum[f[mx[[j]],my[[k]]]*lx*ly,{j,1,20},{k,1,20}]]
Out[7]= 126.163
```

Notice that the **familiar Sum[ ] command is used to create this double sum by inserting two indexing ranges inside its argument.**

By way of comparison, Mathematica's decimal value for this integral turns out to be 127.4384921, as we shall see shortly.

The definition of a double integral can be extended to planar regions other than rectangles as follows. Let $f(x, y)$ be a bounded real valued function of x and y defined over an arbitrary bounded planar region D. We enclose D in a rectangle R $(D \subseteq R)$, and extend the definition of f to all of R by defining $f(x, y)$ to be 0 for all points (x, y) in R outside of D. Then define

$$\int\int_D f(x, y)\, dxdy = \int\int_R f(x, y)\, dxdy.$$

It can be shown that the double integral of a continuous function f over a planar region D, always exists, as long as D is not "unreasonably complicated." Conditions on D which would imply the existence of the integral are discussed in any standard calculus text.

Example 12.2 *Let $f(x,y) = 2x^2 + 5y^2$, and let D be the closed disk of radius 2 centered at the origin. Approximate the value of $\int\int_D f(x,y)dxdy$ with a Riemann sum obtained by partitioning $R = [-2,2] \times [-2,2]$ into 20^2 subrectangles using a grid of 20 equally spaced horizontal lines and 20 equally spaced vertical lines, and by letting (x_j, y_j) be the midpoint of the j-th subrectangle for each $j = 1, 2, \ldots, 400$.*

Since the rectangle $R = [-2,2] \times [-2,2]$ is defined by the same interval in both the x and y directions, we only need to partition one of these intervals to create the rectangular partition of R.

```
In[8]:= f[x_,y_]:=If[x^2+y^2<=4,2x^2+5y^2,0]
```

This extends the given function $f(x,y)$ to be 0 everywhere outside of closed disk D.

```
In[9]:= l=4/20;
```

```
In[10]:= P=Table[-02+j*l,{j,0,20}]
```

Out[10]= $\{-2, -\frac{9}{5}, -\frac{8}{5}, -\frac{7}{5}, -\frac{6}{5}, -1, -\frac{4}{5}, -\frac{3}{5}, -\frac{2}{5}, -\frac{1}{5}, 0, \frac{1}{5}, \frac{2}{5}, \frac{3}{5}, \frac{4}{5}, 1, \frac{6}{5}, \frac{7}{5}, \frac{8}{5}, \frac{9}{5}, 2\}$

```
In[11]:= m=Table[(P[[j]]+P[[j+1]])/2,{j,1,20}]
```

Out[11]= $\{-\frac{19}{10}, -\frac{17}{10}, -\frac{3}{2}, -\frac{13}{10}, -\frac{11}{10}, -\frac{9}{10}, -\frac{7}{10}, -\frac{1}{2}, -\frac{3}{10}, -\frac{1}{10}, \frac{1}{10}, \frac{3}{10}, \frac{1}{2}, \frac{7}{10}, \frac{9}{10}, \frac{11}{10}, \frac{13}{10}, \frac{3}{2}, \frac{17}{10}, \frac{19}{10}\}$

```
In[12]:= S=N[Sum[f[m[[j]],m[[k]]]*l^2,{j,1,20},{k,1,20}]]
Out[12]= 88.9616
```

Mathematica's decimal value for this integral turns out to be 87.96459431, as we shall see next.

Double integrals are computed by Mathematica using the familiar Integrate[ ] command in an expected way. If D is the domain defined by

$$\{(x,y)|p(x) \leq y \leq q(x), a \leq x \leq b\},$$

then

A =Integrate[$f(x,y)$, $\{y, p(x), q(x)\}$]

defines the integral

$$\int_{p(x)}^{q(x)} f(x,y)\,dy,$$

and so,

B =Integrate[A, $\{x, a, b\}$]

defines the integral

$$B = \int_a^b \int_{p(x)}^{q(x)} f(x,y)\,dydx.$$

Of course, both integrals can be brought together in one input statement as

B =Integrate[Integrate[$f(x, y)$, {$y, p(x), q(x)$}], {x, a, b}].

Mathematica has the following abbreviated way of writing this double integral:

Integrate[$f(x,y)$, {x, a, b}, {$y, p(x), q(x)$}]

Notice that the two integration intervals appear in reverse order compared to the first input statement. With this notation, integration is performed first with respect to the outer (or right) variable and then with respect to the inner (or left) variable. This same pattern is used in higher order multiple integrals. Integration is performed first with respect to the variable furthest to the right, and the order of integration proceeds from right to left. This may strike you as such an unnatural way to enter the limits of integration, that you may prefer the first input statement, which is certainly more natural. On the other hand, this abbreviated notation is a real convenience (especially for triple integrals). There is no mathematical reason for choosing one over the other. For a few examples, we will use both methods, but then we shall defer to the convenience of the abbreviated notation.

Mathematica's decimal value for the integral in Example 12.1 and for the integral in Example 12.2, expressed as the iterated integral

$$\int_{-2}^{2}\int_{-\sqrt{4-x^2}}^{\sqrt{4-x^2}}(2x^2+5y^2)dxdy,$$

are computed as follows.

```
In[13]:= f=x*Exp[x*y];g=2x^2+5y^2;

In[14]:= N[Integrate[Integrate[f,{x,-1,2}],{y,1,3}]]
Out[14]= 127.438

In[15]:= N[Integrate[f,{y,1,3},{x,-1,2}]]
Out[15]= 127.438

In[16]:= N[Integrate[Integrate[g,{y,-Sqrt[4-x^2],Sqrt[4-x^2]}],{x,-2,2}]]
Out[16]= 87.9646

In[17]:= N[Integrate[g,{x,-2,2},{y,-Sqrt[4-x^2],Sqrt[4-x^2]}]]
Out[17]= 87.9646
```

The integrals $A=\int_{p(x)}^{q(x)} f(x,y)dy$ and $B=\int_a^b A\,dx$, which together determine the double integral

$$B=\int_a^b\int_{p(x)}^{q(x)} f(x,y)dydx,$$

can also be evaluated separately. By evaluating the integrals separately, the intermediate antidifferentiation processes can be observed. Integrating in this way can be a good learning experience.

Example 12.3 *Evaluate the iterated integral* $\int_{-4}^{1}\int_{3x+7}^{9-x^2} x\sin(y)\,dydx$ *directly, using a single Mathematica input statement. Evaluate the integral again, as a sequence of two separate single integrals. Use indefinite integration so that the individual antidifferentiation processes can be observed in greater detail.*

```
In[18]:= A=Integrate[x*Sin[y],{x,-3,3},{y,3x+7,9-x^2}]
```

Out[18]= $-\frac{\cos[2]}{9}+\frac{\cos[16]}{9}-\sin[2]+\sin[16]$

```
In[19]:= Integrate[Integrate[x*Sin[y],{y,3x+7,9-x^2}],{x,-3,3}]
```

Out[19]= $-\frac{\cos[2]}{9}+\frac{\cos[16]}{9}-\sin[2]+\sin[16]$

```
In[20]:= A1=Integrate[x*Sin[y],y]
Out[20]= -x cos[y]
```

```
In[21]:= A2=(A1/.y->9-x^2)-(A1/.y->3x+7)
```

Out[21]= $x\ \cos[7+3\ x]-x\ \cos\left[9-x^2\right]$

```
In[22]:= B1=Integrate[A2,x]
```

Out[22]= $\frac{1}{18}\left(2\ \cos[7+3\ x]+6\ x\ \sin[7+3\ x]+9\ \sin\left[9-x^2\right]\right)$

```
In[23]:= B2=(B1/.x->3)-(B1/.x->-3)
```

Out[23]= $\frac{1}{18}\ (-2\ \cos[2]-18\ \sin[2])+\frac{1}{18}\ (2\ \cos[16]+18\ \sin[16])$

The expression $B2$ looks slightly different from the answer we obtained by a direct evaluation of the double integral, but careful observation shows that the two answers are the same. To show that two answers are the same in more difficult cases, **subtract one answer from the other and show that their difference is 0**.

As we mentioned in the introduction, the definition of a double integral, as a limit of Riemann sums, is itself a numerical decimal approximation scheme, although not a very good one. More effective methods are available with the familiar NIntegrate[ ] command, which has been used to give decimal approximations to single integrals. Actually, there are several methods one can choose from, but if a method is not selected, the default method will usually be satisfactory. Look at the Help File, if you wish to try another method. The input statement:

NIntigrate[$f(x,y)$, $\{x,a,b\}$, $\{y,p(z),q(x)\}$]
immediately activates a genuine numerical approximation technique, without a search for an antiderivative, regardless of whether $f(x,y)$ has or does not have an elementary antiderivative. By way of contrast, the input statement:

N[Intigrate[$f(x,y)$, $\{x,a,b\}$, $\{y,p(z),q(x)\}$]]

first attempts an exact evaluation by way of antidifferentiation, before it gives a decimal approximation, and it uses a genuine numerical approximation technique only if an antiderivative cannot be found.

For double, and even more so, for triple integrals, the performance differences can be quite dramatic! As you can imagine, numerical approximations techniques for double integrals take considerable time even for simple integrals. It often takes less time for Mathematica to evaluate a double integral by antidifferentiation techniques. Even simple integrals which might take Mathematica a few seconds to evaluate using the command N[Integrate[ ]] might take several minutes to evaluate using the command NIntegrate[ ].

If Mathematica cannot evaluate a double integral exactly, it may still be possible to avoid using Mathematica's (occasionally) slow numerical techniques for double integrals. Try setting up the integral as a sequence of two separate single integrals. If the first integral can be evaluated exactly, then numerical techniques only have to be applied to the second single integral. As we have noticed throughout this manual, numerical techniques on single integrals are usually quite fast.

Thankfully, Mathematica will usually respond to our requests with a quick answer. Only rarely, does it balks, and it is only then that we have to resort to these more delicate maneuvers.

Example 12.4 *Let D be the region in the x,y-plane bounded by the curves $y = 2e^{3x}$ and $y = 7x^3 + 28$. Evaluate both iterated integrals associated with $\int\int_D x^2y^3 dxdy$ and verify their equivalence.*

To set up the first iterated integral, fix x, and collect the terms $f(x,y)dxdy$ with respect to y as y goes from its value on the bottom curve to its value on the top curve. This determines (in a manner of speaking) a vertical "line of values" at x. To finish up the integral we collect these "lines" with respect to x as x goes from the left hand "corner" of the region D to the right hand "corner" of D.

```
In[24]:= f=2Exp[3x];g=7x^3+28;
```

```
In[25]:= Plot[{f,g},{x,-2,1}];
```

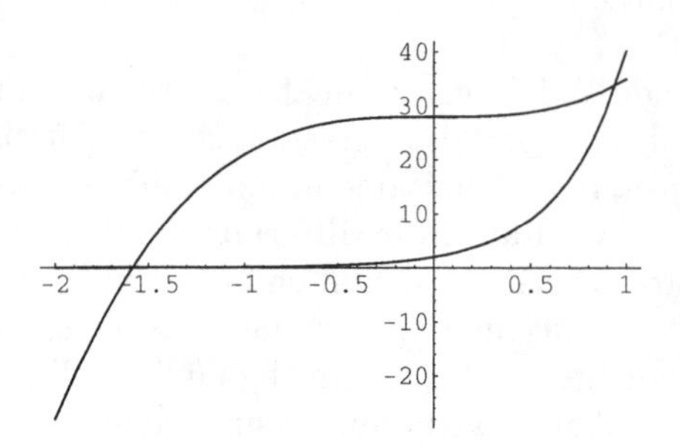


```
In[26]:= s1=FindRoot[f==g,{x,-1.5}]
Out[26]= {x → -1.58708}
```

```
In[27]:= s2=FindRoot[f==g,{x,1}]
Out[27]= {x → 0.943155}
```

```
In[28]:= a=(x/.s1);b=(x/.s2);
```

```
In[29]:= A1=Integrate[x^2*y^3,{x,a,b},{y,f,g}]
Out[29]= 85705.2
```

To set up the other iterated integral, fix y, and collect the terms $f(x,y)dxdy$ with respect to x as x goes from its value on the left hand curve to its value on the right hand curve. To find these x-values, we must solve the equations $y = g(x)$ and $y = f(x)$ for x in terms of y. We denote their values by xl (for x on the left) and xr (for x on the right). This collection gives us (in a manner of speaking) a horizontal "line of values" at y. To finish the double integral, we collect these "lines' with respect to y as y goes from the bottom "corner" of D to the upper "corner" of D.

```
In[30]:= X1=Solve[f==y,x]
Solve::ifun : Inverse functions are being used
    by Solve, so some solutions may not be found.
```

Out[30]= $\{\{x \to \frac{1}{3} \log[\frac{y}{2}]\}\}$

```
In[31]:= xr=x/.X1[[1]]
```

Out[31]= $\frac{1}{3} \log[\frac{y}{2}]$

```
In[32]:= X2=Solve[g==y,x]
```

Out[32]= $\{\{x \to -\left(-\frac{1}{7}\right)^{1/3} (-28+y)^{1/3}\}, \{x \to \frac{(-28+y)^{1/3}}{7^{1/3}}\}, \{x \to \frac{(-1)^{2/3} (-28+y)^{1/3}}{7^{1/3}}\}\}$

This is a major complication! Which solution do we want? Remember that Mathematica will **always evaluate a negative number raised to a fractional power, as a complex number**. Actually, the original equation is not that complicated, so we make a mid-course correction, drop the above approach, and try the following.

```
In[33]:= s=Solve[(g/.x^3->p)==y,p]
```

Out[33]= $\{\{p \to \frac{1}{7} (-28+y)\}\}$

We want the real valued cubed root of this expression. Mathamatica will only cooperate when $(-28+y)$ is nonnegative, that is to say, when $y \geq 28$. When $y < 28$, the cubed root of the resulting negative expression will be complex valued. To avoid this problem, we factor out the negative sign, so that we are computing the cubed root of a positive number. We have had to use this approach before to avoid complex valued output.

```
In[34]:= {xl1=-((28-y)/7)^(1/3),xl2=((y-28)/7)^(1/3)}
```

Out[34]= $\{-\frac{(28-y)^{1/3}}{7^{1/3}}, \frac{(-28+y)^{1/3}}{7^{1/3}}\}$

We will need the y-coordinates of the region's corner points. Either f or g can be used; we evaluate both as a means of checking values.

```
In[35]:= {c=(g/.x->a),d=(g/.x->b),f/.x->a,g/.x->b}
Out[35]= {0.0171101, 33.8728, 0.0171101, 33.8728}
```

The formula for the x-coordinate on the left changes depending on whether $y < 28$ or $y \geq 28$, so we must split the integral into two parts.

```
In[36]:= Integrate[x^2*y^3,{y,c,28},{x,xl1,xr}]+
         Integrate[x^2*y^3,{y,28,d},{x,xl2,xr}]

Out[36]= 85705.2
```

If Mathematica cannot evaluate a double integral after it is set up as one iterated integral, it is at least possible that the integral could still be evaluated by changing the order of integration, and evaluating the alternate iterated integral instead.

Example 12.5 *Evaluate the double integral*

$$\int_0^5 \int_{y^2}^{25} e^{-x\sqrt{x}/10} y^3 \, dx dy$$

We first try to evaluate the integral directly. A glance at the integral would suggest that Mathematica probably cannot find the antiderivative with respect to x of $e^{x\sqrt{x}/10}$.

```
In[37]:= f=Exp[-x*Sqrt[x]/10]*y^3
```

Out[37]= $e^{-\frac{x^{3/2}}{10}} y^3$

```
In[38]:= Integrate[f,{y,0,5},{x,y^2,25}]
Out[38]= $Aborted
```

After several minutes of patiently waiting for an answer, we pulled down the **Kernel Menu** and clicked on **Abort Evaluation**. To set up the alternate iterated integral, we must determine D—the region of integration. There is no need to use Mathematica here. In the integral, as it is expressed in the example, for y fixed, x goes from $x = y^2$ to $x = 25$, and then y goes from $y = 0$ to $y = 5$. It follows that D is the region bounded by $y = \sqrt{x}$, $x = 25$ and the x-axis.

```
In[39]:= Integrate[f,{x,0,25},{y,0,Sqrt[x]}]
```

Out[39]= $\frac{50}{3} - \frac{225}{e^{25/2}}$

When a double integral is transformed into a polar coordinate integral, the variables x and y are replaced by

$$x = r\cos(\theta), \quad y = r\sin(\theta),$$

and the area differential $dxdy$ is replaced by $rdrd\theta$. Of course, the differentials are not inserted inside Mathematica's integration command, but when polar coordinated integrals are set up, the extra factor of r must be inserted in the integrand.

Example 12.6 *Let D be the region bounded by $x + y \geq 0$ and $x^2 + y^2 \leq 4$. Evaluate the double integral*

$$\int\int_D xye^y \, dxdy$$

first as a cartesian integral and then as a polar coordinate integral.

The boundary curves are plotted below. The region D lies inside both the circular disk and the "upper," or "right" half plane formed by the line. The points of intersection between the two curves are $(x, y) = (\mp\sqrt{2}, \pm\sqrt{2})$. From the plot, we can see that the cartesian integral will have to be split into two integrals, depending on whether $x \leq \sqrt{2}$ or $x \geq \sqrt{2}$.

```
In[40]:= f[x_,y_]:=x*y*Exp[y]; eq=x^2+y^2==4;
```

```
In[41]:= <<Graphics`ImplicitPlot`
```

```
In[42]:= ImplicitPlot[{eq,x+y==0},{x,-3,3},AspectRatio->1];
```

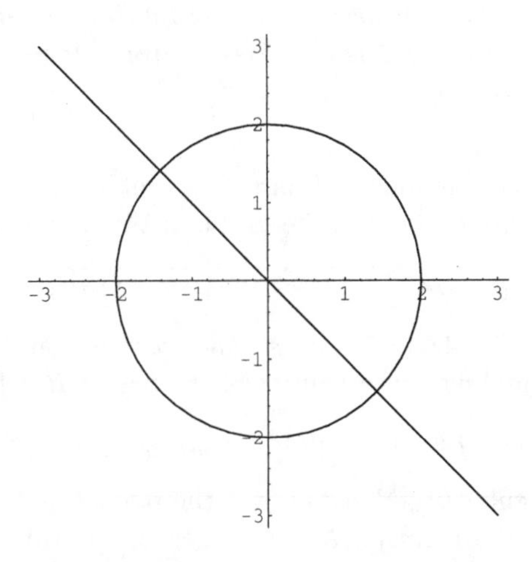


```
In[43]:= cart=Integrate[f[x,y],{x,-Sqrt[2],Sqrt[2]},{y,-x,Sqrt[4-x^2]}]+
         Integrate[f[x,y],{x,Sqrt[2],2},{y,-Sqrt[4-x^2],Sqrt[4-x^2]}]
```

Integrate :: gener : Unable to check convergence

Out[43]= $-2\left(5+3\sqrt{2}\right)e^{-\sqrt{2}}-2\left(-5+3\sqrt{2}\right)e^{\sqrt{2}}$

The letter θ can be used in a Mathematica work session by using \[Theta]. We used the letter t instead, just for the sake of convenience. To set up the limits of integration, think of holding $\theta = t$ fixed, and "collect" differentials terms with respect to r as r goes from $r = 0$ to $r = 2$. This sweeps out a "thin circular sector" at $\theta = t$ of angle opening $d\theta = dt$. Then we collect these "thin sectors" with respect to $\theta = t$ as t goes from $t = -\pi/4$ to $t = 3\pi/4$.

```
In[44]:= pol=Integrate[f[r*Cos[t],r*Sin[t]]*r,{r,0,2},{t,-Pi/4,3Pi/4}]
```

Out[44]= $-2\left(5+3\sqrt{2}\right)e^{-\sqrt{2}}-2\left(-5+3\sqrt{2}\right)e^{\sqrt{2}}$

Mathematica labored over the cartesian integral. The polar integral was clearly easier to set up and easier for Mathematica to calculate.

12.2 Applications of Double Integrals

Area is the most immediate application of a double integral. If D is a region in the x,y-plane, then $\int\int_D dxdy$ is the area of the region D. Obviously, area is also an application of the more familiar single variable integral, and there is no real advantage gained by treating area as a double integral. Volume, on the other hand, is an application that depends, more fundamentally, on double integration.

Example 12.7 *A scenic building in the shape of a circular cylinder with a hemispheric top is built into the side of a mountain. Determine the volume of the building. The mountain is the surface defined by the graph of*

$$z = m(x,y) = 4850 - 0.0019x^2 - 0.0037y^2,$$

where x, y, and z are expressed in feet. The radius of the cylinder is 48 feet. and the central axis of the cylinder (which is vertical, of course) is 83 feet tall from where it strikes the side of the mountain to the top of the cylinder. The axis passes through the point on the mountain corresponding to $(x_0, y_0) = (79, 113)$. This defines the cylinder, and the hemispheric top sits on top of this cylinder.

The hemisphere is the top half of the sphere of radius 48, centered at the point $(79, 113, z_0)$, where $z_0 = m(79, 113) + 83$ Consequently, it is the graph of the function

$$s(x, y) = z_0 + \sqrt{48^2 - (x - 79)^2 + (y - 113)^2}.$$

If D is the closed disk of radius 48 in the x, y-plane, centered at $(x_0, y_0) = (79, 113)$, then the volume of the scenic building is the volume of the region R defined by

$$R = \{(x, y, z) | (x, y) \text{ is in } D, \text{ and } m(x, y) \leq z \leq s(x, y)\}$$

Think of a thin, vertical, rectangular tube located over the point (x, y) in D and extending from the mountain's surface to the hemisphere. The area of the tube's base is $dxdy$, its height is $(s(x, y) - m(x, y))$, and so its volume is $(s(x, y) - m(x, y))dxdy$. We simply add, in the sense of a double integral, the volumes of all of these thin tubes, as (x, y) ranges over D.

There is no need to plot these surfaces to evaluate the volume of R. It would be nice to see (in a manner of speaking) the building, and how it sits on the mountain, and it would be a good opportunity to bring several somewhat incompatible plots together using the Show[ ] command. Unfortunately, this is not an easy task, and since a plot is not critical, we will not provide one.

```
In[45]:= m[x_,y_]:=4850-0.0019x^2-0.0037y^2

In[46]:= z0=m[79,113]+83
Out[46]= 4873.9

In[47]:= s[x_,y_]:=z0+Sqrt[48^2-(x-79)^2-(y-113)^2]

In[48]:= Y=Solve[(x-79)^2+(y-113)^2==48^2,y]
```

Out[48]= $\{\{y \to 113 - \sqrt{-3937 + 158\ x - x^2}\}, \{y \to 113 + \sqrt{-3937 + 158\ x - x^2}\}\}$

```
In[49]:= y1=(y/.Y[[1]]);y2=(y/.Y[[2]]);

In[50]:= {a=79-48,b=79+48,c=113-48,d=113+48}
Out[50]= {31, 127, 65, 161}

In[51]:= vol=NIntegrate[s[x,y]-m[x,y],{x,a,b},{y,y1,y2}]
```

Out[51]= $855744. + 2.10574\ 10^{-46}\ i$

Obviously, the small complex valued term can be dropped. The volume of the building is 855,744 cubic feet.

Let D be a region (a flat plate) in the x, y-plane. Let $\rho = \rho(x, y)$ denote the mass density of the plate at the point (x, y) in D. Then double integrals can be used to determine the total mass of the plate and its center of mass.

Example 12.8 *Determine the mass and the center of mass of the flat plate bounded by the curves*

$$y = x^4 - 14x^3 + 68x^2 - 136x + 86, \text{ and } y = 23 + 12x - x^2,$$

if the mass density $\rho = \rho(x, y)$ at the point (x, y) on the plate is

$$\rho(x, y) = 0.03(8 - x - y)^2.$$

Let D denote the region bounded by the two polynomial curves. Think of a small square of dimensions $dx \times dy$ located at the point (x, y) in D. Its mass (density times area) is $\rho(x, y)dxdy$, its moment with respect to the y-axis (directed distance to the y-axis times mass) is $x\rho(x, y)dxdy$, and its moment with respect to the x-axis (directed distance to the x-axis times mass) is $y\rho(x, y)dxdy$. To find the mass of the whole plate, we add the individual masses, in the sense of a double integral as (x, y) ranges over the region D. The total moments are found in the same way.

```
In[52]:= f=x^4-14x^3+68x^2-136x+86; g=23+12x-x^2; ρ =0.03(8-x-y)^2;
```

```
In[53]:= Plot[{f,g},{x,0,7}];
```

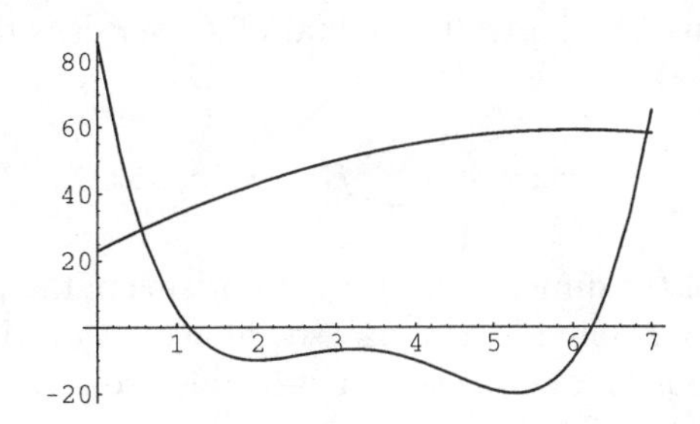


```
In[54]:= s=NSolve[f-g==0,x]
Out[54]= {{x → 0.552795}, {x → 3.25089 - 2.41673 i},
          {x → 3.25089 + 2.41673 i}, {x → 6.94543}}
```

```
In[55]:= a=(x/.s[[1]]);b=(x/.s[[4]]);
```

```
In[56]:= M=NIntegrate[ρ ,{x,a,b},{y,f,g}]
Out[56]= 7249.22
```

```
In[57]:= Mx=NIntegrate[y*ρ ,{x,a,b},{y,f,g}]
Out[57]= 295991.
```

```
In[58]:= My=NIntegrate[x*ρ ,{x,a,b},{y,f,g}]
Out[58]= 33636.7
```

```
In[59]:= CenterMass={My/M,Mx/M}
Out[59]= {4.64005, 40.8308}
```

12.3 Triple Integrals

Before we start this section, it should be pointed out that evaluating a triple integral, almost always starts with a carefully drawn 3-dimensional picture of the region of integration. In order for the plot to help, it must be turned just right, and fine-tuned to get what is needed from the picture. This is not an easy task and usually requires a trial and error process of drawing several plots until a useful one is found. In this manual, only the final plot is shown, but each one is the result of this trial and error process.

The definition of a triple integral is very similar to the definition of a double integral. Here, briefly, are some of the highlights.

Suppose that $f(x, y, z)$ is a bounded real valued function of x, y, and z defined on a solid region R. Enclose R inside a rectangular prism R^*, of the form

$$R \subseteq R^* = [a_1, a_2] \times [b_1, b_2 \times [c_1, c_2],$$

(unless it is already a set of this form) and extend the definition of f by defining f to be zero at points of R^*, which are outside of R. The prism R^* is partitioned, in the expected way, into smaller prisms. A triple indexing scheme would usually be used, but for the sake of simplicity, simply order the subprisms in some way to form a partition $\{P_1, P_2, \ldots, P_n\}$. Let $\Delta x_j \times \Delta y_j \times \Delta z_j$ denote the dimensions of the j-th subprism P_j in the partition, and let (x_j, y_j, z_j) denote a point in P_j. The triple integral of f over R is defined as a limit of of the form

$$\int\int\int_R f(x, y, z)\, dxdydz = \lim \sum f(x_j, y_j, z_j) \Delta x_j \Delta y_j \Delta z_j,$$

where the limit is taken as all of the dimensions of the subprisms in the partition go to zero. The details can be found in any standard calculus text, along with a theorem stating that the triple integral of a continuous function over a "reasonable" solid R always exists.

Think of $dx\, dy\, dz$ as the volume of a infinitesimally small prism ("box") of dimensions dx by dy by dz located at the point (x, y, z) in R. Multiply this volume by the value of f at that same point (x, y, z) to form the term $f(x, y, z)dxdydz$. The symbol $\int\int\int_R f(x, y, z)\, dxdydz$ represents, intuitively, the sum of all the terms $f(x, y, z)dxdydz$ as (x, y, z) ranges over all of the points in R. This interpretation helps immensely in setting up integrals and also in setting up applications.

Mathematica is an excellent environment for constructing and evaluating Riemann sums and effort in this direction may help gain some insight into the definition of a triple integral. Examples 12.1 and 12.2 can be used as models to form similar sums for triple integrals, and a few problems of this sort appear in the exercise set.

The input statements:

Integrate[Integrate[Integrate[$f(x, y, z)$, $\{z, p(x, y), q(x, y)\}$], $\{y, u(x), v(x)\}$],$\{x, a, b\}$]

and,

Integrate[$f(x, y, z)$, $\{x, a, b\}$, $\{y, u(x), v(x)\}$, $\{z, p(x, y), q(x, y)\}$]

are equivalent ways of evaluating the triple integral

$$\int_a^b \int_{u(x)}^{v(x)} \int_{p(x,y)}^{q(x,y)} f(x, y, z)\, dz\, dy\, dx.$$

With triple integrals, the abbreviated notation is even more of a convenience, but again, be forewarned that the **order of integration is entered in a reverse** and somewhat unnatural order in the abbreviated notation. For decimal approximations, the commands

NIntegrate[ ] and N[Integrate[ ]], and various other combinations, are used here as they were used in double integrals. Recall that the commands are not entirely equivalent, and the performance differences between the various combinations is even more dramatic with triple integrals than with double integrals (See page 261).

Example 12.9 *Evaluate the triple integral* $\int\int\int_R 4x^2y^3\,dxdydz$ *where* R *is the region bounded by the planes* $y = 0$, $z = 0$, $z = 4 - x - y$, *and* $z = 8 + 2x - y$. *Evaluate the integral a second time as a sequence of three separate single integrals. Use indefinite integration so that the individual antidifferentiation processes can be observed in greater detail.*

Some thought must be given to how the various plot commands are to be used, in order to get Mathematica to present a useful picture of the region R. If too much of each plane is shown, the region will be lost inside all of the drawings. In the present case a simple devise works well. **Notice the use of the PlotRange option, inside of the Plot3D[ ] command. The bounding box is used to show some of the planar faces** (and some numerical values). **Adding x, y, z labels to the axes** definitely helps set up the integral. Finally, in a key step, **the plot is turned around to see the planar faces $y = 0$ and $z = 0$.**

The plot shows that we **must** integrate first with respect to x if we wish to set up the problem in terms of just one (triple) integral.

```
In[60]:= p1=Plot3D[4-x-y,{x,-4,4},{y,0,16/3},PlotRange->{0,16/3},
         DisplayFunction->Identity];
         p2=Plot3D[8+2x-y,{x,-4,4},{y,0,16/3},PlotRange->{0,16/3},
         DisplayFunction->Identity];
```

```
In[61]:= Show[{p1,p2},DisplayFunction->$DisplayFunction,
         AxesLabel->{"x","y","z"}];
```

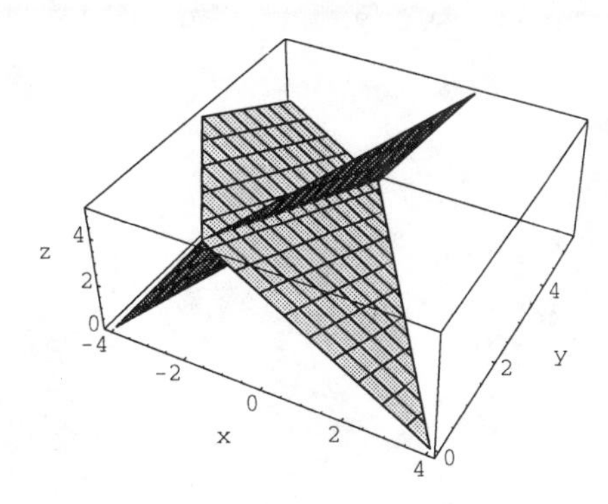

Look at this plot. For an arbitrary y and z, think of collecting terms first with respect to x, as x goes from one oblique plane to the next (from smaller to larger values of x). This forms, at each y and z, a horizontal "line of values" (for lack of a better phrase) parallel to the $y = 0$ plane.

```
In[62]:= {s1=Solve[z==4-x-y,x],s2=Solve[z==8+2x-y,x]}
```

Out[62]= $\{\{\{x \to 4 - y - z\}\}, \{\{x \to \frac{1}{2}\ (-8 + y + z)\}\}\}$

```
In[63]:= x1=(x/.s1[[1]]);x2=(x/.s2[[1]]);
```

Then, collect these "lines" with respect to y, as y goes from $y = 0$ to where it intersects the two oblique planes. This y-coordinate, denoted by $y1$, in the next input statement, is found by setting $x1 = x2$ and solving for y. We now have—to coin another suggestive phrase—a horizontal "plane of values" for each z.

```
In[64]:= s3=Solve[x1==x2,y]
```

Out[64]= $\{\{y \to \frac{1}{3}(16 - 3z)\}\}$

```
In[65]:= y1=(y/.s3[[1]]);
```

Finally, we collect these "planes" as z goes from $z = 0$ to $z = 16/3$, this last value being quite evident from the formula for $y1$.

```
In[66]:= Integrate[4x^2*y^3,{z,0,16/3},{y,0,y1},{x,x2,x1}]
```

Out[66]= $\frac{134217728}{76545}$

```
In[67]:= Integrate[Integrate[Integrate[4x^2*y^3,{x,x2,x1}],
         {y,0,y1}], {z,0,16/3}]
```

Out[67]= $\{\frac{134217728}{76545}\}$

The individual antidifferentiation processes involved in the evaluation of this integral can be observed in the following way.

```
In[68]:= A1=Integrate[4x^2*y^3,x]
```

Out[68]= $\frac{4x^3y^3}{3}$

```
In[69]:= A2=(A1/.x->x1)-(A1/.x->x2)
```

Out[69]= $\frac{4}{3}y^3(4 - y - z)^3 - \frac{1}{6}y^3(-8 + y + z)^3$

```
In[70]:= B1=Integrate[A2,y]
```

Out[70]= $-\frac{3y^7}{14} + \frac{1}{12}y^6(40 - 9z) + \frac{1}{10}y^5(-192 + 80z - 9z^2) + \frac{1}{24}y^4(1024 - 576z + 120z^2 - 9z^3)$

```
In[71]:= B2=(B1/.y->y1)-(B1/.y->0)
```

Out[71]= $\frac{(40-9z)(16-3z)^6}{8748} - \frac{(16-3z)^7}{10206} + \frac{(16-3z)^5(-192+80z-9z^2)}{2430} + \frac{(16-3z)^4(1024-576z+120z^2-9z^3)}{1944}$

```
In[72]:= C1=Integrate[B2,z]
```

$$\text{Out[72]=}\ \frac{159383552\ z}{76545} - \frac{262144\ z^2}{243} + \frac{131072\ z^3}{405} - \frac{5120\ z^4}{81} + \frac{128\ z^5}{15} - \frac{4\ z^6}{5} + \frac{z^7}{21} - \frac{3\ z^8}{2240}$$

```
In[73]:= c2=(C1/.z->16/3)-(C1/.z->0)
```

$$\text{Out[73]=}\ \frac{134217728}{76545}$$

Example 12.10 *Find the volume of the region bounded by $x = 25 - y^2$, $x + 8y + 2z = 56$, $z = 0$ and $x = 0$. Confirm with an alternate calculation.*

It helps to plot the region involved with some care. Notice the method used in this example. **The bounding box is used to show the boundaries $x = 0$ and $z = 0$.**

The parabolic cylinder is plotted using the ParametricPlot3D[ ] command. We can not use the Plot3D[ ] command on this surface, if we wish to keep the z-axis vertical. We think of its equation as determining a point $\{x, y, z\}$, where y and z are arbitrary, and $x = 25 - y^2$. Alternately, write both equations in the form $x = f(y, z)$, and they can be plotted using the Plot3D[ ] command. It is not a critical issue, but the down side of this approach is that the x-axis will be vertical.

```
In[74]:= p1=ParametricPlot3D[{25-y^2,y,z},{y,-5,5},{z,0,60},
         DisplayFunction->Identity];
         p2=Plot3D[28-4y-x/2,{x,0,25},{y,-5,5},DisplayFunction->Identity];
```

```
In[75]:= Show[{p1,p2},BoxRatios->{2,2,2},DisplayFunction->$DisplayFunction,
         ViewPoint->{-2.206, 1.907, 1.718}];
```

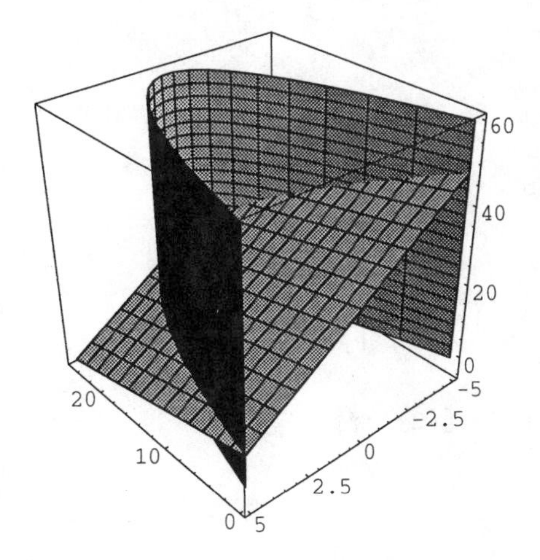

It is necessary to integrate first with respect to z, otherwise two integrals would be required.

```
In[76]:= Integrate[1,{y,-5,5},{x,0,25-y^2},{z,0,28-4y-x/2}]
```

Out[76]= $\frac{11500}{3}$

```
In[77]:= Integrate[Integrate[Integrate[1,{z,0,28-4y-x/2}],
         {x,0,25-y^2}],{y,-5,5}]
```

Out[77]= $\frac{11500}{3}$

To confirm this value, we use a different order of integration. There is only one other convenient order.

```
In[78]:= Integrate[1,{x,0,25},{y,-Sqrt[25-x],Sqrt[25-x]},{z,0,28-4y-x/2}]
```

Out[78]= $\frac{11500}{3}$

Example 12.11 *Determine the volume of the region bounded above by $x^2 + y^2 + z^2 = 25$ and below by $z = \sqrt{x^2 + y^2}/2$, as a triple integral in rectangular, cylindrical, and spherical coordinates,*

```
In[79]:= f=Sqrt[25-x^2-y^2];g=Sqrt[x^2+y^2]/2;
```

We begin with some preparations for the plot and for setting up the integrals. The equation, $f = g$, is the curve of intersection between the surfaces. We enter it without square roots.

```
In[80]:= eq=f^2-g^2==0
```

Out[80]= $25 - x^2 - y^2 + \frac{1}{4}\left(-x^2 - y^2\right) == 0$

```
In[81]:= eq=Simplify[eq]
```

Out[81]= $-\frac{5}{4}\left(-20 + x^2 + y^2\right) == 0$

The curve of intersection is a circle of radius $\sqrt{20}$.

```
In[82]:= s=Solve[eq,y]
```

Out[82]= $\left\{\left\{y \to -\sqrt{20 - x^2}\right\}, \left\{y \to \sqrt{20 - x^2}\right\}\right\}$

```
In[83]:= y1=(y/.s[[1]]);y2=(y/.s[[2]]);
```

```
In[84]:= a=Sqrt[20];
```

```
In[85]:= <<Graphics`ContourPlot3D`
```

```
In[86]:= p1=ContourPlot3D[x^2+y^2+z^2-25,{x,-a,a},{y,-a,a},
         {z,Sqrt[5],5},DisplayFunction->Identity];
         p2=Plot3D[g,{x,-a,a},{y,-a,a},DisplayFunction->Identity];
```

```
In[87]:= Show[{p1,p2},ViewPoint->{1.618, -2.849, 0.845},
         DisplayFunction->$DisplayFunction];
```

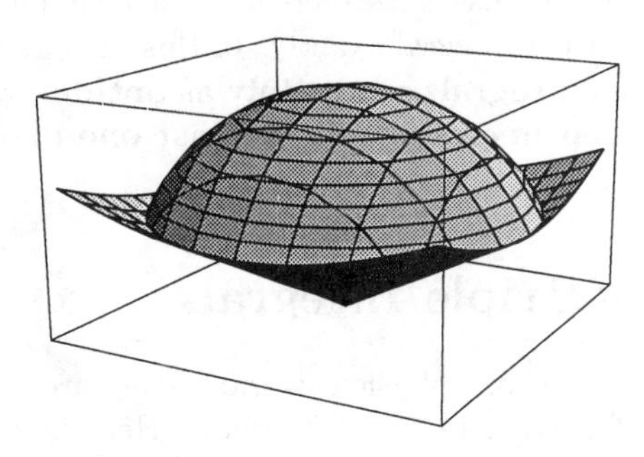

```
In[88]:= Integrate[1,{x,-a,a},{y,y1,y2},{z,g,f}]
Out[88]= $Aborted
```

This calculation was allowed to run for over four minutes, before we conceded defeat, pulled down the Kernel Menu and clicked on Abort Evaluation. Actually, Mathematica would eventually evaluate this integral (probably) if we were patient enough, but we have made our point. The integral is too complicated to compute in Cartesian Coordinates.

In spherical coordinates, it is much easier to compute this integral. The ranges on ρ and θ are straight forward, and the range on ϕ is determined by the angle at the vertex (from the positive z-axis). Elementary right triangle trigonometry shows that this vertex angle, from the positive z-axis, is $\arctan(\sqrt{20}/\sqrt{5}) = \arctan(2)$.

```
In[89]:= angle=ArcTan[2];
```

```
In[90]:= vol1=Integrate[ρ ^2*Sin[φ ],{θ ,0,2Pi},{φ ,0,angle},{ρ ,0,5}]
```

Out[90]= $2\left(\frac{125}{3} - \frac{25\sqrt{5}}{3}\right)\pi$

```
In[91]:= N[vol1]
Out[91]= 144.719
```

Greek letters look nice in input statements, but they are tedious to enter, as you may recall from our previous work. Look this topic up in the index to review this matter or to simplify the typing of input statements, use any three unassigned letters instead of these three Greek letters.

Finally, we compute the integral in cylindrical coordinates.

```
In[92]:= vol2=Integrate[r,{θ ,0,2Pi},{r,0,a},{z,r/2,Sqrt[25-r^2]}]
```

Out[92]= $2\left(\frac{125}{3} - \frac{25\sqrt{5}}{3}\right)\pi$

Before we begin the applications in the next section, and before we get too involved in the more applied exercises at the end of the chapter, something should be said about **numerical approximation processes for double and triple integrals**. The command NIntegrate[ ] is used to initiate a numerical approximation process. As you can imagine,

numerical approximation techniques for double and triple integrals often require a fairly a fairly large amount of computer time.

Frequently, Mathematica is able to use antidifferentiation techniques to compute the first or the first two integrals of a triple integral exactly. In this case, it **often takes much less computer time to do these integrals separately as antiderivatives and to use numerical techniques only when necessary on the last one or two integrals** (see page 261).

12.4 Applications of Triple Integrals

Centers of mass are very important in our physical world. They are needed to determine motion stability, for example, of a wide range of objects. Here is a classic type of an application of triple integrals.

Example 12.12 *The nose section of a high speed locomotive has the shape of the region bounded by the planes $z = 0$, $x = 0$, and the surface $z = 3.248 - 1.657y^2 - 0.1299x^2$, where x, y, and z are in meters. Naturally, a model for density can only be given approximately, but suppose that a reasonable value for the density, in ${}^{kg}/_{m^3}$, at the point (x, y, z) in the nose section is $\delta(x, y, z) = 2400 - 2(x-2)^2 - 3y^2 - 425(z-2)^2$. Determine the mass and the center of mass of the nose section.*

The region should be plotted, so that we see the nose section in a true scale. As you can see, other adjustments of the plot are made below to get a good view.

Because of the symmetry of both the locomotive and the density function, it is obvious that the locomotive is balanced on the $y = 0$ coordinate plane, so that the moment Mxz with respect to this coordinate plane is zero, as is the y-coordinate of the center of mass. We compute this moment anyway, since it is hardly any extra work.

```
In[93]:= f=3.248-1.657y^2-0.1299x^2;
```

```
In[94]:= Plot3D[f,{x,0,5},{y,-1.5,1.5},PlotRange->{0,4},
         BoxRatios->{5,3,4},AxesLabel->{"x","y","z"}];
```

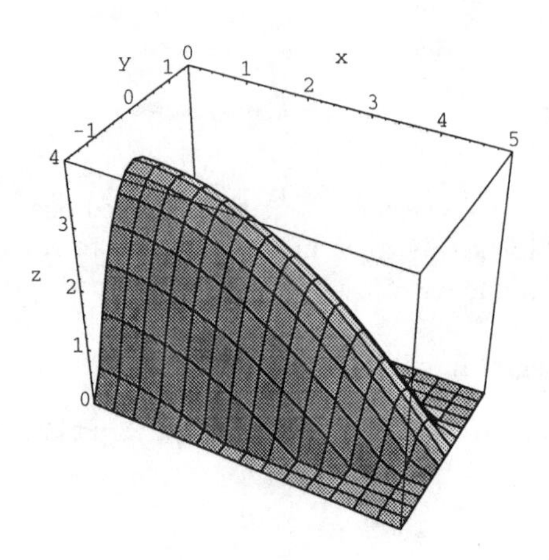


```
In[95]:= δ =2400-2(x-2)^2-3y^2-425(z-2)^2
```

```
Out[95]= 2400 - 2 (-2 + x)^2 - 3 y^2 - 425 (-2 + z)^2

In[96]:= s=Solve[f==0,y]
Out[96]= {{y -> -0.776853 Sqrt[3.248 - 0.1299 x^2]},
          {y -> 0.776853 Sqrt[3.248 - 0.1299 x^2]}}

In[97]:= y1=(y/.s[[1]]);y2=(y/.s[[2]]);

In[98]:= s1=Solve[(f/.y->0)==0,x]
Out[98]= {{x -> -5.00038}, {x -> 5.00038}}

In[99]:= x1=(x/.s1[[2]]);

In[100]:= M=Integrate[δ ,{x,0,x1},{y,y1,y2},{z,0,f}]
Out[100]= 31959.4

In[101]:= Mxy=Integrate[z*δ ,{x,0,x1},{y,y1,y2},{z,0,f}]
Out[101]= 40843.8

In[102]:= Mxz=Integrate[y*δ ,{x,0,x1},{y,y1,y2},{z,0,f}]
Out[102]= 0

In[103]:= Myz=Integrate[x*δ ,{x,0,x1},{y,y1,y2},{z,0,f}]
Out[103]= 51689.7

In[104]:= CenterMass={Myz/M,Mxz/M,Mxy/M}
Out[104]= {1.61736, 0, 1.27799}

In[105]:= Mass=M
Out[105]= 31959.4
```

In our last example, we solve a challenging problem, which requires the use of many of the ideas of multivariate calculus. Call this a celebration of the end of this manual and of the new found power we've acquired by blending mathematics and Mathematica.

Example 12.13 *a new uncharted planet has been detected, which has the typical circular ellipsoid shape of a planet which rotates about a central axis. The distance between the North and South Poles (the central axis length) is 2800 kilometers and the planet's equator is circular with a diameter of 3600 kilometers. Find the volume of the atmosphere, which is 30 kilometers deep.*

Place the origin of an x, y, z-coordinate system at the center of the planet, with the North Pole on the positive z-axis and the South Pole on the negative z-axis.

The volume of the atmosphere is more difficult to compute than it may appear to be at first glance. The ellipsoid is almost spherical. If it were, then the spherical coordinate variable ρ could be used to specify the altitude of a point above the planet's surface. As it stands, radial lines through the origin are not quite orthogonal to the surface. Consequently, if $(\rho_0, \theta_0, \phi_0)$ represents a point on the surface of the planet, then $(\rho_0 + 30, \theta_0, \phi_0)$ represents a point in the atmosphere which is (except for a few special points) not quite 30 kilometers in altitude. Ignoring this complication does not lead to a good approximation, and ultimately it dooms the problem and prevents a direct decimal solution.

Setting up a triple integral with ρ ranging from a point on the planet's surface, say $\rho = \rho_0$ to $\rho = \rho_0 + 30$ would appear to be a reasonable way to approximate the volume of the atmosphere, but it can be shown that it is not very accurate. The approach we use is both simpler and more accurate. Equally important, we are able to demonstrate the accuracy of the models used to approximate the volume.

The equation of the surface representing the upper limits of the atmosphere is very complicated and a simpler equation will be used instead. It would seem reasonable to simply add 30 to the elliptical dimensions of 1800 and 1400 which define the equation of the planet's surface. The resulting equation

$$\frac{x^2}{1830^2} + \frac{y^2}{1830^2} + \frac{z^2}{1430^2} = 1$$

turns out to be a reasonably good approximation of the boundary of the atmosphere.

With such a simple model, an approximate volume of the atmosphere will be easy to compute. Most of the effort in the Maple work session below will be devoted to measuring the accuracy of the model. We will show that the approximation is an under approximation. By adding slightly more than 30 to the dimensions of the ellipsoid, which define the planet's surface, we will form yet another good approximation of the upper boundary of the atmosphere. Using this model leads to an over approximation for the volume of the atmosphere. The actual volume, then, will be in between these two reasonably close approximations.

```
In[106]:= Planet=x^2/1800^2+y^2/1800^2+z^2/1400^2==1
```

Out[106]= $\frac{x^2}{3240000} + \frac{y^2}{3240000} + \frac{z^2}{1960000} == 1$

```
In[107]:= Upper=x^2/1830^2+y^2/1830^2+z^2/1430^2==1
```

Out[107]= $\frac{x^2}{3348900} + \frac{y^2}{3348900} + \frac{z^2}{2044900} == 1$

We plan to show that every point 30 kilometers above the planet's surface is outside of (or on) *Upper*. This will establish our claim that the volume of the region between *Planet* and *Upper* is an under approximation for the volume of the atmosphere.

Because of the rotational symmetry of the surfaces, this matter can be addressed in a simplified setting. We consider instead the ellipse generated by setting $y = 0$ in each of the two equations.

```
In[108]:= Planet0=(Planet/.y->0)
```

Out[108]= $\frac{x^2}{3240000} + \frac{z^2}{1960000} == 1$

```
In[109]:= Upper0=(Upper/.y->0)
```

Out[109]= $\frac{x^2}{3348900} + \frac{z^2}{2044900} == 1$

Let $\vec{r}$ be a vector from the origin to a point (x, z) on the ellipse *Planet0* and let $\vec{r_n}$ be a vector of length 30, which is normal to the ellipse *Planet0* at (x, z). We show that the components of the vector $\vec{r} + \vec{r_n}$ (that is, the coordinates of the head of this vector) are always outside of *Upper0*, the second ellipse. The figure shown below might help, although nothing is drawn to scale.

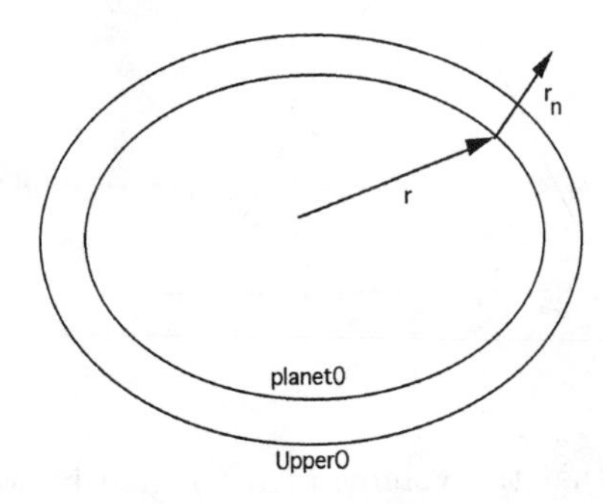

It is convenient to produce the vectors $\vec{r}$ and $\vec{rn}$ parametrically.

```
In[110]:= r={1800Cos[t],1400Sin[t]}
Out[110]= {1800 cos[t], 1400 sin[t]}
```

The vector $\vec{r}$ is the position vector for the ellipse *Planet0*.

```
In[111]:= rp=D[r,t]
Out[111]= {-1800 sin[t], 1400 cos[t]}
```

The vector $\vec{rp}$ is tangent to the ellipse. To create a normal vector $\vec{n}$, simply switch the components of $\vec{rp}$, and change the sign of one of its components to create a vector whose dot product with $\vec{rp}$ is zero. It is a simple matter to decide which component should take the change in sign in order to produce an outer normal rather than an inner normal. Then, we turn $\vec{n}$ into a unit vector $\vec{un}$.

```
In[112]:= n={rp[[2]],-rp[[1]]}
Out[112]= {1400 cos[t], 1800 sin[t]}
```

```
In[113]:= rn=Simplify[(30/Sqrt[n.n])n]
```

$$\text{Out[113]= } \left\{\frac{210\ \cos[t]}{\sqrt{65-16\ \cos[2\ t]}}, \frac{270\ \sin[t]}{\sqrt{65-16\ \cos[2\ t]}}\right\}$$

```
In[114]:= r30=r+rn
```

$$\text{Out[114]= } \left\{1800\ \cos[t] + \frac{210\ \cos[t]}{\sqrt{65-16\ \cos[2\ t]}},\ 1400\ \sin[t] + \frac{270\ \sin[t]}{\sqrt{65-16\ \cos[2\ t]}}\right\}$$

```
In[115]:= f=Simplify[Upper0[[1]]/.{x->r30[[1]],z->r30[[2]]}]
```

$$\text{Out[115]= } \frac{\cos[t]^2 \left(60 + \frac{7}{\sqrt{65-16\ \cos[2\ t]}}\right)^2}{3721} + \frac{\left(140 + \frac{27}{\sqrt{65-16\ \cos[2\ t]}}\right)^2 \sin[t]^2}{20449}$$

To show that the point corresponding to the head of the vector $\vec{r30}$ is always outside of (or on) the ellipse *Upper0*, it is enough to show that $f \geq 1$ for all t. Actually, because of symmetry, it is enough to show this for all t between 0 and $\pi/2$

```
In[116]:= Plot[f,{t,0,Pi/2}];
```

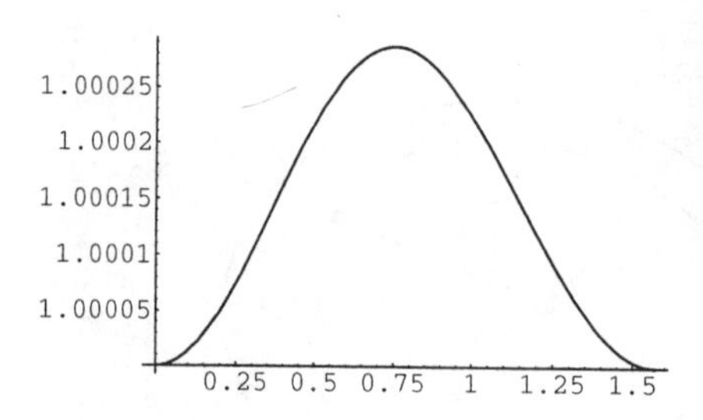

This establishes our claim that the volume of the region between *Planet* and *Upper* is less than the volume of the atmosphere.

Before we compute the volume, we first create a slightly larger ellipse that will contain the entire atmosphere in its interior. This will lead to an over approximation for the volume of the atmosphere.

```
In[117]:= UpperPlus=x^2/(1830+1/3)^2+y^2/(1830+1/3)^2+z^2/(1430+1/3)^2==1
```

Out[117]= $\frac{9\ x^2}{30151081} + \frac{9\ y^2}{30151081} + \frac{9\ z^2}{18412681} == 1$

```
In[118]:= UpperPlus0=(UpperPlus/.y->0)
```

Out[118]= $\frac{9\ x^2}{30151081} + \frac{9\ z^2}{18412681} == 1$

```
In[119]:= f1=Simplify[(UpperPlus0[[1]]/.{x->r30[[1]],z->r30[[2]]})]
```

Out[119]= $\frac{8100\ \cos[t]^2 \left(60+\frac{7}{\sqrt{65-16\ \cos[2\ t]}}\right)^2}{30151081} + \frac{900\ \left(140+\frac{27}{\sqrt{65-16\ \cos[2\ t]}}\right)^2\ \sin[t]^2}{18412681}$

```
In[120]:= Plot[f1,{t,0,Pi/2}];
```

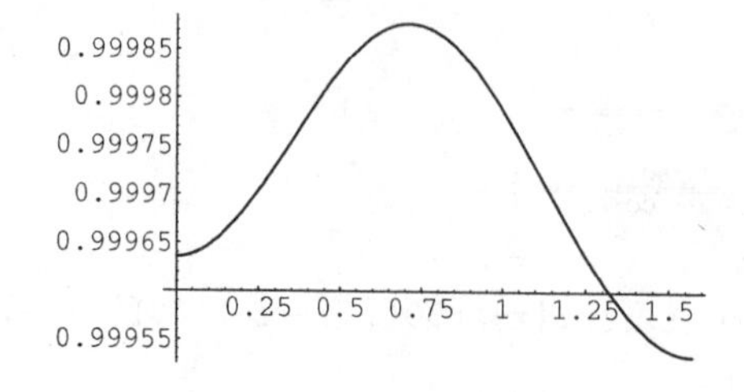

The volume of the region between *Planet* and *UpperPlus* will therefore be greater than the volume of the atmosphere.

All that is left to do is evaluate the two volumes. We use spherical coordinates to set up the integrals.

```
In[121]:= Planet2=(Planet/.{x^2->ρ ^2-y^2-z^2})
```

$$\text{Out[121]=}\quad \frac{y^2}{3240000}+\frac{z^2}{1960000}+\frac{-y^2-z^2+\rho^2}{3240000}==1$$

```
In[122]:= SphericalPlanet=Simplify[Planet2/.{z->ρ *Cos[φ ]}]
```

$$\text{Out[122]=}\quad \frac{\rho^2\ (65+16\ \cos[2\ \phi])}{158760000}==1$$

```
In[123]:= Upper2=(Upper/.{x^2->ρ ^2-y^2-z^2})
```

$$\text{Out[123]=}\quad \frac{y^2}{3348900}+\frac{z^2}{2044900}+\frac{-y^2-z^2+\rho^2}{3348900}==1$$

```
In[124]:= SphericalUpper=Simplify[Upper2/.{z->ρ *Cos[φ ]}]
```

$$\text{Out[124]=}\quad \frac{\rho^2\ (26969+6520\ \cos[2\ \phi])}{68481656100}==1$$

```
In[125]:= UpperPlus2=(UpperPlus/.{x^2->ρ ^2-y^2-z^2})
```

$$\text{Out[125]=}\quad \frac{9\ y^2}{30151081}+\frac{9\ z^2}{18412681}+\frac{9\ (-y^2-z^2+\rho^2)}{30151081}==1$$

```
In[126]:= SphericalUpperPlus=Simplify[UpperPlus2/.{z->ρ *Cos[φ ]}]
```

$$\text{Out[126]=}\quad \frac{9\ \rho^2\ (24281881+5869200\ \cos[2\ \phi])}{555162236258161}==1$$

```
In[127]:= s1=Solve[SphericalPlanet,ρ ]
```

$$\text{Out[127]=}\quad \left\{\left\{\rho\to-\frac{12600}{\sqrt{65+16\ \cos[2\ \phi]}}\right\},\left\{\rho\to\frac{12600}{\sqrt{65+16\ \cos[2\ \phi]}}\right\}\right\}$$

```
In[128]:= s2=Solve[SphericalUpper,ρ ]
```

$$\text{Out[128]=}\quad \left\{\left\{\rho\to-\frac{261690}{\sqrt{26969+6520\ \cos[2\ \phi]}}\right\},\left\{\rho\to\frac{261690}{\sqrt{26969+6520\ \cos[2\ \phi]}}\right\}\right\}$$

```
In[129]:= s3=Solve[SphericalUpperPlus,ρ ]
```

$$\text{Out[129]=}\quad \left\{\left\{\rho\to-\frac{23561881}{3\ \sqrt{24281881+5869200\ \cos[2\ \phi]}}\right\},\left\{\rho\to\frac{23561881}{3\ \sqrt{24281881+5869200\ \cos[2\ \phi]}}\right\}\right\}$$

```
In[130]:= RPlanet=(ρ /.s1[[2]]);RUpper=(ρ /.s2[[2]]);
         RUpperPlus=(ρ /.s3[[2]]);
```

Our integrals are independent of θ, so integration with respect to θ contributes a multiple of 2π.

```
In[131]:= MinVol=2Pi*NIntegrate[ρ ^2*Sin[φ ],{φ ,0,Pi},{ρ ,RPlanet,RUpper}]

Out[131]= 1.05946 10^9

In[132]:= MaxVol=2Pi*NIntegrate[ρ ^2*Sin[φ ],{φ ,0,Pi},{ρ ,RPlanet,RUpperPlus}]

Out[132]= 1.07144 10^9
```

The volume of the atmosphere (in cubic kilometers) is between *Min Vol* and *Max Vol*. On the negative side, it should be said that an approximation technique is of greater value if it can be improved, and the technique used in this problem does not appear to offer much room for improvement.

12.5 Exercise Set

Some New and Some Old Mathematica features and Commands

ContourPlot[ ]	ContourPlot3D	Cross[ ] §
Dot[ ] §	FindRoot[ ]	Get[ ] or <<
<<Graphics`ImplicitPlot`	If[ ]	ImplicitPlot[ ]
Integrate[ ]	NIntegrate[ ]	N[Integrate[ ]]
NSolve[ ]	ParametricPlot3D[ ]	Part[or [[ ]]
Plot[ ]	Plot3D[ ]	Show[ ]
Sum[ ]	Table[ ]	Which[ ]

Plot options

AspectRatio− > 1	AxesLabel− > {"x","y","z"}
BoxRatios− > { }	DisplayFunction− >Identity
DisplayFunction− >$DisplayFunction	Mesh− >False
PlotPoints− > n (a positive integer)†	PlotRange− > { }
ViewPoint− > { } ‡	ViewVertical− > { }

§ Use the infix forms for scalar, dot and cross products: $a\vec{v}$, $\vec{v}\cdot\vec{w}$, and $\vec{v}\times\vec{w}$ can be computed using input statements $a * v$, $v.w$, and $v \times w$ (enter \[Cross] for $\times$).

† the default value for PlotPoints is $n = 25$. Don't make it too big.

‡ Use the **3D ViewPoint Selector...** in the **Input Menu** to insert the ViewPoint option. See page 184 for details.

Use the identity $|\vec{v}|^2 = \vec{v}\cdot\vec{v}$ to compute the length of a vector $\vec{v}$.

1. Let R be the rectangle $R = [2,4] \times [-2,3]$, let n be a positive integer, and let $\mathcal{P}_n$ be the partition obtained by dividing R into n^2 subrectangles using a grid of n equally spaced horizontal lines and n equally spaced vertical lines. Approximate the value of the integral $\int\int_R x^2y^3\,dxdy$ with a Riemann sum over the partition $\mathcal{P}_n$ for the following values of n and

the following choices of the points (x_j, y_j), $(j = 1, 2, \ldots, n^2)$. Compare the approximations with the decimal value of the integral obtained by using Mathematica's integration command. Notice how the approximations get better as n gets larger.

a) Let $n = 10$, and let (x_j, y_j), be the midpoint of the j-th subrectangle $(j = 1, 2, \ldots, 100)$.

b) Let $n = 20$, and let (x_j, y_j), be the midpoint of the j-th subrectangle $(j = 1, 2, \ldots, 400)$.

c) Let $n = 20$, and let (x_j, y_j), be the lower left hand corner of the j-th subrectangle $(j = 1, 2, \ldots, 400)$.

d) Let $n = 50$, and let (x_j, y_j), be the midpoint of the j-th subrectangle $(j = 1, 2, \ldots, 2500)$.

2. Let D be the region bounded by the curve $y = 9 - x^2$ and the x-axis. Enclose D in the rectangle $R = [-3, 3] \times [0, 9]$, and let $\mathcal{P}_n$ be the partition of D obtained by dividing R into n^2 subrectangles using a grid of n equally spaced horizontal lines and n equally spaced vertical lines. Approximate the value of the integral $\int\int_D ye^x \, dxdy$ with a Riemann sum over the partition $\mathcal{P}_n$ for the following values of n and the following choices of the points (x_j, y_j), $(j = 1, 2, \ldots, n^2)$. Compare the approximations with the decimal value of the integral obtained by using Mathematica's integration command. Notice how the approximations get better as n gets larger.

a) Let $n = 10$, and let (x_j, y_j), be the midpoint of the j-th subrectangle $(j = 1, 2, \ldots, 100)$.

b) Let $n = 20$, and let (x_j, y_j), be the midpoint of the j-th subrectangle $(j = 1, 2, \ldots, 400)$.

c) Let $n = 20$, and let (x_j, y_j), be the lower left hand corner of the j-th subrectangle $(j = 1, 2, \ldots, 400)$.

d) Let $n = 50$, and let (x_j, y_j), be the midpoint of the j-th subrectangle $(j = 1, 2, \ldots, 2500)$.

3. Use a single Mathematica input statement to evaluate the double integral

$$\int\int_D \frac{y^3}{x^2} \, dxdy,$$

where D is the region bounded by the curves $xy = 1$ and $10x + 7y = 50$. Evaluate the integral again, as a sequence of two separate single integrals. Use indefinite integration so that the individual antidifferentiation processes can be observed in greater detail.

4. For each of the following functions $f(x, y)$ and regions D, set up and evaluate (if possible) both iterated double integrals for $\int\int_D f(x, y) \, dxdy$. Confirm that the two values are the same. Evaluate each iterated integral a second time, using a sequence of two separate single integrals. Use indefinite integration so that the individual antidifferentiation processes can be observed in greater detail. Notice the advantage of integrating in one order over the other. In some cases, Mathematica may be able to evaluate one iterated integral, but not the other.

a) $f(x, y) = \frac{x^2 y^3}{x^2 + y^2}$, D is the region bounded by the x-axis, $y = x$ and $x = 1$.

b) $f(x, y) = y^2 e^{xy}$, where D is the region bounded by $y = x$, $y = 1$, and the y-axis.

c) $f(x, y) = \sqrt{x^7} \cos(x^2 y)$, where D is the region bounded by $x = y^2$ and $x = 1$.

5. For each of the following functions $f(x, y)$ and regions D, evaluate, as an approximate decimal, the double integral $\int\int_D f(x, y) \, dxdy$.

a) $f(x, y) = \sqrt{5x^2 + 3y^2}$, where D is the closed disk of radius 2 centered at the origin.

b) $f(x, y) = \frac{2x^2}{\sqrt{y+9}}$, where D is the triangular region with vertices at the points (3,5), (3,-2), and (8,14).

c) $f(x, y) = \frac{x^2 - y^2}{x^2 + y^2}$, where D is the region bounded by the curves $y = x^2 - 6x + 5$ and $y = 2x + 4$.

6. Evaluate, as an approximate decimal, the double integral

$$\int\int_D (x^3 - y^3) \, dxdy,$$

where D is the right hand portion of the two regions bounded by the curves

$$4x^2 + 25y^2 + 32x - 150y + 200 = 0, \text{ and } y - x^2 + 2x + 7 = 0.$$

7. Find the area of the region bounded by

$$x = y^4 - 5y^3 + 2y^2 - 13, \text{ and } x = 57 + 10y - 4y^2.$$

8. Find the mass and center of mass of a thin plate bounded by

$$x = y^4 - 5y^3 + 2y^2 - 13, \text{ and } x = 57 + 10y - 4y^2,$$

where x and y are expressed in centimeters, if the mass density at the point (x, y), in grams per square centimeter, is $\delta(x, y) = 4 + 2x + y/2$.

9. Use a single input statement to compute the double integral, in rectangular coordinates, of $f(x, y) = x^2y^2$ over the closed disk of radius 5 centered at the origin. Evaluate the integral a second time, using a sequence of two separate single integrals. Use indefinite integration so that the individual antidifferentiation processes can be observed in greater detail.

Compute the integral again as a double integral in polar coordinates and again as a sequence of separate single polar integrals.

10. Compute the double integral, of $f(x, y) = \sqrt{16 + x^2 + y^2}$ over the region D bounded by $y = x$, $y = -x$, and the semicircle $y = \sqrt{4 - x^2}$. Try to evaluate the integral in both rectangular and polar coordinates.

11. Let R be the rectangular box (prism) $R = [1, 3] \times [-2, 1] \times [-1, 4]$, let n be a positive integer, and let $\mathcal{P}_n$ be the partition obtained by dividing R into n^3 subprisms using a gridwork of n equally spaced planes parallel to each of the coordinate planes. Let (x_j, y_j, z_j), be the midpoint of the j-th subprism ($j = 1, 2, \ldots, n^3$). Using this scheme of choosing the n^3 points (x_j, y_j, z_j), approximate the value of the integral $\int\int\int_R (xye^{zx}\, dxdydz$ with a Riemann sum over the partition $\mathcal{P}_n$ for the following values of n. Compare the approximations with the decimal value of the integral obtained by using Mathematica's integration command. Notice how the approximations get better as n gets larger. This problem should be accessible on most machines, but its computational complexity increases dramatically as n gets large. The partition $\mathcal{P}_{50}$, for example, has 125,000 subprisms.

a) $n = 10$ b) $n = 20$ c) $n = 50$

12. Let D be the closed sphere of radius 2 centered at the origin. Enclose D in the rectangular box (prism) $R = [-2, 2] \times [-2, 2] \times [-2, 2]$, let n be a positive integer, and let $\mathcal{P}_n$ be the partition obtained by dividing R into n^3 subprisms using a gridwork of n equally spaced planes parallel to each of the coordinate planes. Let (x_j, y_j, z_j), be the midpoint of the j-th subprism ($j = 1, 2, \ldots, n^3$). Using this scheme of choosing the n^3 points (x_j, y_j, z_j), approximate the value of the integral $\int\int\int_R (xyz - 4)^2\, dxdydz$ with a Riemann sum over the partition $\mathcal{P}_n$ for the following values of n. Compare the approximations with the decimal value of the integral obtained by using Mathematica's integration command. Notice how the approximations get better as n gets larger.

a) $n = 10$ b) $n = 20$ c) $n = 50$

Computing the Riemann sum corresponding to $\mathcal{P}_{50}$ will require some patience or a machine with a fast processing time. This partition has 125,000 subprisms.

13. Use a single Mathematica input statement to evaluate the triple integral

$$\int_0^2 \int_0^{\sqrt{16-4z^2}} \int_0^{8-4z} dydxdz.$$

Evaluate the integral again, as a sequence of three separate single integrals. Use indefinite integration so that the individual antidifferentiation processes can be observed in greater detail.

14. Set up and evaluate all six iterated integrals for the volume of the region bounded below by $z = 0$, above by $z + 2y = 4$, and on the side by the cylinder $y = x^2$.

15. Compute the triple integral of $f(x, y, z) = x^3(xz - y)^2$ over the region D, which is bounded by the planes $x = 0$, $3x + 7z = 85$, and the cylinder $3y^2 + 8z^2 = 28$

16. Find the volume of the region D, which is bounded by the planes $2x + 3y + z = 8$, $2x + 3y - z = 8$, and the cylinder $y = x^2 - 4$.

17. Find the volume of the region D, which is bounded by $y = 8x^2 + 3z^2 - 10$ and $y = 10 - 2x^2 - 5z^2$.

18. Find the volume of the region which is bounded above and below by $z = 900 - 25x^2 + 24y$, $z = 4y^2 - 250x$, and which lies inside the cup of the parabolic cylinder $x = 8 - y^2 - y$.

19. Find the volume of the region common the the interior of the cylinders $3y^2 + 5z^2 = 84$ and $3x^2 + 5z^2 = 84$.

20. Let D be the region bounded above by the sphere $x^2 + y^2 + z^2 = 900$ and below by the plane $z = 20$. Set up and evaluate the iterated triple integral for the volume of D in (a) rectangular, (b) cylindrical, and (c) spherical coordinates.

21. Find the (exact) volume of the region inside the sphere $x^2 + y^2 + z^2 = 16$ and outside the cylinder $(x - 2)^2 + y^2 = 4$.

22. Find the mass and center of mass of the region bounded by $z = 200 - 0.03x^2 - 0.04y^2$ and $z = 10 - 0.02x^2 - 0.03y^2$, if the density at the point $P(x, y, z)$ in the region is $\delta(x, y, z) = (40 + 0.18z)^2$. Symmetry suggests that the center of mass should be on the z-axis. Confirm this with your calculations.

23. Let D be the region bounded by $x = 0$ and

$$x = 25\sqrt{1 - \frac{y^2}{16} - \frac{z^2}{9}},$$

where the units are in centimeters. Find the mass and center of mass of D under a uniform mass density of 8 grams per cubic centimeter.

24. A fuel tank has the shape of a circular cylinder with a hemisphere at each end. It lays on its side and its position is carefully adjusted so that the axis of the cylinder is horizontal. The depth of the fuel level in the tank naturally determines how much fuel remains in the tank. Determine a formula for the volume of the fuel remaining in the tank as a function of the depth h of the fuel level. Let r denote the radius of the cylinder (hence the radius of each hemisphere), and let l denote the length of the cylinder (without the hemispheric ends).

For some values of h the fuel volume is obvious. Verify that your formula holds for these values of h.

25. A mixture of copper and waste rock is located inside of an immense hill, and mining engineers are attempting to measure the total mass of the pure copper inside the hill, before an attempt is made to buy the mineral rights to the land. A flat map of the region is being studied, and a two dimensional grid has been placed over the map with units x and y in miles. The copper deposits exist within an area defined roughly by the ellipse

$$1.4x^2 + 3.2y^2 = 5.$$

The elevation H, of the hill, in feet above sea level, at the point (x, y) on or inside the ellipse is

$$H = (5 - 1.4x^2 - 3.2y^2)(652y^2 + 248^y + 350) + 43.$$

The copper/rock mixture begins roughly 20 feet below the surface of the hill, and it continues down to a depth defined by the equation

$$U = 950 - 280x^2 - 640y^2.$$

Let z denote the elevation, in feet above sea level, of a point P above the point (x, y) on the map. Engineers have made a "reasonable" conjecture, that the density of pure copper within the rock mixture, at a point $P(x, y, z)$ in the ore field, is proportional to the product of the vertical distances from P to the upper and lower limits, $H - 20$ and U of the field. This means that for some constant of proportionality a, the density formula is

$$\delta = a(H - 20 - z)(z - U).$$

Empirical evidence suggests that a maximum density of 42 pounds per cubic foot (of pure copper) can be expected. Use this to determine a value for the constant a, and then find the total mass of pure copper in the field.

Remember that the elevation z is expressed in feet, and x, and y are expressed in miles. The search for the point $P(x_0, y_0, z_0)$, which maximizes δ, can be simplified. Surely $(H(x, y) - U(x, y))$ will be a maximum, for $x = x_0$ and $y = y_0$. Furthermore, z_0 will be midway between $(H(x_0, y_0) - 20)$ and $U(x_0, y_0)$.

Project: Gravity

Let m denote the mass of a point located at P, and let M denote the mass of a point located at Q. The gravitational force of attraction exerted at the point P between the two point masses is

$$\vec{F} = \frac{GmM}{|\vec{r}|^3}\vec{r},$$

where $G =$ is the gravitational constant and $\vec{r}$ is the vector from P to Q. This is one of the three classic *"Inverse Square Laws"* that govern the behavior of not only gravitational forces, but electrical and nuclear forces as well.

If m and M are in *grams*, $\vec{r}$ is in *cm*, and $G = 6.670 \times 10^{-8}$, then $\vec{F}$ is expressed in *dynes*. The problems in this section can be solved, however, without a value for G. Just carry G along as a constant. Usually, we let $m = 1$ and talk about the gravitational force exerted on a unit point mass located at P.

If a mass-object W (a world) is not a point mass, then the gravitational field created by W is a consequence of the its global shape and size. To determine a formula for the gravitational field for W, we partition W into lots of infinitesimal point masses. At each point mass, the *Inverse Square Law* applies, and the individual forces of each point mass exerted on a unit mass located at the fixed point P can be computed. We simply add up (in the sense of integration) all of the infinitesimal forces.

Force, of course, is a vector quantity. Normally it would have to be decomposed into three real valued components, and the integration would then be performed on each of the three components. The following two problems, however, have a great deal of symmetry. Use it wisely, and the problems can be greatly simplified.

1. Show that the gravitational field created by a spherical planet of mass M and constant mass density, is the same as the gravitational field created by a point mass of size M located at the center of the sphere. Of course, this only makes sense for points far enough from the center to be outside the sphere.

We may as well suppose that the planet is centered at the origin. Let R denote the radius of the planet, and let δ denote its constant mass density. Then the total mass M of the planet is just δ times the volume of the sphere.

Let $P_0(x_0, y_0, z_0)$ be the location of a unit point mass, and let R_0 be the distance from $P0$ to the origin. **Mathematica may have to be told that $R > 0$ and that $R_0 > R$.** If so, try replacing R by a new variable $R = R_0 + S^2$.

2. Have you ever read a science fiction or fantasy book about life inside a hollow planet? Is this possible? Could such a world support a gravitational field which would hold the planet's inhabitants to the "ground"?

Consider a large hollow sphere (planet) of radius R, Suppose that R is quite large (perhaps several thousand kilometers), and that the skin of the sphere (the ground) is made out of a very heavy material of uniform mass density δ (expressed in kilograms per square meter).

Determine a vector valued formula, which describes the gravitational force of attraction exerted by the entire hollow sphere on a unit point mass located at a point $P_0(x_0, y_0, z_0)$ interior to the sphere.

This problem is meant to be done in spherical coordinates ρ, θ and ϕ, with ρ fixed at $\rho = R$. Consider a small patch of area on the sphere as a point mass.

A formula for a "small patch of area" on the skin of the sphere will be needed. Recall that the "infinitesimal cube" in spherical coordinates has dimensions $d\rho$ by $\rho\, d\phi$ by $\rho \sin(\phi)\, d\theta$. The area of the face of this cube which lies on the sphere $\rho = R$ is just the product, $\rho^2 \sin(\phi)\, d\theta d\phi$ of the last two dimensions with $\rho = R$. It follows that the expression $R^2 \sin(\phi)$ represents the area of a small patch of area on the surface of the hollow sphere.

let R_0 be the distance from P_0 to the origin. **Mathematica will have to be told that $R_0 > 0$ and that $R_0 < R$** (see part 1 of this project).

Project: The Atmosphere of a Planet

A new uncharted planet has been discovered, and scientists need to know the total amount of methane in the planet's atmosphere.

The origin of a coordinate system is placed at the planet's center, with units expressed in kilometers. The surface of the planet is defined by the equation

$$\frac{x^2}{5000^2} + \frac{y^2}{5000^2} + \frac{z^2}{4200^2} = 1,$$

and the outer limits of the planet's atmosphere is defined by the equation

$$\frac{x^2}{5030^2} + \frac{y^2}{5030^2} + \frac{z^2}{4230^2} = 1.$$

The density of methane, in grams per cubic meter, is estimated to be

$$\delta = \frac{1.023}{h^2 + 1}$$

at an altitude h meters above the planet's surface. (Altitude is a dimension, which is always orthogonal to the planet's surface.) Determine the total amount of methane in the atmosphere.

If you use NIntegrate[ ] to initiate a numerical approximation process, you will probably have time to enjoy a leisurely dinner while you wait for Mathematica to calculate an answer to this problem. To speed things up, it may help to take advantage of the comments at the end of Section 12.3. Altitude may be difficult to express in terms of spherical coordinates, and an approximation strategy may be necessary. See the solution to Example 12.13 for some insight.

Index